T0205567

Supply Chain

Antonio Rizzi

Supply Chain

Fundamentals and Best Practices to Compete
by Leveraging the Network

 Springer

Antonio Rizzi
Department of Engineering
and Architecture
University of Parma
Parma, Italy

ISBN 978-3-030-95706-3   ISBN 978-3-030-95707-0   (eBook)
https://doi.org/10.1007/978-3-030-95707-0

This Springer imprint is published by the registered company Springer Nature Switzerland AG
The registered company address is: Gewerbestrasse 11, 6330 Cham, Switzerland

This book is in memory of Gino Marchet.

My master, my colleague, my friend

Preface

This book stems from my desire to put together and organize teaching material I have accumulated in 10+ years of supply chain management classes for the Post Graduate Degree in Industrial Engineering and Management at the University of Parma, and to create a textbook to help future students.

The contents combine my academic research in the fields of supply chain management, industrial logistics distribution and retail, with the experience I have gained in these fields thanks to consultancy provided over the 20 years in which I have worked side by side with supply chain and logistics, retail and ICT officers and managers of major companies in the food and consumer goods, textile, clothing and fashion, healthcare and pharmaceutical industries.

The Chap. 1 is an introduction to the book, since it defines the supply chain, its process and its players, as well as a manifold representation from different angles and perspectives: the supply chain can be talked about through the players that make it up, as well as by describing the company's functions which operate therein, or the information that players exchange at different levels, and the service that the supply chain delivers. Each of these aspects makes it possible to highlight the different issues at stake and to deliver specific insights, and is the subject of its own chapter.

Chapter 2 describes the supply chain through the players involved in it. A particular focus is paid to the distribution and retail structure, its players, and the various organizational and operational models. This chapter details best practices in the organization of distribution processes in the food, FMCG and e-commerce sectors. Chapter 2 also delves deeply into how players like Amazon or Dell have been able to leverage and innovate the supply chain to generate a competitive advantage.

Chapter 3 begins from Porter's Value Chain, and sets out the role of the functions involved, illustrating in thorough detail the role of each core and support function, as well as the conflicts that can arise between them. Instead, in the second part of the chapter I address how the various functions involved in the supply chain process must be orchestrated by the supply chain manager, both inside and outside the company boundaries, to achieve a strategic coherence and fit.

Chapter 4 deals with the information flows exchanged by the players in the supply chain, which make it possible to govern the physical flows of products and the related services. Purchase orders and confirmation orders, bills of lading and dispatching advices, point of sales data and inventory data are introduced and dealt with in thorough detail, while the role of information sharing in the management of product flows and service generation are also analysed. A deep dive into identification standards is also presented here.

In conclusion, Chap. 5 describes the supply chain from a final perspective: the service flows that stem from the supply chain process. It first gives a definition of customer service, and tackles the strategic role of customer service in various industries. Then it sets out the factors affecting customer service in a structured framework, addressing in particular those that fall within the boundaries of supply chain processes, such as lead time, delivery accuracy, frequency and flexibility. Last but not least, the chapter analyses how companies can measure and manage customer service to create a competitive advantage.

The book addresses the different topics by combining a quantitative approach typical of engineering disciplines and adopted particularly in the operational parts, with a qualitative approach which fits in the more strategic parts.

This combination makes the volume suitable for both under- and post-graduate degree students in industrial engineering and management, as well as undergraduates or masters students confronting supply chain management from different cultural backgrounds.

Industry operators and practitioners can also find useful insights in this text, since I thoroughly describe industry best practices, especially in the food and consumer goods, fashion and apparel, pharmaceutical and healthcare sectors, stemming from real life experiences.

I wish to thank and congratulate my son Tommaso Rizzi, 17 years old, who illustrated the whole book and hand drew all figures. Kudos Tommy!

My deepest gratitude goes to my teachers in Supply Chain Management and Industrial Logistics. Martin Christopher, Sunil Chopra, Claudio Ferrozzi, Roy D. Shapiro, Giulia Urgeletti Tinarelli, Gino Ferretti, Remigio Ruggeri, and Gino Marchet.

Parma, Italy   Antonio Rizzi

Contents

Acronyms

3PL	Third-party logistics service provider
3PM	Third-party manufacturer
4PL	Fourth-party logistics service provider
A%	Accuracy
ABC	Activity-based costing
ACS	Average customer seniority
AD%	Advance deliveries percentage
AF	Administration and finance
AGV	Automated guided vehicle
AI	Artificial intelligence
AIDC	Automatic data capture identification technologies
AM%	Accuracy of mix
AP%	Accuracy of place
AQ%	Accuracy of quantity
AS%	Accuracy of the state
AS/RS	Automated storage retrieval system
ASIN	Amazon serialized item number
ASN	Advanced shipping notification
AT%	Accuracy of time
AV	Artificial vision
AxB	Ordered day A; delivered day B
AxC	Ordered day A; delivered day C
B2C	Business to consumer
BC	Bar code
BD%	Backlog deliveries percentage
BE	Bullwhip effect
BE	Backend
BL	Backlog
BLT	Buyer lead time
BM	Basic models
BoD	Board of Directors

BoL	Bill of lading
BOM	Bill of materials
BOPIS	Buy Online Pick Up In Store
BOSFS	Buy Online Ship from Store
BOSTS	Buy Online Ship to Store
BP	Basic products
BPMN	Business process modeling notation
BPR	Business process reengineering
BR	Backroom
BR	Breath of range
BSSFS	Buy In Store Ship from Store
BSSTD	Buy In Store Ship to Door
C	Consumables
C	Capacity
CAPEX	Capital expenditures
CAT	Cost adding time
CAT%	Percentage of value adding cost
CC	Consolidation centre
CD	Cross docking
Cdir	Direct costs
CEO	Chief Executive Officer
CFO	Chief Financial Officer
CI	Industrial cost
Cind	Indirect costs
CLE	Customer life expectancy
CLT	Consumer lead time
CLT	Central limit theorem
CLV	Customer lifetime value
CM	Collateral models
CMO	Chief Marketing Officer
COO	Chief Operation Officer
CP	Collateral products
CPFR	Continuous planning forecasting and replenishment
CPO	Chief Procurement Officer
CR	Conversion rate
CR	Continuous replenishment
CRM	Customer relationship management
CRR	Customer retention rate
CS	Cycle stock
CS	Customer service; customer satisfaction
CSO	Chief Sales Officer
CTR	Cost of transaction
D	Demand
D	Distributor
D	Distribution

D	Detractors
D/S	Downstream
DB	Database
DC	Distribution centre
DDP	Delivery duty paid
DESADV	Despatch advice
DET	Data entry time
DF	Delivery frequency
DF%	Delivery frequency percentage
DP	Decoupling point
DPP	Demand penetration point
DRP	Distribution requirement planning
DURC	Documento unico regolarità contributiva
DW	Data warehouse
E	Energy
EAN	European article number
EBITDA	Earnings before interest taxes depreciation and amortization
ECR	Efficient consumer response
ED	Expiration date
EDI	Electronic data interchange
EDLC	Every day low cost
EDLP	Every day low price
EI	Electronic invoice
EMEA	Europe Middle-East and Africa
EOQ	Economic order quantity
EPAL	European Pallet Association
EPC	Electronic product code
ERP	Enterprise requirement planning
ETO	Engineered to order
EXW	Ex works
F	Forecast
FAA	Federal Administration Aviation
FAQ	Frequently asked questions
FBA	Fulfilled by Amazon
FC	Focal company
FE	Front end
FMCG	Fast moving consumer goods
FO%	Percentage of fulfilled orders
FP	Finished products
FTL	Full truckload
FX%	Flexibility of (X=mix; quantity, place, time)
G	Inventory on hand
GLN	Global location number
GLP	Good logistics practices
GMP	Good manufacturing practices

GOM	Gross operative margin
GP	Gross profit
GR	Gross requirement
GRAI	Global returnable asset identifier
GTIN	Global trade item number
GXP	Good x practices
HR	Human resources
HU	Handling unit
ID	Identification number
IPV	Items per version
IR	Rotation index
IT	Information technology
JIT	Just in time
K	Capex; equipment
KPI	Key performance indicator
L	Labour
LAN	Local area network
LN	Lot number
LSP	Logistics service provider
LT	Lead time
LTG	Lead time gap
LTmin	Minimum lead time
LTT	Less than truckload
M	Unit margin
M	Manufacturer
M2M	Many to many; Machine to machine
MAPE	Mean absolute percentage error
MBO	Management by objectives
MES	Manufacturing execution systems
MH	Material handling
MMC	Market mediation cost
MRO	Maintenance, repair and operational materials
MRP	Material requirement planning
MTO	Make to order
MTS	Make to stock
NAD	Number of actual deliveries
NC	Number of customer
NCAT%	Percentage of non value adding costs
NOEX	Number of order with errors under dimension X
NOLEX	Number of order lines with errors under dimension X
NP	Net profit
NPS	Net promoter score
NR	Net requirement
NRC	Number of retained customer
NRS	Number of references in stock

NRVXF	Number of request for variation (X=mix, quantity, place, time) fulfilled
NRVXR	Number of request for variation (X=mix, quantity, place, time) received
NSD	Number of scheduled deliveries
NT	Normal trade
NVAC	Non value added cost
NVAC%	Non value added cost percentage
NVAT	Non value adding time
NVAT%	Non value adding time percentage
O	Order
O	Operation
OC	Order confirmation
OCT	Order cycle time
ODF	Optimal delivery frequency
ODL%	Orders delivered late percentage
ODOT%	Orders delivered on time percentage
OEE	Overall equipment efficiency
OFMA%	Orders fulfilled with mix accuracy percentage
OFQA%	Orders fulfilled with quantity accuracy percentage
OFT	Order forwarding time
OH	On hand inventory
OIA%	Orders delivered in advance percentage
OIT	Order issuing time
OLFMA%	Order lines fulfilled with mix accuracy percentage
OLFQA%	Order lines fulfilled with quantity accuracy percentage
OMD	Organized mass distribution
OND%	Orders not delivered percentage
OUTL	Order up to level
OOS	Out of stock
OP	Order planning
OPEX	Operational expenditures
OPT	Order preparation time
OR	Order release
OTC	Order to cash
OTD	Order to delivery
OTD	On time delivery
OTIF	On time in full
OTIF%	On time in full
P	Selling price
p	Selling price
P	Procurement
P	Promoters
PC	Product category
PC	Personal computer
PC	Physical cost
PG	Physical goods

PKL	Pick list
PL	Private label
PLC	Programmable logic controller
PO	Purchase order
POS	Point of sale data
Q	Quantity, volume
R	Retailer
R&D	Research and development
RECADV	Receiving advice
RECT	Receiving time
RFID	Radio frequency identification
RFP	Request for proposal
RI	Reorder interval
RM	Raw materials
RO	Retail outlet
ROE	Return on equity
ROI	Return on investment
RPC	Reusable plastic container
RPV	References per variants
S	Stores
S	Supplier
SAR	Stock availability ratio
SC	Supply chain
SCFT	Supply chain flow time
SCLT	Supply chain lead time
SCM	Supply chain management
SD	Scheduled delivery
SF	Sales floor
SGTIN	Serialized GTIN
SKU	Stock keeping unit
SLA	Service level agreements
SLAM	Scan label apply manifest
SN	Serial number
SQR	Stock quantity per reference
SS	Safety stock
SSCC	Serial chipping container code
SSR	Stock service ratio
SU	Shipping unit
SVR	Service vendor rating
T	Turnover
TAC	Time for administrative checks
TCP	Time for consolidation and packing
TCU	Time to customize items
TD	Transport document
TDOQ	Time to define the order quantity

TDP	Time for delivery planning
TIC	Time for inventory checks
TIT	Time for internal transmission
TLP	Time for picklist generation
TM	Transportation management
TOC	Time for order confirmation
TOF	Time for order fulfillment
TOF	Total orders fulfilled
TOLF	Total order lines fulfilled
TOP	Time for order processing
TOT	Time for order transport
TP	Transit point
TPI	Time for picking items
TPK	Time for picking
TRT	Transport time
TSO	Time for sorting
TSOS	Time for sorting/staging
TTR	Time for transfer
TVE	Time for verification
TVL	Time for vehicle loading
U/S	Upstream
UI	User interface
ULT	Unloading time
VAC	Value added cost
VAC%	Value added cost percentage
VAS	Value added service
VAT	Value adding time
VAT%	Value adding time percentage
VICS	Voluntary interindustry commerce standards
VLT	Vendor lead time
VMI	Vendor managed inventory
VPBM	Versions per basic model
VPBP	Variants per basic product
W	Wholesaler
WIP	Work in progress
WMS	Warehouse management system
WS	Web services
WTU	Worktime units
YFP	Yearly frequency of purchase

Chapter 1
Introduction to the Supply Chain Concept

1 What is a Supply Chain?

If we want to talk about Supply Chain Management (SCM) we ought to start by explaining the literal meaning. Given that, for non-English native speakers at least, this is a concept with no direct correspondence. So let's analyse the two terms:

- Management: easily understood. But what is it that is being managed? Furthermore the term *management* omits to mention the aspects of an *integrated management of a system*, typical of SCM, that will be thoroughly tackled in the following chapters.
- Supply Chain (SC): an Anglo-Saxon term, and quite literal.

So just what does Supply Chain actually mean?

An SC can be seen as a macro-process, aimed at achieving a final result which means satisfying a consumer's need for a product and a service, for which the consumer is willing to pay a price.

Using black-box modelling, the description of this process is contained within the control surface in Fig. 1: it is a complex of activities aimed at satisfying the end customer's needs via a product and a service.

Indeed, the end consumer is an integral part of this process in the sense of an expressed need to be satisfied. Consequently, by extending the control surface to this player, a full description of the SC, of which the customer is an integral part, is obtained. This concept is detailed further below.

As mentioned above, the macro-process by which the supply chain is defined brings the end consumer two things: a *product* and a *service*.

- **Product**: any object, more or less complex, available on the market, which meets the needs of the end consumer. The consumer's needs may be tangible, that is to say, a need which can be satisfied by the functional characteristics of a product (e.g. a drug able to cure a disease), or intangible, in which the need is satisfied by the immaterial characteristics of a product, typically by the added value of a brand. The meaning of *product* is exceedingly generic, since it can refer to any type of

Fig. 1 The SC process; an
SC also includes the end
consumer

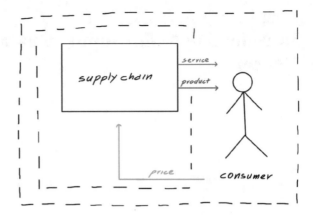

product: (i) fast-moving consumer goods, (ii) foodstuffs, (iii) capital goods such
as an industrial plant, or (iv) durable goods, e.g. a garment, a household appliance,
a furniture item, a car, and so on.

- **Service**: the term service will be discussed in more depth later in Chap. 5, however,
  it is a good idea to introduce a first notion. When end consumers purchase a
  product, they are paying for its technical functionality and intangible value, or for
  the purchase experience connected with the brand, but they are also purchasing
  ancillary services which accompany the product itself. A typical example of a
  service which might be requested/provided is: (i) the instant when the product
  becomes available (known as *lead time*), but also (ii) the fact that the consumer
  can go to a physical place at a certain time and leave that physical place with the
  available product, or instead, have the product delivered to the home or a locker
  hub a few days later. Suffice to think of purchasing electronic products (e.g.
  smartphones), or any specific good (e.g. a book) purchased at a retail chain, or
  from an online store. The product purchased from both of the channels described
  is the same, but what changes is the service, interlinked with the point in time
  (immediately in the case of a retail purchase, within the space of a few days in the
  case of an online sale) along with the place where the product is made available.
  The fact of a product being available in different formats, or the presence of
  pre/post-sales assistance are other forms of service. In short, customer service is
  defined as that set of intangible services, whether expressed or implicit, which
  accompany a product, and for which the consumer is willing to attach a value and
  therefore pay a surcharge on the product. In recent times, the concept of service
  has seen a considerable increase in interest on the part of the business players,
  and above all of the end consumers, who, from one day to the next, increase
  their service requirements in terms of demand for any product, in ever more rapid
  timeframes, to be delivered anywhere. Consequently, in order to compete, also a
  supply chain must be able to meet this demand for higher service levels, and this
  increases the complexity of the problem to be dealt with.

End consumers can be reached through various SCs which offer specific combinations of products and services. End consumers are willing to pay a different purchase price depending on the value they attach to the combination of product/service. Typically, as the technical performance of the product and service increases, the purchase price increases as well. The purchase price is therefore the total value which the consumer is prepared to attach to the process before making a purchase (*price acceptance*). This sum of money is the "reward" for the system which made this process possible, and which the players who make up the system can divide amongst themselves to cover their costs.

The system which makes this process possible, through which the end consumer is reached, is that object which we improperly call "Supply Chain". Improperly, as said before, since the supply chain involves not only the players who make the process possible, but also the end consumer, who is an integral part of it. In fact, a supply chain exists to satisfy a demand for a product and/or service on the part of an end consumer, and in this sense the end consumer is actually the driver of it. Without the end consumer there would be no reason for the process and therefore the supply chain to exist.

Meanwhile, it can be observed that any product or service which meets the needs of a customer is backed up by a more or less complex SC. The end consumer is, therefore, the final link of this chain which we call Supply Chain, and which it is possible to envisage by retracing the history of a product.

By imagining that we are able to get onboard a good purchased by an end consumer (for example, a product purchased in a physical shop), we can watch the object as it moves and is intercepted by different players, in different places and at different times.

By way of example, imagine "rewinding" the tape of a hypothetical video camera placed on a packet of pasta sold in a supermarket, from the moment it is added by the end consumer to his or her shopping trolley. The resulting description is described in Table 1

The lexicon used and the meaning of each activity will become clearer in Chap. 2, where we present a thorough physical description of the supply chain.

Let's now introduce the concepts of SC depth and breadth. The *depth* of an SC means the number of levels involved.

Instead, the *breadth* of an SC is used to identify the number of players per level. As is clear, from the bottom up, the base of the pyramid widens more and more, and the number of players involved in the process at each level increases exponentially.

As the depth and breadth increase, the complexity of the system also increases from a management point of view, with the result that it tends to take the form of a network rather than a chain. In fact using the term *supply networks* rather than supply chains stresses precisely this aspect of complexity in the relationships between players within a single layer and between layers.

It is not unusual to find supply networks with hundreds of players, naturally also distributed across a dozen levels.

It's worth pointing out that the entire process is governed by an interaction between various corporate functions. In a "push" approach, supplies of raw materials are

Table 1 Rewinding the SC tape of a packet of pasta replenished on a store shelf

Location	Activities	Date time	Description
Retail outlet	• Sale	04/03 19.00	The consumer buys the product
	• Replenishment	04/03 14.00	The supermarket store associate refills the shelf in the sales area, by bringing a box of products from the shelf in the backroom area, opening it, and placing it on the shelf
	• Receiving	02/03 06.00	The box of products is unloaded from an articulated lorry which has arrived from a Distribution Centre, in a metal roll container or on a pallet, together with all the other products to replenish that particular retail outlet on that day
Transportation			An articulated lorry carries out multi-drop deliveries to different retail outlets
Retailer's distribution centre	• Shipping	01/03 22.00	Loaded onboard the articulated lorry, the box leaves the centre's shipping dock
	• Packing & marking	01/03 18.00	The multi-reference load unit (roll container/pallet), containing the box of products bearing a shipping label is made stable
	• Picking	01/03 16.00	A forklift truck driver collects the box of products from the picking bay to create a multi-reference unit load
	• Storage	16/02 10.00	The box containing the packet of pasta is put away in a storage area with shelving, on a single-lot, single-reference pallet

(continued)

Table 1   (continued)

Location	Activities	Date time	Description
	• Receiving	15/02 08.00	The single-reference single-lot box enters the Distribution Centre from a receiving dock, together with other single-lot single-reference pallets, which have arrived from the manufacturer's full mix Distribution Centre, all of them of the same brand
Transportation			Direct delivery: the articulated lorry carries out a direct delivery from manufacturer to distributor
Manufacturer's full mix distribution centre	• Shipping	14/02 18.00	Similar process to the Retailer's Distribution Centre
	• Retrieving	14/02 14.00	The forklift truck driver (or an Automated Storage/Retrieval System – AS/RS) picks up the single-product single-batch pallet from the warehouse shelf (automated)
	• Storage	07/02 Noon	Similar process to the Retailer's Distribution Centre
	• Receiving	07/02 11.00	Similar process to the Retailer's Distribution Centre. The material arrives directly from the production lines
Manufacturer's production lines	• Materials handling	07/02 10.00	The single-lot pallet is delivered to the full mix Distribution Centre warehouse by means of an MH system
	• Tertiary packaging—palletizing	07/02 9.55	A palletizer generates a single-product single-lot pallet by stacking boxes of products identified by an SSCC barcode logistic label

(continued)

Table 1 (continued)

Location	Activities	Date time	Description
	• Secondary packaging—boxing	07/02 9.50	The packet of pasta is boxed in a secondary packaging bearing an EAN 128 label
	• Primary packaging—packing	07/02 9.45	A packing machine generates each sales unit, consisting of product + packaging and identified by an EAN 13 label
	• Production processes		
	– Drying – Drawing – Kneading – Mixing	06–07/02	The production lines generate the finished product from the raw materials, using a transformation process
	• Supply lines		The raw materials (RM) are brought from the RM warehouse to the production lines to make the production lot
Manufacturer's RM warehouse	• Procurement of RM from top-level suppliers • Procurement of services		The RM and services are provided by first-tier suppliers
Transportation			
First-tier supplier	• Distribution, production, procurement		The first-tier supplier provides raw materials and components, and produces a finished product which represents the raw material for manufacturing The processes described above are repeated
Transportation			
Second-tier supplier	• Distribution, production, procurement		The second-tier supplier provides raw materials and components, and produces a finished product which represents raw material for the first-tier supplier The processes described above are repeated

generated by a projection based on sales forecasts made by retailers (sales business function), which are suitably consolidated and exploded by the company Material Requirement Planning (MRP) (operations), determining both the total requirements of (i) raw materials (RM), components, and other materials which contribute to producing the lots that need to be procured and supplied at a certain time and place (procurement), (ii) production and logistical capacity (operation and logistics/distribution) (Figs. 2 and 3).

In short then, on the basis of these premises, we can make some concluding remarks:

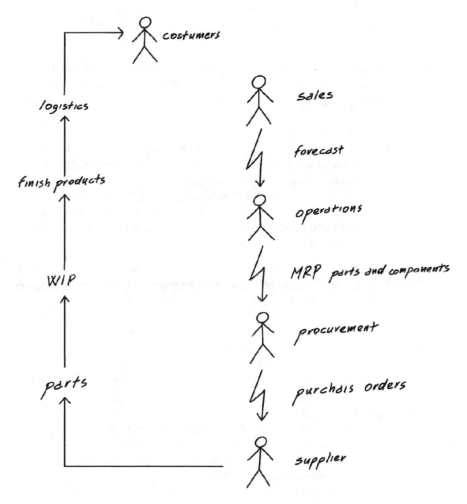

Fig. 2 Analysis of an SC from the perspective of the corporate functions which indirectly contribute to generating flows of materials

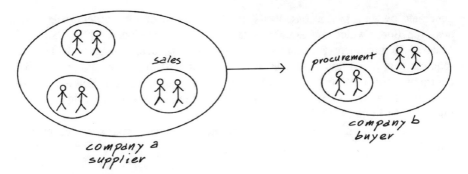

Fig. 3 Intra- and extra-business interactions

- The end consumers are a part of the supply chain, they represent the final link and are the "driver" of the entire process. This process would make no sense if there were no consumer who needed a product and service, and who, to satisfy that requirement, must be willing to (i) attach a value and (ii) pay an amount of money equal to the selling price.
- Any product which comes into the hands of the consumer with a given service has behind it a more or less complex multi-stage SC, relatively broad or deep, of which the consumer is a part and is also the driver.
- The same product can reach the end consumer through different SCs.
- The process is controlled and managed through different business functions of different supply chain's players, that interact each other to streamline the flow of goods through the supply chain to the end consumer.

The last activity, which finalizes the entire process, is the economic transaction, the ultimate end of the process and the only reason for its existence. On the one hand, the SC process is carried out by a series of players who, through a series of phases and individual activities, and using an aggregate of resources and controls, interact and provide the end consumer with a product and a service. On the other, the consumer attaches a value to this product and service; a value represented by the selling price which the consumer is willing to pay the players who made this process possible, and which represents the ultimate goal of the process and also the "reward" which, once all the costs have been covered, generates the profit which the various players then apportion.

The term *Demand Network* is often used instead of SC, precisely to underline two concepts: *consumer demand* is the trigger and driver of the macro-process, rather than this being an end-to-end process which pushes the product forward (supply), while it is the *network* which responds to the complexity of the network (depth and breadth) as well as of the process management, shifting from a "chain" concept to a "network" concept.

2 The Supply Chain—Different Perspectives

It is now possible to delve into the details of the supply chain, trying to describe it by highlighting how the process illustrated in the previous paragraph can be broken down.

The supply chain can be described in different ways, from different perspectives and viewpoints, each of which is useful to highlight miscellaneous aspects, and make some complementary considerations.

2.1 Physical Description: The SC as a Series of Players

A first way of describing the SC in its entirety, therefore, is to consider it as a series of players and physical structures involved in the generation of a flow which can satisfy the needs of an end customer, such as shops, distribution centres, production and service plants. It includes systems for storage and sorting, but also for handling and transportation, computerized networking and calculation and auto id technologies. This approach stresses the physical aspects underlying the system, with an emphasis on the hardware side. This approach is used to underline and design the structure of the network/chain, typically when it comes to aspects of *network design*, i.e., where to locate plants and warehouses, and what type of (i) production system facilities need to be set up (for the local or global market), and (ii) what warehouses need to be built (e.g. including storage spaces, distribution centres, and transit points). Once each single element of the network/chain has been defined, it is possible to identify the characteristics of the hardware for a more detailed design. This type of description highlights the more physical side, the one dedicated to satisfying customer demand.

2.2 Functional Description: The SC as a Set of Corporate Business Functions

The SC is made up of a series of players/infrastructures and hardware, but of course the whole process is organized and managed by *people*, in turn allocated to corporate business functions, e.g. Product Development and R&D, Marketing and Sales, Procurement, Production, Logistics and Distribution, Information Technology, Administration and Finance. All of these functions combine to bring the product into the hands of the end consumer. This was Michael Porter's typical description, using the "Value Chain" concept, which emphasizes the organizational aspect of the SC. In fact, this description is typically used for management purposes, to highlight the interactions between the various corporate functions and between one player and another. These functions, in fact, interact with one another, both within the same company (i.e. intra-company interactions between internal customers and internal suppliers) and with the functions of other companies (e.g., extra-company interactions between internal customers and external suppliers, and vice versa)., Therefore they need to be organized and orchestrated to streamline the process, since the way in which these functions interact has a profound impact on both the effectiveness and efficiency of the process, as explained below.

While efficiency and effectiveness may have a similar meaning in common language, differences are meaningful when applied to the supply chain process. The definitions are detailed below.

2.2.1   Effectiveness and Efficiency

We define SC effectiveness as the ability of the SC to deliver the product and service to the end consumer, i.e. to generate the required performance in terms of product and service, e.g. ability to take a fast-moving consumer good to the required place (on the shelf of the retail outlet), within the time required, in the right state, making it available to the consumer who goes to the retail outlet to shop (i.e., effectiveness of the distribution process, Fig. 4).

Efficiency is instead meant to be the ability to minimize the resources used and therefore the overall cost incurred. With reference to the example described before: how many resources were needed to put that packet of pasta on the shelf? What was the total cost—including shrinkage and wastage—that is, products that were non-compliant, rejects, had expired, were obsolete or left unsold—incurred by all the supply chain players to put that product on the shelf, as well as lost sales opportunities—due to inability to fulfil a customer request due to lack of product (out of stocks). The purchase price which the consumer will be willing to attach to the value of this process will need to cover this cost too, otherwise the process will run at a loss.

Efficiency is a process yield which compares the process output to the resources needed for that process. Therefore In the SC process we put as a numerator—the conforming product obtained and sold, represented by the overall output of the SC process minus the shrinkage (scraps and expired products) and the lost sales opportunities), while the sum—as a denominator—of all the resources used for each i activity carried out by each player j, in terms of labour/manpower L, energy E, consumables C, and equipment K (since equipment is a Capital Expenditure or CAPEX, it is necessary to consider the part of the depreciation of the capital goods.

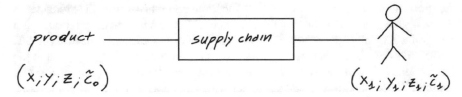

Fig. 4 SC processes: effectiveness understood as the ability to reach the end consumer with the right product and the right service (quantity, place, time, state, etc.)

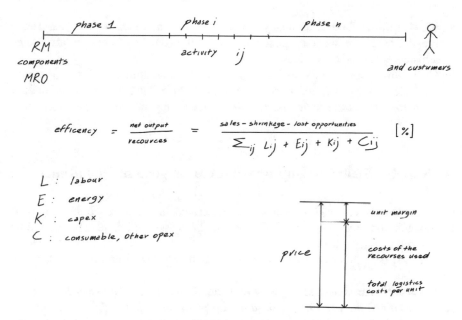

Fig. 5 SC processes: efficiency of the SC process optimizing resources used at parity of output

The efficiency of the process will be at a maximum if, at parity of output, the resources used are minimized, i.e., if the numerator is maximized for the same resources, e.g. if waste and lost opportunities can be minimized (Fig. 5).

2.3 Flow Description: The SC as a Set of Flows

Finally, the SC can be represented by focusing attention on the set of flows through it, to highlight what it is that "flows" inside the SC. These flows can be mono-directional, i.e. forward or backward and therefore from upstream (U/S) towards downstream (finished product FP), or from downstream (D/S) towards upstream (U/S) (returns). They can also be bi-directional (typical of SC), in other words, flows moving in both directions. In this specific case, the flows are multiple and can include:

- **Product flows**: typically forward-moving, these describe the physical flows of RM, Maintenance Repair and Operation (MRO) components, FP, and packaging, but also backward flows like returns, end of life products and assets.
- **Information flows**: a description of the information exchanged between the various players, which "triggers" product flows, and enables governance and management of the system geared to effectiveness and efficiency. Typical information flows include purchase orders (PO) and order confirmations (OC), transport documentation (TD), despatch advices (DESADV) receipt advices (RECADV), inventory, sell-in/sell-through/sell-out, and retail outlet point of sale data (POS).

- **Service flows**: these accompany the product and increase its value, e.g. range breadth, lead time, accuracy, punctuality, flexibility, delivery frequency, after-sales assistance, and other services.
- **Cash flows**: Between the end consumer and the penultimate SC ring (the only input flow to the SC), and from there towards the SC players upstream, both at the same level of the chain and at different levels, e.g. between a producer and a third party.

3   Supply Chain and Supply Chain Management Definitions

Once the supply chain process has been described in detail, some definitions of SC given in the literature, and which are also the subject of this paragraph, will be introduced and discussed.

The first definitions of supply chains are relatively recent and date back to the early 1990s.

In 1992, Martin Christopher gave a first definition of SC, as *"a network of organizations involved through links upstream and downstream, in the different processes and activities that produce value in the form of products and services for the end consumer."* (Christopher 1992). This definition emphasizes the concepts of a network rather than a chain, the processes of supply, production, and distribution which characterize the SC, and the ultimate goal of the supply chain, namely, the creation of value for the end consumer, the target of this process.

Again in 1992, Lee and Billington (1992), gave the following definition of SC: *"A supply chain is a network of facilities that procure raw materials, transform them into intermediate goods and then final products, and deliver the products to customers through a distribution system,"* in which we can note the emphasis on the physical process and the organizations involved in carrying out such a process, which will be discussed in detail in Chap. 2.

Ganeshan and Harrison (1992), defined SC in these terms. *"a network of facilities and distribution options that performs the functions of procurement of materials, transformation of these materials into intermediate and finished products, and the distribution of these finished products to customers."* The theme of the *network* contrasted with the concept of a *chain*, the three fundamental processes (procurement, operation and distribution), is also emphasized. Above all, for the first time, the concepts of the presence of different autonomous bodies (companies) are introduced, which contribute to this process and therefore share common objectives.

La Londe and Masters (1994), also emphasized the aggregation between independent companies in a single process, talking about SC as *"A set of interdependent firms that pass raw material forward, manufacturing materials into final products."*

It was towards the end of the'90s and the beginning of the 2000s that certain key concepts began to be added to the definition of SC. First and foremost, that of the information associated with the flow of products. Parunak and VanderBok (1998), talked about SC as "*a set of organizations through which material and associated information flow in producing a manufactured product and delivering it to the end user.*"

While Mentzer et al. (2001), went further in 2001 and defined the SC as "*a set of three or more companies directly linked by one or more of the upstream and downstream flows of products, services, finances, and information from a source to a customer,*" which summarizes many of the SC concepts described above, and introduces in particular the concepts of forward and backward flows which criss-cross the SC.

By piecing together the various definitions, it can therefore be concluded that an SC is made up of at least 3 players, one of whom—the one downstream—is the end consumer, the ultimate target of the whole process. At the opposite end is the source. In the middle is a more or less extensive network, with many or a few independent organizations (companies), united by the same objective, namely, creating value for the end consumer through a product and a service, and criss-crossed and interconnected by flows of products, information, services and monies.

Christopher's definition underlines that the goal of a supply chain is not merely related to the logistic concept of the forward flow of product and service but is also correlated to the flow of information and knowledge derived from each stage. Each process, activity and partners involved in the chain has to be correctly managed and integrated in all areas of the value chain to reach global efficiency and effectiveness of the system.

Supply Chain performance depends on effective and efficient management of information about inventory (raw materials, components, finished goods, and replenishment), assets (resource capacity, labour, and tools), and location of goods in transit and forecasted, actual end user-demand and movement of finances (Gale et al. 2009). It is precisely these definitions which will be used below for the description of the SC and the structuring of the whole discourse.

On the other hand, supply chain management (SCM) involves the integration of all key business processes across the whole chain of processes and stakeholders.

Mentzer et al. (2001), define a broader view of SCM as: "*the systemic, strategic coordination of the traditional business functions and the tactics across these business functions within a particular company and across businesses within the supply chain, for the purposes of improving the long-term performance of the individual companies and the supply chain as a whole.*"

In this sense, SCM represents a way for supply chains to manage their activities to extract maximum value for end users and stakeholders under four macro-process (material stock, inventory, distribution and information flow).

The process of delivering a product to an end customer involves monitoring and control of products from production facilities to retail stores. Production schedule streamlining, inventory management, pinpointing of bottlenecks, improvement of order response time and supply chain lead-time compression are among typical SCM goals. To achieve the above aims, businesses need to render processes more automated and have updated real-time data about processes and activities in the supply chain to be shared among the parties involved.

Chapter 2
Physical Representation of the Supply Chain

1 The Macro Areas of the Supply Chain

When we attempt to describe the SC using a physical representation, on the one hand we want to highlight the players involved, and on the other the physical structures with which these players operate.

Using black box modelling for (i) the players concerned (box), (ii) the product flows (solid arrows), and (iii) the information flows (broken arrows), the SC can be characterized as in Fig. 1. Each macro area features a certain number of levels which, in turn, are characterized by a certain number of players operating within it, who will be described in detail later in this chapter.

Notice that this representation is an exploded diagram of the first representation proposed in the introductory chapter, in which the system delimited there by Control Surface 1, that is the supply chain, can be split into three interconnected macro areas:

- **Procurement**: which includes all the players involved, from the initial raw materials supplier, to components, auxiliary materials (or MRO—maintenance repair and operations), consumables, and ending with the focal company that is usually in charge of operations.
- **Manufacturing/Operations**: manufacturing includes the players who manage the flows from raw materials to the finished product through semi-finished products (or WIP—Work in progress), and in a "traditional" SC such as that of consumer brand products, for example, this area generally corresponds to the focal company. A definition of the term 'focal company' will follow soon.
- **Distribution**: this is the part of the SC that has to do with the finished product, and in a "traditional" SC it ranges from the focal company to the final consumer, and includes all the intermediate players who contribute to delivering the finished product to the consumer.

A. Rizzi, *Supply Chain*,
https://doi.org/10.1007/978-3-030-95707-0_2

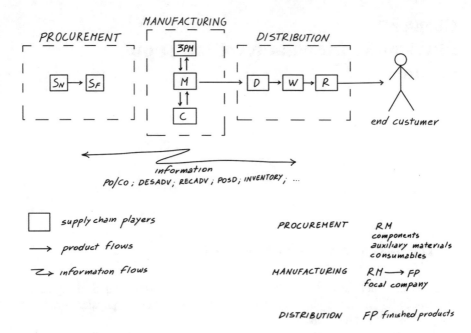

Fig. 1 A physical representation of the supply chain

1.1 Procurement

The Procurement part includes all the players who supply the manufacturer with the following types of supplies:

- **Raw Materials**: all those parts of the Bill of Materials (BOM) of the product subject to a transformation process (chemical and/or physical), and therefore no longer distinguishable in the finished product (e.g. flour for pasta and biscuits, tomato for vegetable preserves).
- **Components**: unlike raw materials, these are not subject to a transformation process but are processed by parts. Typically, these are processes to modify form and size (e.g. machining, chip removal) or involve assembly, and are easily identifiable in the finished product (e.g., the components of a telephone, a car, or the fabrics and accessories of a garment, such as the buttons, studs, tags and labels). This category also includes packaging, or at least the primary kind—used for a product.
- **Auxiliary materials/consumables**: including MROs. This category includes all the materials needed for the production process, which are not found in the finished product but without which the process could not have taken place. They are typically materials for maintenance and repair (e.g., detergents, lubricants, spare parts) but also operating materials that serve the production process (e.g. catalysts, paper, consumables for electronic equipment, labels, etc.), and secondary and tertiary packaging materials (cartons, pallets, stretch/shrink film, etc.).

1.1.1 Levels of Procurement

The players responsible for procurement are typically organized in levels. In fact, we speak of Tier I supplier, Tier II supplier, Tier III supplier, etc., to highlight the level of the supplier, and therefore its distance from the focal company and/or from the operations side.

The ith + 1-tier supplier is therefore a Tier 1 supplier to the i-th tier supplier, as shown in Fig. 2.

The more complex and deeper the SC, the more tiers are involved. Each of these (i-th level) features production plants, processing and service plants, handling and storage systems, which serve for the transformation of raw materials, components and MROs arriving from upstream (level $i + 1$) into finished products, which in turn are raw materials, components and MROs for the level downstream (level $i - 1$). Sometimes, if the i-th tier supplier is just a distributor, it simply stocks and handle products without any transformation.

Level I suppliers are those that interface directly with the focal company and supply it with the raw materials, components, and MROs necessary for operations, aimed at creating the finished product which is the object of the SC. For example, in the case of the FMCG (Fast Moving Consumer Goods) industry, which is detailed in ANNEX A, the first-level packaging supplier provides the primary packaging, the raw material supplier (mill) provides the flour; in the case of telephony manufacturers, the tier one supplier provides the Wi-Fi module or touchscreen of a phone; in the case of the automotive sector, the first level supplier provides the catalytic converter, the electronic control unit, or the cooling unit. In the case of the clothing sector, a Tier I supplier provides the accessories, fabrics or labels.

Upstream of these, however, Tier II suppliers operate, which are therefore first-level suppliers for the above Tier I suppliers. For example, in the case of the consumer goods industry, the second-tier packaging supplier delivers the paper for the production of primary packaging; that of raw materials, wheat for the production of flour (a mill); in the case of a telephone, the Tier II supplier provides the Tier I supplier with the electronic components to assemble the modules; in the case of cars, the first level supplier provides raw materials and components for the production of the dashboard

Fig. 2 The structure of the levels of procurement

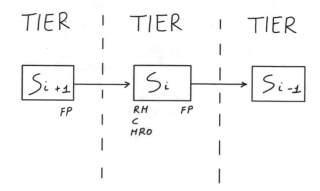

or the electronic control unit, or aluminium for the production of radiators; in the case of a garment, the Tier II supplier provides the yarn or inks for weaving and printing fabrics, and the paper or RFID tags for the labels.

And so on and so forth, throughout the depth of the supply chain for higher level suppliers, up to the nth-level supplier.

In Fig. 3, the block diagram shows the procurement process.

In Table 1 we can see once again the example of pasta production, listing some suppliers of the operations and their level in the SC. For each level, the supplier and the type of supply are identified, as well as the type of procurement for the focal company's operations.

It should be noted that each of these players is necessary and instrumental to the realization of the product by the manufacturer/focal company, and must therefore be arranged and managed within the SC process.

Fig. 3 The procurement process

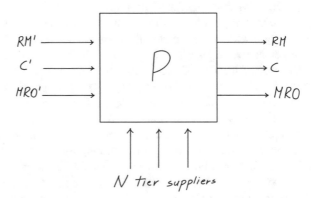

Table 1 The depth of the SC: example of a pasta factory

Tier III level	Supply	Tier II level	Supply	Tier I level	Supply	Operation
Seed industry	seeds	Farm	durum wheat	Mill	Semolina	RM
Forest industry	wood	Paper mill	carton	Box manufacturer	Paper packaging (boxes)	C
		Ink manufacturer	ink			
Chemical industry	Plastic in granules	Plastic film manufacturer	plastic	Packaging	Plastic film wrapping (bobbins)	C
		Ink and paint manufacturer	paints			
	wood	Paper mill	paper	Labeller	Labels	MRO

Table 2 Summary of the physical description of Procurement.

Procurement			
Players involved	Products involved	Structures	Objective
Tier I suppliers	RM, C, MRO	Facilities	Send the flow of RM, C, and MRO to the focal company in the mix, with a quantity and timeframe suitable to meet production schedules
Tier II suppliers		Production lines	
		Manufacturing lines	
Tier N suppliers		Assembly lines	

If, on the one hand, the increase in network complexity, expecially in terms of depth, increases the management complexity and the risk of system failure (disruption), on the other the system itself benefits in resilience from breath (i.e., its ability to absorb unexpected events), precisely because of the possibility of finding an alternative route in the event of a branch breaking down. In fact, the supply network can be compared to a hydraulic circuit or a vascular system, in which the interruption of one branch could compromise the ability of the circuit to provide the service downstream. However, if it is possible to activate a bypass and make the fluid flow through another branch (i.e., supplier and/or alternative process), the fluid is able to avoid the blocked branch and reach the users (i.e., the final consumer).

It is therefore possible to summarize the players, products, structures and objectives of the procurement area as in Table 2.

1.2   Operations

The part of the SC identified by the term Operations is that assigned to the transformation of raw materials, components and MROs, and to the production of the finished product which the distribution system will then deliver to the final consumer.

By describing the macro areas through the physical flows of materials, we can characterize as Operations that process which starts from the RM, Components (C), and MRO made available on the procurement side, and realizes the finished products, passing via a series of intermediate states, represented by semi-finished products (WIP = work-in-process) according to the process outlined in Fig. 4.

Fig. 4 The operations process

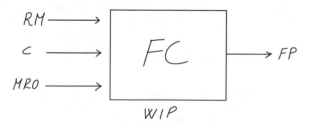

As mentioned in the previous paragraph, the transformation from raw material/component/MRO into a WIP and subsequently into the finished product can take place in different ways.

- assembly and machining:, i.e., a non-obligatory technological cycle, transformations using parts, machining and/or assembly operations. These transformations modify the form of the components, through a work-in-process (WIP), to transform them into FPs through machining and/or assembly lines.
- Process transformations, i.e., an obligatory technological cycle, and transformations that change the chemical-physical nature of the RM/WIP into an FP. We can think of the dairy sector where milk, through appropriate operations, is chemically and physically transformed into cheese, yoghurt, butter, cream, and other dairy products. These transformations take place by means of processing plants, and the raw materials are no longer recognizable in the finished product

Since the focus of this paragraph is a physical representation of the SC, the operations make use of:

- Production plants for the transformation of the form of RM into FPs, e.g., machining using machine tools. The RM will be irreversibly transformed into an FP.
- Assembly lines to assemble different components and semi-finished products in FP. From the FP it is possible to trace the parts it is composed of.
- Processing plants that perform a chemical-physical transformation into FP of RM, MRO, auxiliary materials, catalyst agents, and all the materials necessary for the transformation.
- Handling and storage systems, both for (i) RM, components, MRO, and (ii) WIP and (iii) finished products).
- Facilities, such as service systems (i.e. general system services) but also infrastructures (e.g. Information Technology IT, both hardware and software).

1.2.1 The Focal Company

Very often, the Operations part of the SC identifies the focal company of the SC itself.

The focal company is that subject, central to the whole SC, to which the consumer attributes the functional performance of the product and which consumers identify as the entire SC, regardless of the number and role of all the other players involved. The Focal company identifies itself in most cases with the owner of the product brand (e.g. Barilla, Coca Cola, Max Mara, Apple, DELL).

To identify the focal company of a SC it is therefore necessary to ask ourselves this question:

> Which company do consumers perceive as the SC, and attribute the functional performance of the product to?" Or: "Who do consumers blame in the case of a product that does not correspond to their expectations, especially from a functional or safety point of view, but in some cases also that of service?

The second aspect is particularly important to define the role and responsibility of the focal company (FC) within any SC. This becomes obvious in the case of a product being withdrawn from the market—the withdrawal does not involve the final consumer—but even more so in the case of a recall—which does involve the consumer. In the event of a recall, the FC is the one that suffers most from the related damage, even if this was not its responsibility. Consider, for example, the recall/withdrawal of products from the market in the automotive sector, or the withdrawal of essential products in the food sector, described respectively in the original articles in ANNEX A (Case 1 and Case 2 respectively):

- The recall in 2013 by Nestlé of Buitoni filled pasta because it had been contaminated with horse meat. In this case, the blame for the contamination fell on a supplier who delivered a batch of raw material (meat) containing traces of horsemeat, but the company that suffered the most damage was Nestlé, through a drop in sales of its Buitoni brand, which the consumer perceived as the focal company (repubblica 2013).
- Toyota recalled one million Corollas in 2014 for an airbag problem caused by a Tier II supplier. Also in this case the consumer perceived Toyota as the focal company and attributed the cause of the malfunction to the manufacturer (Sicurauto 2014).

Generally, as mentioned, the FC is the one that owns the brand and produces the product, especially in the case of consumer goods (see Fig. 5), but also in the automotive or electronics sectors (suffice to think of the smartphone industry).

In the FMCG industry, it is amazing to notice how few companies share the vast majority of brands, as depicted in (Fig. 5).

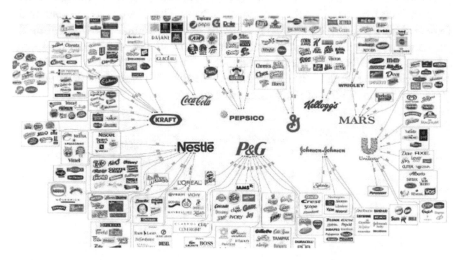

Fig. 5 The top 10 focal companies which own the main brands in the FMCG sector (Fast Moving Consumer Goods) from https://www.smartweek.it/ecco-le-10-aziende-piu-grandi-del-mondo/

1.2.2   Private Label—The Distributor's Brand

However, the focal company is not always the manufacturer of a product.

In many sectors, the FC is not the manufacturing company but the retail company. Think, e.g., of IKEA (see Fig. 6) or Amazon (for the latter see ANNEX C— (Distribuzione moderna 2019), which have become focal companies by exploiting their sales force to produce and distribute a private label product (e.g. Amazon Basics).

It would be more proper to speak of Private Labels.

In the FMCG market, branded products have also become extremely common. These are products, often similar to branded ones but usually of a slightly lower quality. These products feature the brand of the Organized Mass Distribution retailer and are distributed by it (e.g. Coop, Conad, Esselunga—Esselunga products but also Fidel and Smart) and sold at competitive prices.

In this way, companies can enhance their commercial strength through product positioning, also leveraging the reduction in research and development, and pooling of marketing and promotion costs. At the same the unit purchase price of raw materials (in reality, the purchase price of raw materials is also often lower since the focus is on economic convenience rather than on the functional or organoleptic characteristics of the product), distribution companies do not bear the costs for research and development of new products, in addition to the already mentioned marketing and promotion costs, even if in recent times some distributors have undertaken advertising campaigns for their branded products (e.g. Coop, Conad). Furthermore, with this mechanism, these players can obtain particularly favourable conditions for co-packers who work exclusively for them thanks to the volumes they can guarantee.

A mechanism of this type ensures a competitive advantage to the distributor who owns their own brand, since there are lower unit costs (thanks on the one hand to the above said factors, and on the other to the production volumes), and therefore the possibility of offering lower prices with the same margin, or gain higher margins at the same price. Ultimately, therefore, an increase in profit due to both an increase in the margin on the product (thanks to the reduction in unit costs) and sales volumes (thanks to the possibility of practising lower prices).

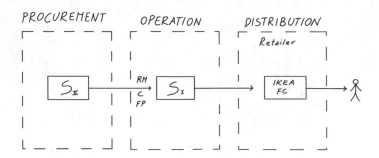

Fig. 6 Retailer's IKEA SC

The image in Fig. 7 highlights the price formulation mechanism and the higher margins in the private label case than in the case of the proprietary brand.

This advantage comes down to: (i) lower direct unit costs—thanks to the reduction of procurement costs thanks to slightly lower quality raw materials (ii) lower indirect unit costs, thanks to lower research and development costs and pooling of logistics and marketing costs. It follows that the Industrial Cost (CI) for a PL product is by far lower than the industrial cost of a branded product, and consequently the Unit Margin (UM) is greater for a PL product than for a branded one, despite the fact that the PL product is sold at a lower price. Since product price and placements are a main driver for FMCG, private labels products are always displayed on the front shelves at bargain prices, compared to branded ones, which means that the volumes (Q) of the PL product sold are often higher than those of the branded product, with a consequent increase in the Gross Operating Margin (GOM) of the private label compared to the branded product due to the combined effect of higher volumes and higher margins.

We talk about co-packers (FMCG), façon (fashion sector), or simply third-party production 3PM—Third-Party Manufacturers (furniture and durable goods sector), to identify those players who carry out the entire production cycle and supply the FC with the finished product. Suffice to think of the aforementioned co-packers who make branded products for OMD Organized Mass Distribution; but also the Chinese, Bangladeshi or Vietnamese façons of fashion or the ready-to-wear products of the districts of Carpi and Prato in Italy, which make all the products which are then resold by the main luxury houses; or to Eastern Europe or Italian suppliers who make furniture for IKEA or sofas for the Poltronesofà company, respectively.

Sometimes, companies that are FCs in an industry with their own brand become co-packers themselves for competing companies, a fairly common phenomenon in the FMCG industry. To cite one example, Conad branded milk is produced by Granarolo which is a premium brand for fresh milk, while coffee is produced by Lavazza (Verde Azzurro notizie 2018), and so on for many other product categories.

It is therefore possible to define three possible combinations of types of co-packer:

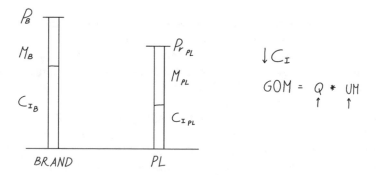

Fig. 7 Price, cost and margin for a branded product and for a distributor's brand

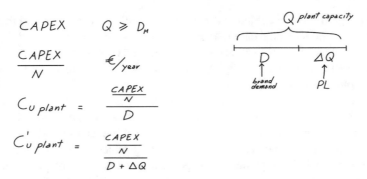

Fig. 8 The saturation of the production capacity of the plant as 3PM

- Manufacturers exclusively for a PL brand, who do not own branded products themselves
- Manufacturers not exclusive to the PL brand, who do not own branded products
- Non-exclusive manufacturers who do own branded products.

In relation to the latter case, the operation at first glance might seem a contradiction in terms. In reality, it is based on a strong economic base, as illustrated in Fig. 8. In fact, it may happen that a manufacturer of branded products ends up with unfilled production capacity, so that by increasing the production volume, it is possible to saturate the plant. This also lowers the unit cost of the branded product. Let's take as an example, a milk producer who owns a potential plant Q with a plant cost of C, amortized over n years. The annual amortization charge at rate 0 based on constant capital repayment is equal to $\frac{C}{n}$ and if D is the demand of the brand product, the share of the plant cost for each unit produced with the brand created is equal to $\frac{\frac{C}{n}}{D}$. Faced with a drop ΔD in demand D, there is an increase in the unit cost of the plant, which can only be lowered by saturating the residual production capacity with a private label PL, that is, by bringing the total production $D - \Delta D + PL$ closer to the Q.

In other words, decreasing the difference $\Delta D = Q - D$. Which is why companies that produce branded products decide to "rent out" a part of their production capacity and also to work as a co-packer for the production of private label products for Organized Mass Distributions (OMD) or discount stores.

Note the different roles of the same companies for different products. In this scenario, the producer is a FC for its own branded product and a co-packer for the private label product, while the OMD company is a FC for the private label product and a retailer for the branded product.

In some cases, a 3PM does not carry out the entire production cycle but only certain phases. This is very common in some sectors such as apparel and fashion (e.g., third parties that deal with cutting/weaving, assembly, dyeing, finishing, ironing, packaging), where there is therefore a continuous flow of semi-finished products between the FC and the façon (Fig. 9).

Fig. 9 3PMs in the textile sector: the façons

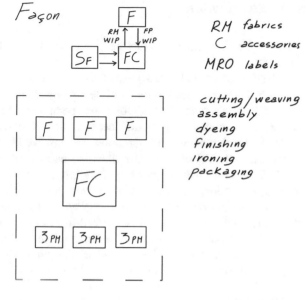

Fig. 10 3PMs in the metalworking sector: work on behalf of third parties

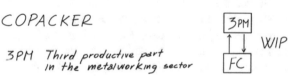

Also the metalworker uses this outsourcing process, where some phases of the production cycle are managed externally: for example, third parties who carry out particularly complex mechanical processing on behalf of third parties (sheet metal processing, plastic moulding, forging, additive manufacturing), with a continuous flow of plastic and metal bins of parts and WIP between the producer and the third-party production (Fig. 10).

It is therefore possible to summarize the players, products, structures and objectives of the Operations area as in Table 3.

Table 3 Summary of the physical description of operations

Operations			
Players involved	Products involved	Structures	Objective
Focal Company	RM, C, MRO	Facilities	Transform the RM/C into FP (1), according to the programmed production plans (2)
Co-packer	WIP	Production lines	
Façon	FP	Manufacturing lines	
3rd party manufacturer		Assembly lines	

1.3 Distribution

Finally, the physical infrastructure of the third area of the SC is described, the Distribution part.

The distribution part is the one that is studied in greatest detail, also from organizational and process points of view, while the organizational aspects of operations and procurement are typically the subject of operations management courses. However, further procurement insights and deeper information will be given in Chap. 3.

While Procurement and Operations were involved in a flow of materials that were subject to a change of state/form, the distribution part is responsible for a constant flow of FPs and their forms of aggregation, i.e., at the level of sales units and/or primary/secondary/tertiary packaging (transport units).

The Distribution part involves the players and related infrastructures that deal with physically and temporally transferring FPs from the production site at a given moment (e.g. Barilla production plant in Pedrignano, Parma, on day xx) to the place and moment of consumption, linked to the purchase by the final consumer (for example, the Coop Consumatori Nord-Est store in Via Sani, Reggio Emilia, on day $x + \Delta T$).

In fact, the production of goods and the request by a final consumer do not take place either in the same physical place or in the same instant in time, but are spatially and temporally out of phase. But for 3D printed objects that are spawned at the place and time of demand/consumption and therefore do not need a SC, Production occurs over time t_0 in a physical place (x_0, y_0), and demand also manifests itself over time t_1 in a physical place (x_1, y_1). The objective of the SC and in particular of the distribution part is precisely to bridge this gap and to deliver the product to the physical place at the instant in time requested by the consumer, aligning the SC and the Operations offering with the final consumer demand. All of which generates sufficient economies of scale both in procurement and production as well as in logistics (distribution side— transport, handling and storage) to ensure that costs are kept as low as possible and consequently the margins for the same prices are always as high as possible. A block diagram of the distribution part of the SC is shown in Fig. 11.

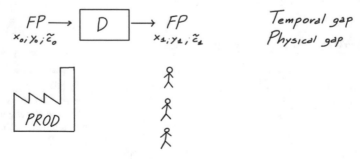

Fig. 11 Block diagram of the distribution process

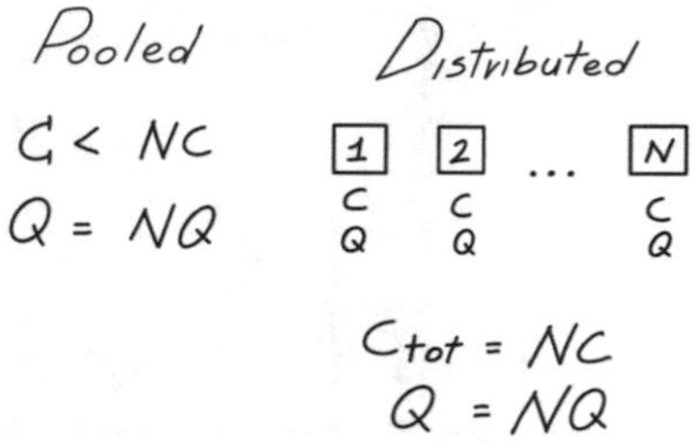

Fig. 12 Pooled versus
distributed
capacity—economies of
scale and scope

$$Pooled \qquad Distributed$$

$$C < NC \qquad \boxed{1} \ \boxed{2} \ \dots \ \boxed{N}$$

$$Q = NQ \qquad \begin{matrix} C \\ Q \end{matrix} \quad \begin{matrix} C \\ Q \end{matrix} \qquad \begin{matrix} C \\ Q \end{matrix}$$

$$C_{tot} = NC$$
$$Q = NQ$$

But why are the physical and temporal gaps created? The physical gap derives from the need to pool production in a single physical location in order to gain economies of scale and scope,[1] both in terms of infrastructure (i.e. a single plant, a single production line, a single location, a single plant of facilities) which will serve different markets, and in terms of operations, i.e. resources necessary for production (energy, manpower, operating costs). This means that the economy of scale concerns both investments in Capex [€] and operating costs in Opex [$\frac{€}{year}$] which are far less than proportional to the sharing of production capacity in N plants, how far depending on the specific industry (see Fig. 12).

Instead, the time gap derives from the need to produce in advance of the emergence of demand.

This is due to various causes, such as:

- Unavailability of the raw material, as depicted in Fig. 13, due in turn to:

  - Seasonal availability of the raw materials: it is necessary to produce in advance (campaign) to meet the annual demand (e.g. tomatoes for the canned vegetable industry).
  - Seasonal demand: it is produced over a medium-long time horizon to cope with a demand limited in time, i.e. recurring events (e.g. Christmas and Easter cakes, and ice cream, whose demand peaks in summer).

- Technological constraints, i.e., the inability to produce the product within the time required by the end customer, so that the product must be available in stock. For example, pasta must be on the supermarket shelf, it cannot be made to order and delivered to the consumer. In this case, it is possible to identify two further types of limits/constraints:

  - Constraints linked to cycle times: for example for the maturation or seasoning of food products.
  - Inability to produce Just In Time: the SC times are not compatible with the waiting times "granted" by the final consumer—in this case we can speak of a Lead Time gap LTG, as fully described in the relevant paragraphs of Chap. 3.

---

[1] We speak of economy of scale when there is a less than proportional increase in costs in relation to the production volumes of the same mix—A + A produced; economy of scope when the less than proportional increase in costs is linked to an increase in volumes but to a variation in the production mix—produced in addition to A B.

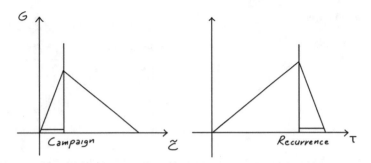

Fig. 13 Inventories in case of seasonal availability and demand

- Inability to carry out a continuous process: e.g., in the pharmaceutical industry, it is necessary to proceed with the validation of production batches, so that production cannot be Just in Time.
- Attempting to contain production and distribution costs through:
  - economies of scale or scope in procurement—linked to procurement volumes. More than instantaneous demand is procured to generate economical lots, in order to optimize reorder and stock maintenance costs, taking advantage of any possible quantity discounts.
  - economies of scale or scope in production—linked to volumes or the mix: productions are grouped into economical lots larger than the instant demand, reducing start-up waste but, on the other hand, incurring storage costs which are more than compensated.
  - economies of scale or scope in transport, Full Truckload (FTL) transport is preferred to dedicated Less Than Truckload (LTT) due to lower unit costs.

Similarly to what we have done with the procurement and operations part, we can therefore summarize the players, products, structures and objectives of the distributions area as in Table 4.

Table 4 Summary of the physical description of distribution

Distribution			
Players involved	Products involved	Structures	Objective
Distributor	FP	Transportation systems Warehouse	Bridging the time and space gap and delivering a product/service to the final consumer
Wholesaler		Distribution centres	
Retailer		Transit points	Generating economies of scale and scope in distribution and transport
3PL		Stores	

Since the focus of the discussion is physical, there are three categories of assets which the Distribution part uses to bridge the physical and temporal gap between production and consumption: goods transport systems, warehouse systems, and retail stores.

1.3.1 Transportation

The transport systems are responsible for the functionality of the distribution system to bridge the physical gap, making the FP available at one point or another

Transport systems are extensively described in industrial logistics classes, so here only a very brief mention will be made for completeness with respect to the discussion of the distribution system.

Different means and modes of transport are therefore used to bridge a physical distance:

- Sea—ship
- Air—cargo plane
- Rail—train
- Intermodality—containers, swap bodies
- Road—tractor-trailer truck, articulated lorry, van.

Each of these features different Lead Times LT, unit Costs (with full truckload saturation) C, and Flexibility F (in terms of point-to-point connection capacity).

The choice of one method over another depends on the availability of infrastructure and the Lead Time, as well as the distance to be covered, according to the rationale described in the following image (Fig. 14).

From a distance point of view, air and sea transport are typically suitable on international and transoceanic routes, where other modes of transport cannot be employed, and are therefore adopted to cope with very long distances. Rail and rail/road intermodality, on the other hand, cover continental transport, over medium to long distances. Finally, road transport is typically adopted locally for last mile deliveries, or on long-haul transport in the case of an impossibility to saturate rail

Fig. 14 Trends in costs, Lead times and distances for different modes of transport

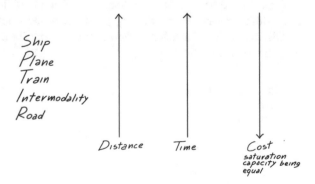

transport, with the need to protect a load by means of dedicated transport, or due to lack of infrastructure.

As for the unit cost of transport, the large cargo capacity of a ship means that this unit cost is particularly low. At the opposite extreme is road transport, since unit costs rise as the vehicle's load capacity decreases. The case of rail is intermediate. In the case of air transport, the absolute cost of air travel and the limited carrying capacity of a cargo plane mean that the unit cost is again high.

Finally, as regards Lead Times LT, on intercontinental routes, transport by ship is certainly the slowest, generally with LT of weeks or months, while air transport allows times in the order of days. At a continental level, road is certainly the preferable means of transport for fast deliveries, while the train is burdened by loading and unloading times on the one hand, and inadequate infrastructures on the other (both nodes and network infrastructures) that increase lead times. To overcome the first drawback, intermodal transport tends to be used, which consists in the use of two or more different modes of transport for the same load unit, typically a container for transport which also includes the mode by ship or a swap body in the case of rail and road transport only. For example, a garment made in China is loaded onto a container and taken to a port of origin and after travelling by ship, received at a port of destination. After import customs clearance—which in reality can take place in a bonded warehouse at any level even further downstream—the container is directly loaded onto a train to reach a freight village, where it is loaded onto a road vehicle to be delivered to a distribution centre.

Intermodal transport combines the benefits of rail on long journeys, reducing the preponderant part of the transport cost, with those of road for the first and last legs.

1.3.2 The Warehouse System

Distribution structures such as warehouses are used to bridge the time gap between operations and demand.

A warehouse can be represented by a block diagram (see Fig. 15), with instant f_{in} inbound flows and outbound flows f_{out}.

Gaps in flows are both temporal (inbound and outbound flows may occur at different times) and qualitative-quantitative (outbound flows f_{out} at the same instant t' are different from those entering f_{in} at the same instant).

Obviously, in a sufficiently broad horizon, the integral inbound and outbound must be the same:

Fig. 15 Conceptual representation of the warehouse system

Fig. 16 Time gap and inventory for a warehouse, distribution centre and transit point

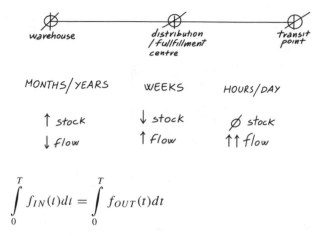

$$\int\limits_0^T f_{IN}(t)dt = \int\limits_0^T f_{OUT}(t)dt$$

As seen above, the time gap is generated by various contributing causes, e.g., limited availability of raw materials, limited duration of demand, technological constraints, and control of production and transport costs.

In relation to the time and mix/quantity phase shift gap, a fundamental distinction is made between warehouses, distribution centres DCs, and transit points TP or hubs (see Fig. 16).

While a warehouse mainly creates a time phase shift, and therefore the flow balancing horizon is extremely long (months, years), the transit point generate only a qualitative-quantitative phase shift between inbound and outbound flows, and the balancing horizon can be daily or even hourly. The distribution centre falls in the middle of the warehouse and transit point, and the balancing horizon is typically weeks/days.

From a process point of view, a warehouse is therefore characterized by a predominant storage process, while inbound/outbound flows are somewhat limited. Conversely, a transit point is characterized by zero stock and extremely significant receiving and shipping processes, so that its receiving, sorting and shipping capacity (flow) predominates.

The distribution centre DC is placed in an intermediate situation, in which there are both storage and inbound/outbound flows, and therefore both functions (warehousing and flow) must be ensured. Recently, in relation to the e-commerce model, the term *fulfilment centre* has been introduced for a DC, to emphasize the role of the warehouse system: satisfying customer orders through stock availability (warehousing) and order fulfilment capacity (flow).

The layouts of these structures (Table 5) also reflects their function. In a warehouse, the storage part is predominant, and the processes are limited to storage, receiving, and shipping.

A transit point has only receiving and shipping docks and a free area flow, since its function is that of sorting. The hub's processes are receiving, sorting, packing, and shipping.

Table 5 Layout, processes and typical infrastructure of a (**a**) warehouse, (**b**) distribution centre, and (**c**) transit point

Processes	Infrastructure	Processes	Infrastructure	Processes	Infrastructure

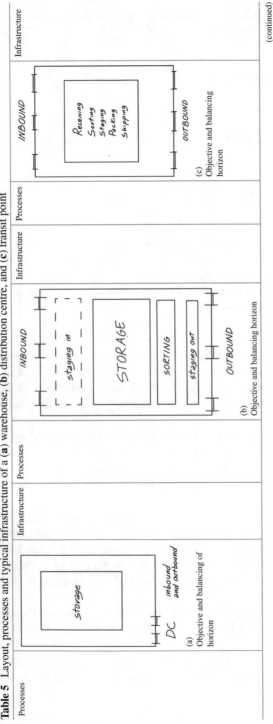

(a) Objective and balancing of inbound and outbound horizon

(b) Objective and balancing horizon

(c) Objective and balancing horizon

(continued)

Table 5 (continued)

Storage, limited receiving & shipping capacity	KEEPING the FP in stock years/months	Storage facilities, docks for loading and unloading (common)	Storage stowing flow receiving, picking/sorting packing shipping returns	To fulfil orders quickly; minimum stock to disconnect demand with supply weeks/days	(Many) loading and unloading docks (differentiated or common for in/out flows if they occur in different time slots), structures (and equipment) for storage and handling (e.g. shelving, sorters, roller conveyors)	Flow receiving, sorting packing, shipping	Sorting the flows in/out day/time	Loading and unloading docks (differentiated by in/out flows), structures (and equipment) for handling and picking

A distribution centre/fulfilment centre has both the part relating to inbound and outbound flows, and the storage part, since its function is the fulfilment of an order. The processes are receiving, storage (stowing), picking, retrieving, sorting, packing, shipping, returns.

1.3.3 Transit Point—Functionality

A transit point/hub can perform three functions:

1. Consolidation
2. Sorting
3. Mixed consolidation/sorting, i.e., cross docking.

The consolidation function, as depicted in Fig. 17, sees to grouping several incoming flows into one or a few outgoing flows. This structure is typical of collection platforms: n flows entering the transit point, very often in a multi-picking logic, are consolidated into a number of outgoing flows $n^* < n$, most often in a single one. With reference to the "express courier" function as an example, we can think of m incoming flows by vans, and $n^* < n$ outgoing flows with fully loaded heavy vehicles, i.e., FTL.

The sorting function (see Fig. 18) mirrors the consolidation function, and provides for both n^* incoming and $n > n^*$ outgoing flows, very often managed in a multi-drop logic.

With reference to these two functions, a TP can have a dual operation depending on the time window in which it operates. We can think of the operating logic of the "express courier" in the example above, whose platform typically performs the sorting function in the morning, and the consolidation function in the afternoon.

Fig. 17 Transit point—consolidation function

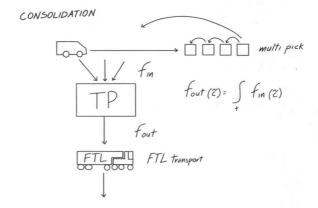

Fig. 18 Transit
point—sorting function

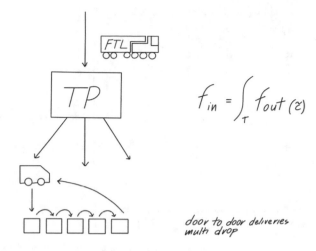

$$f_{in} = \int_T f_{out} (z)$$

door to door deliveries
multi drop

1.3.4 Transit Point—Cross Docking

The Cross-Docking Technique

There is also a third model, different from the case of the express courier seen above, in which the transit point combines both the consolidation and sorting functions. This is the case in which the transit point is used to apply the cross docking technique.

Cross docking is a distribution technique introduced by Wal Mart in the late 1990s to optimize the supply of FMCG stores. It will therefore be detailed with reference to the FMCG and food industries, where it is predominantly applied for fresh food distribution today.

We can think of a structure with M suppliers and N Retail Outlet RO (stores) to be served; the goal of cross docking is to supply the RO every day, guaranteeing fresh produce, therefore operating with zero stocks in both the DC and the RO, to reduce the number of trips and in any case carry out optimized FTL transportation.

Assuming a specific time reference ΔT (typically daily), each store RO_i, seen as the output of the TP function, issues product orders k to different (if not all) suppliers. Be S_j the generic supplier, and q_{ijk} the quantity of specific product k-th requested by the store RO_i from the j-th supplier S_j in the reference timeframe ΔT. The origin/destination matrix depicted in Fig. 19 can be fulfilled for every timeframe ΔT.

The totality of the quantities Q_i ordered by the individual RO_i, and the totality of the orders Q_j fulfilled by the single supplier S_j are calculated summing by rows and columns, as in Eqs. (1) and (2). The totality of the orders fulfilled, over a given programming horizon ΔT by the TP is given by Eq. 3 shown below.

$$Q_i = \sum_j q_{ij} \qquad (1)$$

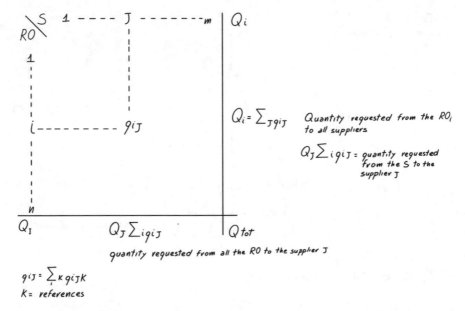

Fig. 19 Cross docking: origin destination matrix

$$Q_j = \sum_j q_{ij} \tag{2}$$

$$Q_{TOT_{T=\Delta T}} = \sum_{i=1}^{n} \sum_{j=1}^{m} \sum_{k=1}^{k} q_{ijk} \tag{3}$$

One of the major benefits of cross docking is the reduction of the number of daily trips between suppliers and stores. Absent cross docking (direct delivery) there would be a direct connection between N suppliers and M stores, for a total, if anything, of $V = N * M$ non-optimized trips, since a supplier delivers to each store only the products ordered and is therefore forced to use smaller vehicles in order not to incur the cost of LTT (Less Than Truckload) trips.

By applying the cross docking technique and then using the hub as a consolidation and sorting centre, we can have both N inbound and M outbound trips, and the total number of trips will be $V = N + M$. In this case, it will be possible to use vehicles with a greater capacity and therefore optimized costs, saturating the load capacity of the vehicle (optimized FTL journeys)., since each load groups together all the orders of all the RO for products provided by a generic supplier. The situation is summarized below (Fig. 20).

In addition to reducing and optimizing the number of trips compared to direct delivery, the application of the CD enables a further fundamental benefit. The RO can be replenished every day by zeroing the stocks both in the distribution system (the Transit Point fills and empties within a day) and on the RO (which keeps a maximum

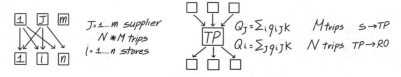

Fig. 20 Cross docking reduction of the number of trips from N × M low capacity trips to N + M high capacity ones

of one-two demand days as safety stock to protect against peak demand/non-delivery for one two days at most).

The cross docking process is described in relation to the case of fast moving consumer goods, the industry in which it was invented and where it is widely applied by the main OMDs for the daily distribution of all fresh and perishable products (dairy products and cured meats, fruit and vegetables, meat and fish, etc.) of which they do not wish to keep stock. The specific case of cross docking for express couriers and the related peculiarities will be dealt with in a dedicated paragraph. Obviously, the process is also used in other sectors (e.g., fashion, consumer electronics, spare parts for branches, etc.), where a smooth, stock-less flow of goods is required.

FMCG Cross Docking—The Processes

The cross docking processes in the case of FMCGs can be broken down into single phases as follows, following the temporal order in which they occur:

- the stores place orders on day A
- the suppliers fulfill orders on day A
- the hub/transit point receives, sorts, and sends out using cross-docking on day $A(or\,B)$
- the stores receive orders on day $B(or\,C)$.

The process Lead Time ΔT is therefore $\Delta T = A + 1 \div 2[gg]$; We talk about process (or orders) as $A \times B(or\,C)$. In some cases, the flow can be swifter, especially if the distances to be covered by transport are not particularly great, and can be managed within 48 h $A \times B$; orders are issued on the morning of A and are received in the afternoon/evening of B.

Going into the details of the individual phases of the processes described above, we have:

• At the customer—store RO_i (at A)	
(a) Collection of N orders	10.00 am
(b) M consolidations $Q_j = \sum_i q_{ijk}$, one for each supplier S_j	Noon

(continued)

(continued)

• At the customer—store RO_i (at A)	
(c) Issuing of the relevant order to the j-th supplier S_j	Before 2 pm

Day A

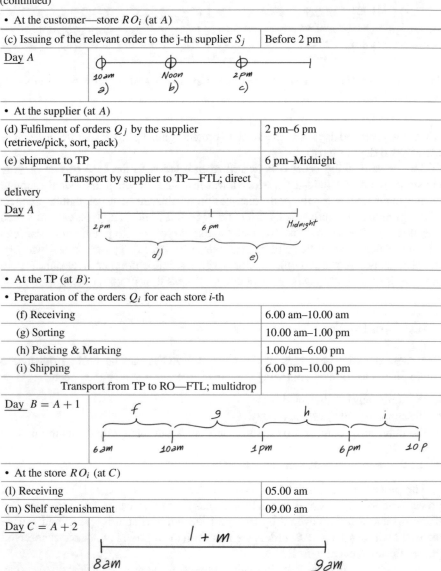

• At the supplier (at A)	
(d) Fulfilment of orders Q_j by the supplier (retrieve/pick, sort, pack)	2 pm–6 pm
(e) shipment to TP	6 pm–Midnight
Transport by supplier to TP—FTL; direct delivery	

Day A

• At the TP (at B):	
• Preparation of the orders Q_i for each store i-th	
(f) Receiving	6.00 am–10.00 am
(g) Sorting	10.00 am–1.00 pm
(h) Packing & Marking	1.00/am–6.00 pm
(i) Shipping	6.00 pm–10.00 pm
Transport from TP to RO—FTL; multidrop	

Day $B = A + 1$

• At the store RO_i (at C)	
(l) Receiving	05.00 am
(m) Shelf replenishment	09.00 am

Day $C = A + 2$

The individual activities/processes that take place at the transit point (processes f, g, h, and i) are described in detail in the following paragraphs.

Transit Point—Receiving

Receiving is the process of verifying the quantities of each product k (with respect to the complete mix K) required by all stores from the j-th supplier. In other words, each supplier S_j sends a certain subset of K references to fulfil the order of all the stores RO_i.

Therefore, (4) is valid, where the quantity q_{ijk} represents the product quantity k requested from the supplier j by the store i.

$$Q_j = \sum_i q_{ij} = \sum_i \sum_{k=1}^{k} x_{jk} * q_{ijk} \tag{4}$$

$x_{jk} = 1$ if the product k is provided by the supplier j
$x_{jk} = 0$ if the product k is not provided by the supplier j

Once it has been quantitatively detailed with (4), as will be seen in detail in the chapter relating to the representation of the SC through information flows and in particular in relation to the Bill of Lading (BoL) document, the receiving process can be described as follows, with reference to the black box of Fig. 21.

The three-way consistency check process is emphasized in Fig. 22, including:

- physical goods received (PG).
- the Bill of Lading (BoL):
- the purchase order (PO).

The operator receives the Bill of Lading/delivery note and enters it into the warehouse management system (WMS) which identifies the delivery, verifying the

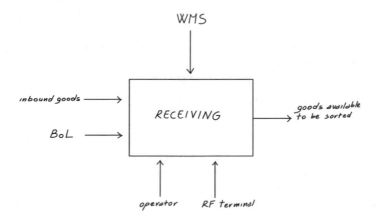

Fig. 21 The receiving process black box

Fig. 22 Receiving process: cross checking purchase order PO, bill of lading BoL and physical goods PG

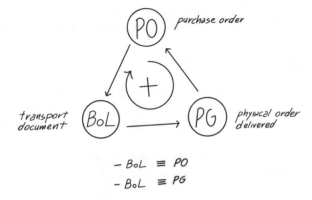

delivery note and the PO. For each delivery, there must be all the products K that represent the consolidated (4) of all the orders of all the N stores for that supplier j. Form this moment on, the Bill of Lading is therefore available in Radio Frequency (RF) to the warehouse operator in charge of receiving, who checks the mix and quantity (i.e., each reference in the right quantity), of the physical delivery compared to the BoL/PO, typically using a barcode reader (BC) or RFID technology.

Given the manual complexity of taking reads using a BC reader compared to the automation possible with RFID, the latter requires less labour for receiving activities, while ensuring greater accuracy. Readings with RFID technology are faster, fully automated, and do not require line of sight between the tags and the reader; if anything, they can take place concurrently with other activities if we think of an unloading area with gates/portals positioned at the entrance dock, where the pallet tags or packages can be read during the unloading of the vehicle. RFID pilots and deployments in the FMCG sector at RFID lab University of Parma have proven that unloading times with RFID can be reduced by up to 50% compared to BC technology.

If the cross check is positive, then we proceed with sorting, otherwise a non-compliance management process is begun. In both cases, the receiving process of the single order (i.e. incoming flow) and for all incoming flows, ends after verification of the mix/quantity for each product. The physical and information flows of the process are described in the swim line flow chart in Fig. 23.

At the end of the receiving process, the product is taken to a staging area to await sorting.

Transit Point—Sorting

Sorting is the activity of assigning goods to their respective destination stores, which takes place once goods have been received. Sorting can be of two types, (i) manual—typical in FMCG—or (ii) automated—more common in sectors where the goods to be handled are uniform in size, shape and weight (e.g., express couriers, textiles and clothing, e-commerce).

Fig. 23 Flow chart of receiving process using BC technology, and advantages of RFID technology versus BC technology

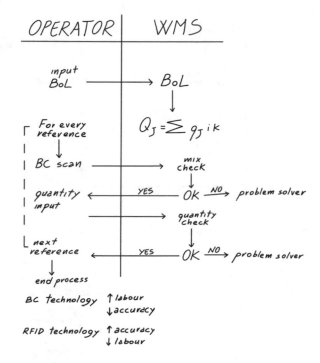

Manual Sorting

This includes a preparatory phase for single-reference Handling Unit (HU). In other words, the incoming flow Q_j received and loaded into the system is divided into K products, each with their relative quantities q_{ijk}.

Usually, this activity is already done upstream by the supplier who prepares the Handling Units (HUs) in single-reference layers/pallets (sometimes also single-lot to guarantee traceability) precisely to facilitate the sorting process at the transit point. During the receiving phase, each single-reference/mono lot layer is identified by the retailer with a unique serial (SN), linked to products, quantities and lots (in Fig. 24 for example, two single-reference single-lot pallets and are shown in A and B). After this phase, the sorting activity can be carried out.

The sorting area is divided by store (RO) and product category (PC) as depicted in Fig. 22. At each point a sorting pallet or roll is positioned (usually pallets are used for large surfaces—hypermarket or superstore; roll for small surface shops—supermarket), which therefore will contain all the products of a specific product category destined for a specific store ROi.

Each HU sorting pallet/roll is identified through a unique serial number (SN) ID (typically a BC/RFID label) which links it with it the information of all the mix and quantities q_{ijk} assigned to fulfil the orders of the specific RO_i. The radio frequency sorting process leveraging BC technology takes place according to the following swim line flow chart (Fig. 25).

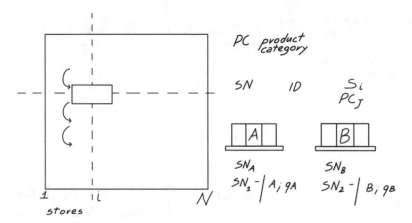

Fig. 24 Manual sorting

Fig. 25 The BC RF sorting
process

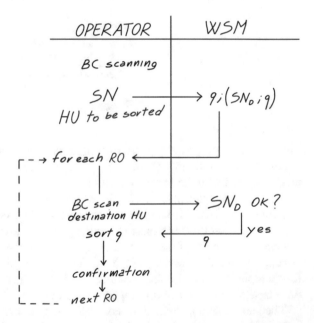

Operators first identify the pallet that they are sorting (by reading the HU SN via the radio terminal) and then the destination sorting HU (again reading the SN relating to the destination HU via the radio terminal), and the system informs them of the quantities to be sorted (typically secondary packages) for that pallet/roll to fulfil the ROi store order.

Therefore the resulting pallets/rolls are consolidated multi-product, multi-supplier, but single-destination. The destination HU is divided by product category, reflecting the store displays in each aisle, so that at the RO it can be taken directly to the aisle to restock the shelf.

The situation is depicted in Fig. 26.

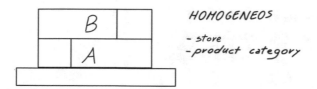

Fig. 26 Sorting process—preparation of the handling units by product category to facilitate the store replenishment

The layers themselves, aside from weight and crushing constraints, represent the sequence of references on the shelf, so that the store associate at the store minimizes travel times during the replenishment of a shelf, thanks to finding the products in sequence. In theory, the store associate can start from the beginning of the shelf and replenish the product on top, ending at the opposite side of the aisle, replenishing the last product on the bottom.

Automated Sorting

Automated sorting is typical of situations where the volumes to be processed are very high, i.e., hundreds of thousands of packages a day. Automated sorters are therefore used, whose investment costs are extremely high, from hundreds of thousands up to several million Euro; and also for this reason they are justified only in the face of very high and stable volumes.

Automatic sorters are not very popular for FMCG due to the extreme heterogeneity of the secondary level packaging to be handled, because of the type (packages, trays, bundles, etc. ...) and the materials, weights and dimensions. Conversely, they are widespread in the case of the clothing market (for hanging or folded garments), e-commerce retailers, or for express couriers, which deal with standard cardboard boxes and weights.

In analysing a sorter from a conceptual and functional point of view (see Fig. 27), it consists of: (i) one or more points of entry—infeed areas, in which the packages of the references to be sorted in the system are input (ii) n exits—out, unloading bays, one for each RO and each PC, from which the packages exit for the specific store for that product category PC.

In the case of an automatic sorter, it is no longer strictly necessary to have single-product pallets from the supplier, since it is the machine which carries out the sorting in an automated manner. The process is as follows.

The infeed operator identifies the product k (e.g., by reading the BC of the product and by manually entering the shelf life or lot numbers for traceability purposes) or by reading RFID tags—in which case it is not necessary to input neither the shelf life nor the lot since it is obtained thanks to the unique SN of the tag Electronic Product Code EPC. The products are moved using a materials handling system (e.g., roller conveyor, conveyor belt trays or tilt tray—trays which each carry a package and can swing from one side to the other). In this way, the system is informed of the package/tray combination, the latter uniquely identified by the system.

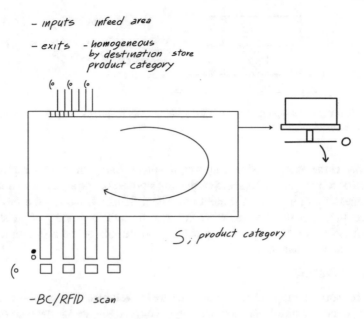

Fig. 27 Conceptual scheme of a sorter

The outputs of the sorter are dynamically linked to a store and a product category, so that the sorter decides how to sort the package introduced into the system according to the RO and the PC for which it is intended.

A highly automated process of this type guarantees high capacity and productivity of hundreds/thousands of packages per hour.

In addition, the operators' work is limited to supervision of the process, the release and loading of outgoing deliveries, or the handling of infeed packages, so that the use of manpower is reduced (one per infeed bay at most). The levels of accuracy that can be obtained using an automated sorter are also almost 100%, and are limited only by identification errors in the feed.

On the other hand, as already mentioned, the CAPEX associated with such investments is extremely high (hundreds of thousands or even millions of euros) so that it is necessary to precede the investment with a careful economic evaluation (ROI feasibility study). The fundamental parameter is the volumes and their stability over the medium to long term to allow the investment to be amortized. Low volumes do not saturate the capacity of the system with a consequent explosion of the unit costs of sorting linked to the CAPEX part.

In addition to such quantitative factors as hourly productivity, accuracy, the cost of labour employed and the higher cost of energy, generally speaking, qualitative factors such as flexibility must also be considered. On the whole, a manual system is more flexible in terms of volumes (just add staff), in terms of the variability of the packaging characteristics, and of the boundary conditions in general.

Figure 28 depicts the main quantitative and qualitative factors to be taken into account for the automated sorting feasibility study.

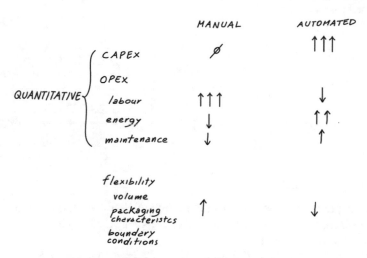

Fig. 28 The factors to consider in an ROI feasibility study

Transit Point—Packing/Marking

Once the references have been sorted, and, as a result, homogeneous multi-reference units have been prepared by store and product category, the next process is to pack them.

This process sees the load units in input (e.g. pallets/rolls) prepared by sorting (HU—handling unit), and the Shipping Units (SUs) in output, ready for shipment. The packing and marking process black box is shown in Fig. 29 below.

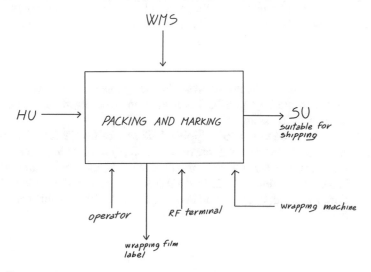

Fig. 29 Cross docking—packing and marking process black box

Fig. 30 The link between
SU, HUs and product
quantities

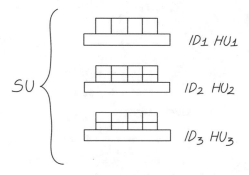

The process can include:

(i)   a sample or complete mix/quantity check of HUs prepared for stores;
(ii)   the formation of Shipping Units SUs by stacking the HU pallets to make them
       suitable for shipping;
(iii) a wrapping activity, with plastic film shrink wrap, to make the SUs stable; and
       finally.
(iv)   shipping unit labelling, using an SSCC (Serial Shipping Container Code),
       which links each SU to a serial number and therefore to the products and
       quantities it contains.

The phase to identify the SU is particularly important to reconstruct the (1) SU
$- (n)HU$ link in the information system, i.e. the Serial Shipping Container Code
(SSCC), and the ID of the HU serials. In fact, each SU is made up of one or more
HUs (multi-product), and each SU corresponds to a SSCC SN identifier (typically a
BC or RFID tag, consisting of a prefix which identifies the company plus the serial
of the SU), which uniquely identifies the particular shipping unit. In this way, it is
possible to use an SSCC code to trace the various HU codes of the sorting platforms,
and for each of these the mixed products and quantities q_{ik}, as shown in Fig. 30.

Transit Point—Shipping

The shipping phase closes the entire cross-docking process at the transit point, in
which each SU identified by an SSCC is loaded onto the means of transport and
shipped to the destination store.

This process is carried out in the shipping dock area of the TP (Fig. 31), divided
into (i) a loading bay in front for vehicle parking and (ii) the area to prepare the SUs
to be loaded (staging area). In the case of palletized SUs, the staging area typically
consists of 3 rows and 11 lines of standard 1200 × 800 mm wooden pallets (SSCC),
compatibly with the width (2.50 m) and depth (>11.3 m) of the loading trailer of the
truck—see Fig. 31.

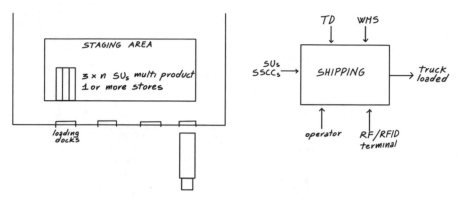

Fig. 31 Layout of shipping area and shipping process black box

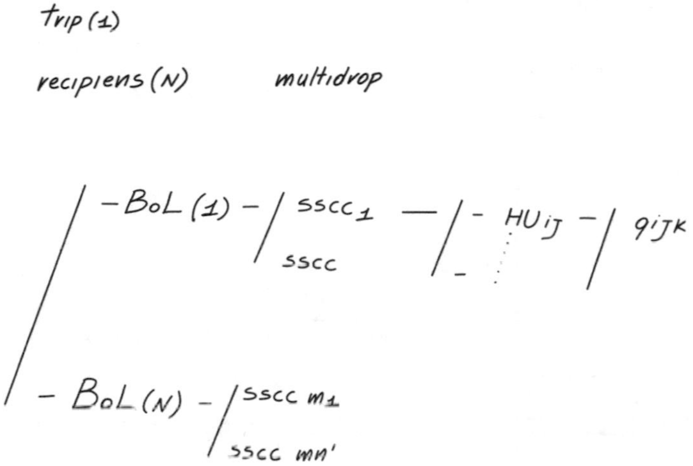

Fig. 32 Logical tree structure of a bordereau

The process flowchart is illustrated in Fig. 33 and involves taking charge of the n shipping SSCCs to be loaded on a specific dock door, associated with a carrier and n BoLs (Bill of Ladings) according to a specific bordereau, as represented in Fig. 32.

Each outgoing SU is verified by reading a BC or by passing it through an RFID gate, as shown in the flowchart of Fig. 33. Also in this case, the check involves a test block which is followed by a non-compliance management phase should the test not confirm a match between the SU loaded onto the vehicle (BC reading or RFID) and the SSCCs listed in the bordereau.

With reference to the shipping process described, further benefits of the use of RFID technology can be added to those introduced previously in the description of the receiving process, due to an increase in accuracy resulting from the elimination of inversion errors and the increase in productivity.

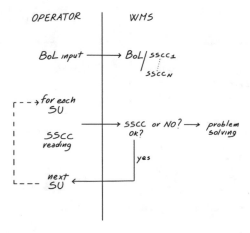

Fig. 33 Flowchart of the shipping process

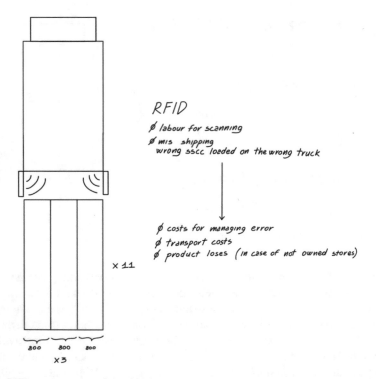

Fig. 34 RFID management of the shipping process

Let us consider a loading area with an RFID gates/portals for automated tag reading (Fig. 34). As far as accuracy goes, the first major advantage which RFID technology guarantees compared to BC technology is that of zeroing the inversion errors (i.e. the loading of an incorrect SSCC onto the wrong dock), since the phases to control the matching of the SU to the bordereau and the loading of the goods are done simultaneously. As an immediate consequence, the inversion errors are eliminated— meaning an incorrect SU loaded onto the wrong vehicle. These errors can instead occur with BC technology, when the check is usually made before loading, while during loading an operator can make a mistake and take a pallet from the adjacent staging area, not directed to that particular store, or overlook one.

Consequently, the related costs of managing inversion errors (i.e. returns) are annulled. But above all—on the store side—eliminating inversion errors means eliminating market mediation costs related to stockouts—both present and future. The lack of a product at a store due to lack of availability of the goods after an inversion error, in addition to generating immediate zeroing of sales due to the stockout, is interpreted by the store automatic reordering system as a reduction in demand and therefore gives rise to a reduction in forecast. Future orders and related sales are therefore negatively affected.

As far as productivity is concerned, a second major advantage of RFID is that of carrying out the identification operations while the trailer is unloaded, with the result that an operator does not need to physically take any BC readings, because the SSCC is read while the pallet jack/fork lift truck passes through a gate (see Fig. 34) to take the pallet to the staging area.

Time required to receive a full truckload of 33 pallets with RFID can be half of the respective time with BC technology.

Critical Issues of a Cross-Docking System

If, on the one hand, cross docking allows work within the logic of zero stocks and a reduction in the number of transports as well as an increase in the transport saturation, thanks to FTLs, on the other, the main critical issues of the process remain the following:

(i)   Need for synchronization of flows in transit: the lack of a purchase order can block the entire cross-docking flow. If it is decided to process and close the flows anyway without waiting for the goods, the day's orders plus the previous day's backlog can be sorted on the following day.

(ii)   Zero stocks: the lack of stocks at a store and in the TP in addition to representing a strength can also prove to be a weakness. Without stocks at the transit point, a non-delivery upstream by a supplier will cause unavailability of the references in all the stores with a consequent stockout, since the stocks are also minimal at the store (the system is indeed engineered for daily deliveries, or in any case at a high frequency, so that also at a store the stocks are minimal, if not zero).

(iii) Rigid timing: the time window in which to perform the activities is severely limited since all the activities are rigidly sequenced to guarantee the flow, and unforeseen circumstances or delays in a process (e.g. due to unavailability of labour or of the IT system) propagate in a cascade on all subsequent ones, without the possibility of being absorbed.

Cross Docking for an "Express Courier"

The cross docking technique is also adopted daily by express couriers—transport companies specializing in express shipments, such as DHL, UPS, Amazon Logistics, and FedEx.

In this case, the TP is configured as a consolidation/sorting hub, which sees the consolidated inbound by origin, and the sorted/sorted outbound by destination (Fig. 35).

Typically, the hubs of express couriers are located at airports, and the primary transport is by air (i.e. the covering of large distances in a short time). Air transport is justified from an economic point of view thanks to economies of scale; large volumes being handled which allow the cargo to become saturated and therefore split the cost of the means and reduce, or at any rate contain, the unit cost of delivery within acceptable limits.

$$C_{u,min} = \frac{C_{trailer}}{Q_{loaded}}$$

Road transport is typically at case level, not at pallet level, given that the truck trailer are saturated by volume, since an express courier normally handles light, bulky goods.

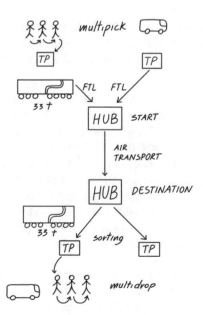

Fig. 35 Cross-docking scheme for an express courier

The main hubs are then connected to the local hubs (which are TPs operating locally) in the afternoon, in a logic of consolidating multipick deliveries, and the following morning, in a logic of sorting and preparation of multidrop deliveries.

The cycle is extremely fast, it takes place within a few hours at night, and is described in detail in Table 6.

At the local Transit Point (branch) of origin, the packages to be shipped are collected, in a multipick perspective, using small-capacity vans (3.5 tons). The packages are taken to a TP (branch) where the processes are typically manual, given the limited volumes of a local branch. These processes are:

- Receiving/Inbound
- Manual consolidation
- Shipping/Outbound.

The FTL long vehicles then travel to the departure hub, which operates nationally, where the processes are:

- Receiving/Inbound: At the hub, the inbound vehicle is directed to the receiving area. The package, identified by a unique serial number printed on the shipping label provided by the courier, is already present in the system and is identified by scanning the BC or by reading the RFID tag on the label. Also in this case, there is a decision block, so that any non-conformity of the label is appropriately managed, otherwise the package is conveyed to the sorting area.
- Automated sorting (at large Hubs): As mentioned, sorting is typically manual at small branches, while it is highly automated at the main hubs. In the case of

Table 6 Cross docking at an "Express Courier"; processes and timeline

Day	Time	Activities	Where
A	• Noon	Multipick collection	Sender
		Receiving Consolidation Shipping	TP origin
		FTL Road transport to Hub	
	• 20.00	Receiving/inbound	Hub origin
		Sorting	
		Shipping/outbound	
	• 24.00	Air transport	
	• 09.00	Receiving/inbound	Destination Hub
		Sorting Shipping/outbound	
	• 11.00	FTL Road transport	
		Receiving/inbound	TP destination
		Sorting	
	• 15.00	Multidrop delivery	Recipient

manual sorting, an operator simply sorts the packages received by shipping dock, according to the outbound routes. In the case of automated sorting, the operator loads the packages onto a sorting system, and the sorter automatically directs the packages directly to the appropriate output and thus shipping dock, again using the unique BC RFID serial on the shipping label.

- Shipping/Outbound: Each sorted package is finally loaded onto the (shipping) vehicle using extendible roller conveyors which enter the loading compartment to bring the packages from the sorter outputs to the operators loading them inside the trailer. Before being loaded, the package can be read again by a BC scan or RFID read of the SN, confirming the correct exit position, and updating the event real time on the courier web tracking portal.

At this point, the main route (e.g. air transport) is completed, to arrive at the destination hub where the processes duplicate those of the departure one, and therefore:

- Receiving/Inbound
- Sorting
- Shipping/Outbound.

The loads thus consolidated by destination arrive at the local hub where single parcels are manually sorted (mirroring what happens at the departure hub) to be delivered the next morning by means of multidrop van deliveries. Consequently, the local hub works in the afternoon as a consolidation centre for deliveries departing from the hub, and in the morning as a sorting centre for deliveries arriving from the main hub.

1.3.5 Distribution Centre

An intermediate structure between a transit point and a warehouse is the distribution centre, at which there is both the time phase shift of inbound/outbound flows of the warehouse system (the balancing horizon is in any case limited to weeks), and the mix/quantity phase shift of inbound/outbound flows as a typical transit point.

Unlike the transit point, a distribution centre holds a minimal stock (a few days/weeks of demand, depending on the shelf life of the specific reference), which allows for decoupling of demand and supply, without having to work with a "rigid flow" as in the case of a transit point.

The typical processes and flows of a distribution centre are shown in Fig. 36. The flows are therefore: cross docking (blue), retrieving of full pallets (yellow), picking of cases (order—violet, and batch—green); the processes are receiving, inbound staging, cross docking (full pallets), storage, picking, sorting, packing & marking, outbound staging, shipping.

The process that all flows pass through is the *receiving process*. The receiving process is conceptually similar to that seen both at the warehouse or TP, and therefore the process by which the inbound goods are checked and uploaded into the system.

Fig. 36 The processes and flows of a distribution centre

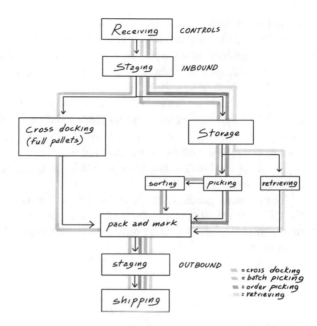

Only after the receiving process are the goods available for subsequent processes. During receiving, there is a three-way cross-check, between physical, transport documentation, and purchase order. As will be seen in detail below (in the chapter relating to information flows), receiving can take place in different ways: from manual to fully automated (if a transport document in electronic format is available, the so-called 'despatching advice' or "advance shipping notification").

In any case, the goods received are placed in a staging area awaiting subsequent processing.

A first flow is that of cross-docking (pale blue flow). In fact, the Distribution Centre can also operate in a Transit Point logic. This occurs in the case of loading units already prepared and destined for a store, which can therefore be transferred in a flow logic directly to the shipping docks to be forwarded.

Normally, however, the load units received are transferred to the storage area for stowing. These can be *stacked warehouses*, in the case of loading units not subject to picking or, more typically, ones which feature double-sided shelving in which the zero level of the shelving is used for picking, while the upper levels are used to store entire pallets.

Starting from the storage, there are three possible flows: retrieving of whole pallets (*yellow flow* in the figure.) These are withdrawals of whole pallets to fulfil an order from a store. Generally, these will be high-rotating products, or products sold in secondary packages—for example, mineral water or paper, and which can then be sorted by whole pallets. Promotion orders are also retrievable. In this case, given the high sales volumes expected in the store following a promo, the picking mission consists of an entire pallet, which is picked up directly in the storage area to be taken to packing & marking (if it requires a logistics label), or directly to outbound staging if the logistics label is already present (most likely).

For all products that are picked, it is possible to adopt either order picking (*purple flow*) or batch picking policies, the latter requiring a downstream sorting process, typically manual (*green flow*).

Once the packages have been picked, and sorted, it is necessary to check and consolidate the load units to make them suitable for shipment by means of the packing & marking process. The processes in these phases take place exactly as seen in detail in the case of the transit point.

The Shipping Units thus formed are left in a staging area facing the shipping dock, awaiting shipment.

The shipping process completes the Distribution Centre cycle and takes place at the DC shipping dock doors. It consists in loading the shipping units onto the truck in accordance with a loading bordereau, as thoroughly described in the transit point paragraph.

1.3.6   Stores

The distribution part is now complete, and to all intents and purposes the retail infrastructures, i.e. the stores, are a part of it as the last point of the distribution chain.

Stores typically fall within the retail or sales business function, even if they are actually the pivotal element of the distribution system, and therefore the supply chain. The store is a fundamental element of the SC, since it represents the interface with the final consumer, in which the product itself is brought to the final consumer in a physical place and at the instant when the consumer is disposed to purchase (at least in the traditional function of a physical brick & mortar store). This moment, therefore, closes and finalizes the entire SC process. The centrality of this link in the chain is twofold, since (i) it finalizes the process, to the extent that all the players have made the product and service available so that the consumer can complete his/her purchasing process, and (ii) it puts the final consumer in the physical condition— product and consumer in the same place when the need arises—and in the mental and psychological predisposition (the importance of the store in terms of generating a shopping experience we shall examine in detail in the following paragraphs) to make his/her purchase, and with this to remunerate and close the circle of the SC process.

To fulfil the core sales function, a store is typically divided into two areas (see Fig. 37):

- The sales floor where the product is visible and available to be purchased, but also to be tried. Suffice to think of large FMCG retail store areas. The sales floor is divided into departments, containing self-service shelves and assisted sales counters which the consumer moves between. In this area it is therefore essential that the product is displayed on the shelves, on the headboards, and in the promo

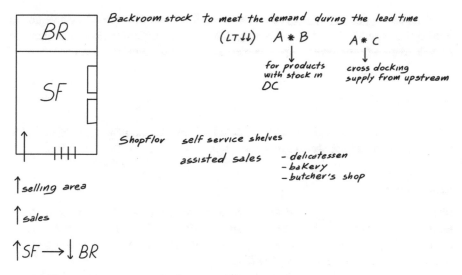

Fig. 37 The two areas of the store—backroom and sales floor

areas, and is available to be picked up and placed in the consumer's shopping trolley or basket. In the case of apparel and fashion stores, the function of the sales floor is slightly different since the sales are generally assisted by sales assistants. The sales floor typically functions as a showcase in which there is a certain stock of products, e.g. only one model/variant, while all the other variants/sizes are in the back room and the product sold is the one kept in the backroom, unless unavailable.

- The back room: which in OMD typically performs the function of inventories; in this way there is no need to restock every day (however, for fresh products this is actually the case), so that the store can be restocked less frequently (e.g., two or three times a week) keeping a minimum of stock available at any one time. reorder time depends on the nature of the upstream distribution point (DC or cross-docking). The backroom area (BR) inventory can also absorb unexpected peaks in demand during the delivery lead time, linked to, e.g., meteorological phenomena. In the case of apparel and clothing, on the other hand, the backroom fulfils the task of meeting the customer's order while availability in the sales area is only exceptionally affected.

Retail often uses the Sales per Square Meter as a reference KPI (Key Performance Indicator) for a store:

$$KPI = \frac{sales}{total\ square\ meters} \frac{[\texteuro]}{[m_{SF}^2 + m_{BR}^2]}$$

Since there is a linear correlation between the surface of the sales floor (SF) and the turnover of the store, in order to increase the sales per m², stores try to maximize the SF surface at the expense of that of the BR.

In addition, due to factors such as an increase in surface costs and inventories holding costs, stores tend to increasingly downsize the backroom area more than the sales area in order to have a larger display area and ensure greater turnover.

Store Models—Apparel

Before analysing in detail the store models for FMCGs, the possible models in the case of fashion and apparel are touched upon.

Retail Stores

Retail stores are chain-owned stores located on the up-market streets of large metropolitan cities (flagship stores) or in town centres. They have a small surface area (ranging from 100 to 1.000 m^2), a wide assortment (more or less the whole range for the season—especially the flagships).

Retail stores are owned by a chain with which they are vertically integrated in a so-called "closed loop" supply chain. In vertically integrated SCs, the brand owner also owns the store, manages the leases, owns the equipment and IT infrastructure, and the store staff are waged employees. The brand owner decides on the assortments, stock levels, visual merchandising policies (what to display and how), restocking and discount sale policies, and returns policies. Point of Sales Data (POS Data or POSD) are available in real time, to manage replenishment and other value chain processes.

Wholesale Stores

These are non-owned stores, where the customer can still find items from the current collection, as in retail stores.

They may be single-brand franchises, or small multi-brand stores managed entirely by third parties.

In the first case, a franchisor, usually the owner of the trademark, exclusively grants an independent entrepreneur (franchisee) the right to sell the products and exploit the trademark for commercial purposes. The franchisor asks the franchisee for an economic consideration, the so-called *entrance fee*, upon signing the contract, and royalties, since a package is provided together with the product to be marketed that may include other products or services such as furniture, IT infrastructure, transport and storage of goods by third parties, plus return of unsold garments at the end of the season. The franchisee must comply with the management and marketing standards and models (e.g. prices, discount sales) which are decided by the franchisor, but is independent, within certain limits, in deciding what and how much to order for each season and in establishing the supply schedules. The staff are waged employees of the franchisee.

As for their size, they can range from large wholesale chains (*Saks Fifth Avenue* or *Macy's*, *La Rinascente* in Italy) to a small shop in a single town or city. In the first case, they are usually present within the chain of owned or franchised corners.

The presence of different brands all in the same store operates as a lever to attract more customers to that store. POS Data can be both shared or not, depending on the business deal.

In the case of a small multi-brand shop, on the other hand, the owner of the shop is an independent entrepreneur, who negotiates the assortment with the brand at the beginning of the season, and therefore which models/variants/sizes to buy, as well as the delivery schedules. In this case, there is/there is not obligation at the end of the season on the part of the owner of the brand to collect the unsold items, which are usually sold off to stockists. The shop owner employs his own staff, and is completely independent from the point of view of the choice of equipment, IT systems, services, and so on. Usually, POS Data are not shared.

Outlets

Outlets were born towards the end of the 1990s. They are typically medium/large stores where the previous season's returns (wholesale or retail) are sold off.

They are usually located in fashion villages. The presence of different stores and brands in the same place acts as a catalyst to attract the final consumer, who travels to find dozens or hundreds of fashion stores in a single physical location, as well as services such as bars and restaurants, and playgrounds and nurseries for children.

As in the case of retail, these are owned stores, medium to large, located in structures where a real estate entrepreneur makes the investment to create the outlet pole and rents the walls of the shop to the owners of the brands. Precisely due to an outlet's ability to attract an extremely high flow of consumers and consequently to generate high turnover (at least in a pre-Covid scenario), rents are particularly high. Many brands that enjoy a strong reputation have therefore decided to move their outlet store outside the fashion district area, into large commercial spaces usually located on the main thoroughfares or close to expressways, easily accessible but with much lower rental costs.

The distribution scheme of an outlet channel is represented in Fig. 38.

Garments from retail and wholesale stores unsold at the end of the season are returned and sent to the DC. Normally, the sell-out part represents 55–65% of the store's sell-in, therefore the return flows are about 35–40% of the related inflows to retail and wholesale. The following processes are performed at an outlet DC:

- Receiving—physical check expected returns
- Quality control: the condition of the garment, the presence of tags, accessories, and everything necessary for marketing are checked
- Conditioning: a series of steps necessary to make the garment suitable for marketing; this may be done internally or by third parties

  - Washing
  - Ironing
  - Unbranding and labelling certain brands: removing the brand labels and rebranding the product with a neutral name
  - Labelling

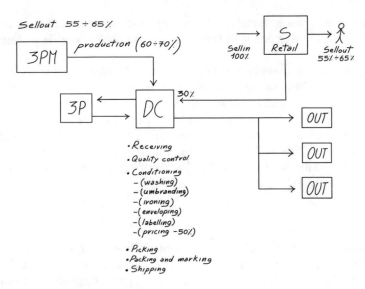

Fig. 38 Distribution scheme of an outlet channel

- Pricing—normally the outlet price offers a discount of 50% upwards compared to the retail store price

- Picking
- Packing & Marking
- Shipping.

Alongside the return flows of the previous season, the outlet channel also sells production garments, made by façons, and necessary to complete the assortment and/or fill the gaps in sizes. The volume ratio may be around 60–40%.

FMCG—Store Models

There are basically 4 +1 store models in the fast moving consumer goods sector, which differ in (i) size of the store and subdivision of the back-store area, (ii) number of product categories and references by category, (iii) stock level:

(i)   Hypermarket
(ii)   Superstore
(iii) Supermarket
(iv)   Traditional Proximity Shops/convenience stores
(v)   Automated Proximity Shops.

Hypermarket

In Itay, the hypermarket model was imported by *Iper, La grande* from the United States in the late 1970s, in view of the success of the Wal Mart superstores. This model was particularly in vogue in the late 1990s until the early 2000s

The hypermarket is a store characterized by an extremely wide range of references (up to 100,000 references or more) divided between a large number of departments and product categories (food, non-food, health and beauty care, multimedia, bazar and homeware, car, pharmaceutical, etc.). The sales model is mainly self-service, even if the sales counters are practically all assisted (butcher, fishmonger, bakery, delicatessen, pharmacy, multimedia, etc.)

The surface areas are also significant ($10,000 \text{ m}^2$ or even more), divided between a large sales area and a particularly important backroom area, as is the amount of inventories present in the reserve, which ensures coverage for several days. The hypermarket was in fact designed to be an autonomous system, with direct deliveries from manufacturers, and able to operate for several days even without fresh supplies.

For each product category, the number of references in the assortment is exceedingly high (20–30 references) so that the customer can find a wide choice for each kind of product. On the contrary, customers go to the hypermarket because they know that there they will find everything they need, with a wide assortment, and also that they can buy enough items to cover the needs of the whole family for a week or more. In the original hypermarket model, the prices were also the lowest, with extensive promotions advertised through leaflets, precisely to stimulate consumers to use a car to make their purchases, and once at the store to buy large quantities that could cover their needs for several days.

A hypermarket is typically located within a shopping centre in a large town or a metropolitan city, where consumers can optimize travel times and combine shopping with family leisure; they go to the shopping centre as a family where they can find other kinds of shop to make other purchases and while away an entire afternoon. The owner of the hypermarket is the name brand (Coop, Auchan, Finiper), while the staff (a few hundred people are needed to run a large hypermarket) are waged employees.

A hypermarket is in all respects a medium-large company with a turnover of hundreds of millions a year and employing hundreds of people.

Supermarket

A diametrically opposite model is that of the supermarket.

The supermarket is characterized by a location in the centre of a town or city, serving a limited number of people represented at most by a small town or a neighbourhood. The supermarket display area is limited (a few hundred m^2) while the backroom area is practically non-existent.

In fact, the supermarket model envisages daily supplies from the DC, while at the store itself there are very few days' demand.

The assortment of a supermarket is extremely limited, 10–15 k references, as is the number of categories and departments. The range is a few references per product:

a premium brand, a private label (PL) and 1–2 additional brands. In a supermarket there is space for food, very few non-food products (typically health and beauty) and some housewares.

Sales are typically self-service with one or at most two counters (i.e., butcher and baker) for assisted sales.

A supermarket is not characterized by any additional service, not even a car park if the supermarket is located in an urban centre. The consumer goes to the supermarket near home for small daily purchases, mainly fresh and short shelf life products, and is willing to spend a little more for a proximity service and to have fresh produce every day, in quantities commensurate with their own consumption.

The supermarket may be owned by a name brand or managed as a franchise by a franchisee, as is the case of Conad or CRAI or Selex, in which the members of the cooperative are the owners of the stores, while Conad itself is a consortium that deals with the purchasing and distribution. CRAI and Selex act as a purchasing consortium, to negotiate purchase prices for all associates.

Some supermarket offer a home delivery service. Consumer buys products and the store offers a home delivery service, so that the customer does not need to carry bulky products.

Superstore

Lying between the hypermarket model and that of the supermarket is a third model, the one known as a superstore. Examples of superstores are the Esselunga or Conad retail outlets whose signs bear the word "Superstore".

The superstore represents an intermediate solution between the hypermarket and supermarket. It is typically located in a populous town or quarter, and serves the neighbourhood at large. It is surrounded by some basic services, e.g. a pharmacy, bar, and parking area, because those who go to a Superstore to shop typically use their car. However, a superstore has small surface areas in the order of a few thousand square metres, divided between the sales area and the backroom area, which guarantees coverage of a few days' demand.

The superstore is stocked at least weekly for dry goods, and daily or at most 3–4 times a week for fresh goods. The assortment is large, superior to that of a supermarket, but not as broad as the hypermarket. The product categories are food, non-food, health and beauty care, and very little else (some consumer electronics references, little homeware).

The sales model is a mix of self-service—predominantly—and assisted sales. In a superstore there is usually space for a butcher, fishmonger, delicatessen and baker. Typically, in a superstore there will be 30–40 k references, divided by category into 1 ÷ 2 market leader brands, one or two private labels of different ranges (e.g. premium and intermediate),. The assortment is completed by other 2 ÷ 5 intermediate brands depending on the product categories.

The superstore is at the service of consumers who enter to make a purchase that can cover the needs of a few days, at most a week. They are willing to use their car to take advantage of medium–low prices, both continuous and promos, since a superstore has prices virtually in line with a hypermarket and makes extensive use of promos.

The consumers benefit from some minimal ancillary services (bar, pharmacy, laundry) but are not disposed to waste time stopping at the store, unlike the case of a hypermarket to visit a shopping centre. In the case of a superstore, the outlet may be owned or franchised.

Introduced into Italy by Esselunga and Conad, this is the model which, in recent years, has given the greatest results in terms of growth, turnover and margins, and for this reason it is the one which is gradually spreading to other brands as well.

Traditional Proximity Shop

Finally, the proximity shop, still very popular in Italy, especially in the centre-south and in suburban areas, and in any case still representing an extremely significant slice of FMCG retail—in Italy at least.

The proximity shop is family run, with an extremely small surface area, in the order of dozens or a few hundred m², and an extremely limited assortment, a few thousand references at most.

For dry goods, supply is organized directly by the owner, who visits wholesalers to buy the products in cartons, while for fresh goods the proximity shop is typically served by distributors using the "attempted sale" model.

The *attempted sale* operation means that the self-service counter is replenished daily by a wholesaler/distributor by a refrigerated vehicle, which first takes care of the inventory and the collection of returns/expired goods; based on the previous day's stock, by looking at the difference, the shop's sales are calculated for each reference. Known as "t", a typical day involves the following relationship:

$$\text{sold } (t) = \text{inventory } (t - 1) - \text{inventory } (t) - \text{returned } (t) + \text{replenished } (t).$$

The main service of a proximity shop is represented, as the name itself says, by its proximity. Typically, consumers go to a traditional proximity shop on foot, to buy products for daily consumption. The sale is typically assisted by the owner who follows the customer in all phases of the transaction.

Proximity shop usually offer a home delivery service for loyal customer.

FMCG Store Models—Conclusions

The following table summarizes the main characteristics of the various store models analysed by type (see Table 7).

The Automated Proximity Shop

Lastly, we have the automated proximity shop, which represents the evolution of the traditional proximity shop.

Table 7 Organized mass distribution summary

Type	Total surface area and characteristics	References	Range	Package Proposal
Hypermarket *BR* *SF*	10,000 m² significant reserve weekly restocking shopping centre with car park and several stores	+100,000	↑↑ +15 brands per reference	Food, non-food, bazaar, homeware, car, household appliances, hobby items, multimedia, pharmacy Self-service assisted sale
Superstore *BR* *SF*	5,000 m² limited reserve ↓ stock ↑ frequency of restocking car park and some ancillary services (tobacconist, bar, pharmacy, laundry)	50,000	↑ 8–10 brands 1 premium 1 ÷ 2PL (one PL and one generic brand) 4 ÷ 5 intermediate	Food, non-food, health and beauty care; housewares some electronic references (in promo) Self-service limited assisted sales
Supermarket *BR↓* *SF*	100 ÷ 1,000 m² reserve → 0; daily restocking only SF; limited BR LT 1 day	15,000–20,000	Limited range 1) Premium: 1 PL 2 ÷ 3 intermediate	Food, non-food, health and beauty care mainly Self-service Home delivery
Proximity shop *SF*	100 m²	1,000	Limited range	Food, very limited non-food, health and beauty care Assisted sales; limited self-service Home delivery

Recently, in order to try and reduce labour costs which, as will be seen in the next paragraph, represent the main cost item of a store, and to improve the shopping experience by eliminating check-out times (which represent one of the main barriers to the purchase itself), some large-scale retailers, as well as the online sales giant Amazon, have been experimenting with new models of proximity stores, character-ized by small or medium-small surface areas, only self-service, an absence of staff except for supervision, and a very fast and easy shopping experience in which the consumers do not waste time checking out but enter and exit freely, paying quickly and automatically for whatever they buy.

Typically, these shops are located close to large office areas, serving people who work and who have little time to devote to shopping, but who need a point where they can take a break and at the same time buy a few basic necessities, or pick up what they have been purchased online.

In the case of Esselunga, some concept stores with the "Esse" sign have recently opened in Milan, created to offer a new shopping experience for consumers. Halfway between an Esselunga store and a Starbucks, Esse is a proximity shop equipped with a café plus kitchen for breaks and a shop for daily purchases, or to quickly collect online shopping from the lockers, choosing from over 15,000 available self-service items.

This supermarket occupies a sales area of about 1,000 m^2 spread across few aisles. These stores offer delicatessen, bakery and pastry departments and self-service fruit, vegetables, meat and fish.

In some concepts, Esse is experimenting with RFID technology (Moroni 2020). All products are equipped with RFID tags, both for self-service and the labels printed by the scales for fruit and vegetables, or the bakery counter. At the time of checking out, consumers place their trolley inside a self-check-out kiosk, where an RFID reader reads the tags inside the basket, and sends the readings of the purchased references to the check-out application, where the consumer can complete the transaction and pay by card or cash, just like at a normal superstore. This eliminates the time associated with the bar code reading of references that are necessary in a traditional store, by the employee in the case of traditional sales or by the consumers themselves in the case of self-scanning applications,[2] however, here the check-out process is even faster, for an innovative shopping experience, in line with a quick purchase of just a few references, made in leisure time or during a break.

A similar proximity store concept conceived and designed for the same shopping experience is that of "Amazon Go", the Amazon proximity store open for the moment in the United States only.

Amazon Go stores are located in large urban centres, near offices. Amazon Go is a small shop of around 500 m^2 with a few thousand references. Again midway between a shop and a café, where users can make their purchases during a break and pick up purchases made online from Amazon Lockers.

The shopping experience is totally innovative, and the control of the entire purchasing process is based on the *Computer Vision* system. The customer enters the store and is recognized by an app by placing his/her phone near a reader with NFC technology or QR scanning. From that point on, an *Automatic Vision System* based on cameras placed on the ceiling and Artificial Intelligence and Machine Learning for image processing, is able on the one hand to recognize a consumer on his/her way around the store, and on the other recognize what is being taken from the shelves

---

[2] Self-scanning means the process in which consumers, through their smartphone using an app, through a radio terminal provided by the brand, or at a dedicated check-out station, scan the bar code of all the products in the trolley or basket.

and placed inside the trolley. The system is extremely sophisticated and can also recognize if the consumer removes something from the cart to put it back on the shelf, updating the receipt accordingly.

In this way, the customer can leave the store without having to perform any operation, since the receipt is automatically updated and, at the time of exiting, the account is debited to the credit card associated with the account with which the customer entered.

All of this makes the flow and the whole shopping experience even faster and easier. Also in this case, the staff are limited to supervision operations and only intervene to provide information or in case of need (youtube 2016).

Organized Mass Distribution—Store Costs

Distinguishing the costs of the store between Capex [€] and Opex Opex $[\frac{€}{year}]$, between the Capex of a store (as usual with particular reference to the FMCG case) we have:

- Purchase of the real estate surface (if owned)
- In-store equipment:

    - Counters for assisted sales
    - Cash desks
    - Shelves for self-service
    - Warehouse equipment (storage and handling)
    - Other equipment

- IT structure:

    - Software
    - Scales
    - Scanners
    - Other retail software.

While the most significant Opex include:

- Property rental (if not owned)—for a hypermarket particularly relevant given the dimensions
- Operating costs:

    - air conditioning and heating
    - electricity for refrigerated counters
    - water
    - lighting
    - other services

- Staff costs—particularly significant for a hypermarket.

Fig. 39 Trend of revenues and costs according to the surface area of the store and the related store model

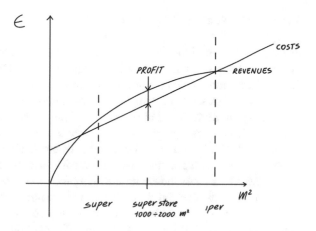

The graph in Fig. 39 shows the trends in turnover and costs as a function of the surface area and therefore of the store model.

While there is a direct correlation between operating costs and surface area, the same cannot be said for turnover. As the surface area increases, costs increase linearly, while turnover follows a logarithmic trend.

The situation of optimal, maximum profit, is that of the Superstore, therefore an area of 2,000 m^2 (up to a maximum of 5,000 m^2).

Which is why, in recent years, brands that have invested most heavily in the development of hypermarkets have been suffering. In the face of extremely significant costs, especially in terms of space costs—Capex and Opex—and staff costs (as mentioned, a few hundred people work in a hypermarket), these brands have seen their turnover decrease significantly, also due to a reduction in sales associated with a difficult economic situation.

Those who enter the hypermarket store are increasingly unwilling to buy large quantities of goods, and therefore in times of economic crisis, models of speculative purchases have also hit a crisis.

Conversely, brands that have invested in small to medium-sized stores have seen a significant development. In fact, consumers tend to go to these stores, close to them, where they can still find prices in line with those of the hypermarket, and where they buy only what they need, wasting less time shopping (time is a decisive variable, fewer and fewer people are willing to devote time to the purchase of basic necessities). In addition, they do not unconsciously find themselves after doing their shopping with trolleys that are "significant" from a price point of view, as can happen at a hypermarket.[3]

---

[3] Some people, and market research has confirmed this, have said that they stopped going to the hypermarket because when they had finished shopping they found that they had spent too much without meaning to.

Retail—The Non-core Function of a Store

Alongside the traditional core function of a store, as the final point of the supply chain where supply and demand meet and where the transaction with the final consumer takes place, there is also a *non-core*, or auxiliary function, which provides the final consumer with a series of complementary *value-added services*.

Among the more banal non-core functions we have seen the proximity of other shops, which allows consumers to optimize travel (e.g. the shopping centre or outlet models), while attracting a higher number of consumers than the single store model.

Another non-core function, particularly important in retail apparel stores but also in the FMCG sector, is the *emotional shopping experience*. Put another way, the shop prepares the consumer psychologically and emotionally to make his/her purchase.

Those buying products such as fashion, high-end food or electronic items, pay not only for a product but for a shopping experience. Shops are increasingly becoming places to enter the world of the product and transfer this experience to the consumer. This experience may be visual—in fact, the technical term normally used is *visual merchandising*—sensory (e.g. Nespresso shops) or the way in which the product is displayed, e.g. corners where a certain look is created, with the presence of videos of a campaign or a fashion show, music, and furnishings in line with the product.

Suffice to think of Apple stores, or Apple corners in household appliance and electronics chain stores. The consumer stays in the store to try the products, talk to the staff to share information, and feels part of a world containing people who love the Apple world as much as he/she does.

The in-store shopping experience is therefore increasingly both psychological but also physical, in the sense that more and more stores are becoming points where the consumer is given the opportunity to try a product (clothing stores where it is possible to try running shoes on a mini-track), or taste it (e.g. In Nespresso stores you can have your coffee brewed, the Eataly chain, where you can buy the product but also eat it according to different models. In a Signorvino or Eataly store, in addition to a shop where you can buy wine or a typical Italian product, there are served and/or self-service restaurants, pizzerias, sandwich bars, or spaces where you can meet sector figures to ask for advice on a particular product.

This function has become increasingly important in recent years, as the store increasingly becomes a point of product promotion, where the consumer goes to try a product and enjoy an experience featuring the product. To then not necessarily buy the product in the same store. Especially thanks to the development of new internet technologies, the consumer could decide to buy from another platform.

In this context, the expression *"The Store Is Media And Media Is The Store"* finds its meaning. Until 3–5 years ago, a product was advertised in the media to bring customers inside a store to make a purchase. The spread of e-commerce has meant that stores have evolved from a being a *Point of Purchase* to an element that increasingly performs the function of the media, in which to let the consumer try a product or the world connected with the product (something which is impossible online). The consumer then buys the product indifferently in a store or via the media, in a shopping experience that is identical to a physical purchase made in-store. In a case like this we speak of *omnichanneling*.

Omnichanneling

After getting to know a product or physically trying one in a shop, thanks to the internet, the consumer approaches a product and buys it by means of a multimedia platform while doing other things, on a website on a desktop at home, sitting on the sofa while watching TV, or using a mobile app on the way to work.

Before the advent of the internet, the only way to reach customers was to bring a product close to them by building a shop. This is why large so-called 'brick & mortar' retail chains have continued to expand their SC by opening stores in an effort to reach more and more customers with an increasingly extensive SC.

However, by doing so, they also increased all the SC costs, above all, those associated with the physical stores, both in terms of Capex to open it, and Opex for the presence of stock and staff. The advent of the internet and the online world meant that it was no longer necessary to physically bring a product close to the final consumer through the SC, since this function was now being performed by the web.

In this new scenario, the store becomes a retail outlet for customers who physically go to the store, but also a distribution centre to fulfil online orders. In this new perspective, the store is to all intents and purposes a distribution centre, with picking, packing, shipping and return management functions, for the exchange of goods or returns.

When it comes to a physical channel versus an online channel it is helpful to distinguish between cases of *single channel*, *multichannel* or *omnichannel* retail.

- Single channel: this is where there is only one model in the retailer's SC: either physical or online. The consumer can therefore buy a product either exclusively on the internet, receiving it at home, or in-store by physically going there. Many players adopt this scheme to avoid competition between the two channels (e.g. Conad only physical; Dell only online).
- Multichannel: we use this term when there are two distinct channels, physical and online, but each is managed by a separate SC, with different players and systems that are not interconnected. The physical side is managed through the chain of stores, in some cases run by third-party franchisees, while the online side is managed directly by the retailer who owns the brand through a dedicated DC. In this case, since there are no links between the two channels, consumers cannot purchase a product online from the store's stock, just as they cannot return a product bought online to the store; product prices are also often different, which generally creates confusion for the consumer as well as for the franchisee, who sometimes needs to be compensated for online retail competition.
- Omnichannel: the physical and online sides merge in a single channel. The product is the same, the price is the same or similar, the supply chain is the same, and the consumer's order is fulfilled indifferently by a store or a DC, just as consumers

can return a product either by sending it to the DC or by returning it to any shop. The sale and costs are correctly attributed to the player who originally generated the hit, and this remains transparent for consumers. Consumers receive the same shopping experience, regardless of whether the sale takes place physically in a store, via a website, or via an app.

In an omnichannel scenario, the store's inventories then become stock with which to fulfil an order in line with traditional models (the customer goes to the store and physically leaves with the product), or where the consumer purchases via a mobile or PC, and the order is then prepared and reaches the consumer following different models, as described below.

- BOPIS: *Buy Online Pick Up In Store*, the store prepares the order issued online using its stock, and the consumer then goes to the store to collect his/her order— also known as "*click and collect*". In this case, processing times are normally in the order of hours. Examples of BOPIS are "click and collect" orders collected at the pickup points at an Esselunga or La Rinascente store. The order is prepared and fulfilled via the shop's stockroom; orders are placed in lockers or in pickup areas for collection.
- BOSTS: *Buy Online Ship To Store*. In this case, an order placed online is collected from a particular store, but the order is fulfilled by a DC or any store. The BOSTS procedure makes it possible to pool inventories between all the stores of a chain (physically in a DC or virtually between stores), and, in the case of picking carried out in DC, to reduce the operating costs of order preparation and the space required.
- BOSFS: *Buy Online Ship From Store*. This is a variant in which the online order is fulfilled by a store, but is shipped to the home and not collected from a store by the final customer. Consequently, it is the shop that does the picking, packing and shipping, sending the order by express courier to the consumer's home (e.g. *Esselunga Home Shopping*).
- BSSFS: *Buy In Store Ship From Store*. These are often retail shops with a high tourist flow (typically the consumer is not local but a tourist with high spending power). The order placed at a store is shipped to the customer's home. This is also the service some FMCG supermarkets or traditional stores offer.
- BSSTD: *Buy In Store Ship To Door*. Concept stores in which the store actually represents a media showcase of the product, and in which the product can be tried and touched (usually there is only $1 \times$ model or $1 \times$ model/variant), but the product is not physically available to be purchased, and the order, issued on an online platform by the shop assistant or the customer via an app, is fulfilled in this case by the DC (e.g., Poltrone & Sofà armchair & cars dealers).

Precisely by combining omnichanneling and the shopping experience, the function of the store is evolving from the end point of the purchase to the point of product promotion and aggregation of the customer, so as to let the customer enjoy an emotional experience connected to the product.

Consumers go to the store not necessarily to buy a product, but to live an experience linked to the product. Consumers will not necessarily conclude a purchase but will eventually complete it later on another channel (e.g. an e-commerce site via PC or smartphone, a third-party e-commerce site via PC or smartphone, social media channels, or others).

As mentioned, the paradigm is reversed: the store becomes part of the media, in which to advertise and make the product known through an emotional experience; while at the same time the media (social channel, website, app) become the store, in which to buy the product tried in-store.

In this sense, and following the development of omnichanneling and the new promotional function acquired by the store, traditional metrics such as the conversion rate $\left(CR = \frac{sales}{numer\ of\ entries} \right)$ or the sales KPI m^2 seen previously may prove obsolete, since consumers are not necessarily buying in-store.

Metrics such as the number of accesses, number of trials, and customer loyalty become important (how many times the same customer returns to the store; just as it becomes essential to be able to follow the consumer on all channels, to understand, for example, if the experience lived in a store or on the social networks (YouTube, Instagram) then turns into a purchase on one of the online channels (app, website, social network, shop, or others).

Think for example of the shopping experience depicted in (Fig. 40). a consumer who is commuting by train to a large city; he/she accesses Facebook or YouTube and is offered a video related to the product; 15 min later the consumer logs on to the e-commerce site via an app and checks the availability in the closest brick and mortar store in the city and price of that product. On arriving at the destination, passing in front of the store of the brand, he/she decides to enter and try that product or others; but does not conclude the purchase, he/she goes home and that same evening or after a few days, he/she concludes the purchase through web access to the e-commerce site while sitting on the sofa. The product is then delivered to the home via BOSFS (Fig. 40).

In an example like this, both the store and the promotion on social media have played a fundamental role in triggering the sale, but a metric such as the conversion rate does not recognize this role at all, on the contrary, the customer contributes to lowering its value. Even the sales per m^2 does not take into account the role of the store, while all the sales are attributed to the online channel which in reality had a merely finalizing role in the process, while the trigger and promotion were enacted respectively by the social network and the physical store.

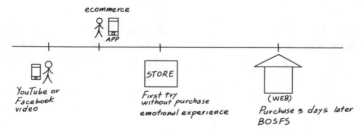

Fig. 40 An example of a purchasing process in an omnichanneling scenario

For details on th evolution of retail in an omnichanneling perspective, see (Mason and Knights 2017; Stephens 2017).

1.3.7   Retail 4.0

In this context, the term Retail 4.0 means a brand-new retail model based on the technologies of Industry 4.0.

The predominant technologies are Artificial Intelligence, often combined with computer vision, Big Data and Cloud Computing, and the Internet of Things enabled by RFID technology. The goal of Retail 4.0 is to make the shopping experience increasingly simpler, smoother, and unmanned.

In traditional retail, the customer visits a store, where the supply chain has had the product arrive, to make a purchase. Assisted by a cashier, he or she buys the product and leaves the shop carrying the product. The model is complex precisely because it requires the consumer to move and therefore the purchase frequency is low, while the quantities purchased tend to be high. Suffice to think of the purchase at a large consumer outlet where the customer travels by car, enters the store to make some purchases, pays at a checkout barrier assisted by a cashier and leaves with the product.

In the e-commerce model, the customer places the order online and receives the product at a predetermined point, normally his or her home, in the shortest time possible. In this case, the tendency is to make the purchase experience as simple as possible precisely to facilitate the impulse purchase of even small quantities, but constantly. We will thoroughly detail, for example, how Amazon through its Amazon Prime programme guarantees free deliveries in 24 h even for a single order line.

In Retail 4.0, the consumer returns to the store, or at any rate to a physical place, where the product is available, but the purchase experience is so fast and smooth that the purchase is continuous, and a high frequency predominates at the expense of quantities.

Unmanned stores, self-service retail, and automatic home replenishment are all included in the Retail 4.0 framework and will be tackled in the following subparagraphs.

Unmanned Stores

These are unattended stores where the customer can enter to buy the product and exit extremely quickly without any hindrance. Typically they are small, located near large office blocks, and designed for people taking a break or for the purchasing of lunch or dinner. One such unmanned store model is Amazon Go, whose operation has been extensively described in he Automated Proximity shop paragraph before.

As mentioned, Amazon Go stores are based on the technologies of Industry 4.0, and in particular computer vision and artificial intelligence, i.e., an extremely complex and sophisticated camera system, which knows the products and recognizes the customers thanks to artificial intelligence, that can distinguish what is taken from the shelves, what is put back and by whom, and automatically calculates and debits the transaction.

In this case, what is predominant when it comes to the Capex is the infrastructural investment in the computer and machine learning systems. Even works to adapt sales surfaces can represent a substantial cost, unless inside an existing shop and not built from scratch, in which case optimizations are obviously possible.

Another model of an unmanned store is the one piloted in 2020 by Esselunga in Italy, or under development in Japan, as part of an economic development programme of the Japanese Ministry of the Economy which aims to transform the 100,000 convenience stores present throughout the country into unattended shops 4.0.

In this case, the enabling technology is RFID. The products are equipped with a UHF radiofrequency tag, which allows them to be uniquely identified. This tag can be used in-store, primarily for inventory counts, since the RFID technology enables accurate and real-time inventories, using fixed or mobile equipment. The former may require significant Capex investment costs connected, for example, to smart shelves or to RFID reader infrastructures installed on the ceiling which cover certain surfaces where stock control is desired, typically the surface of the sales area or some areas of the stockroom. Meanwhile, mobile readers can offer maximum flexibility since the investment costs are reduced to the cost of the reader and the inventory can be carried out at any point of the store according to need. In this case, the modulation of the reader power is the key to having localized readings (low-power) in the case of an inventory by area, or mass readings (high-power) in the case of a complete inventory of the store, with productivity reaching even 50,000 items/hour.

Thanks to RFID, it is possible to distinguish between the stock in the storeroom and the stock in the sales area thereby preventing the out of stock phenomenon, meaning a lack of the product on the shelf, either because it is not in stock, or because the product is present in the back of the shop, but the sales area has not been replenished.

In the first case the OOS problem is typically due to a misalignment between the actual stock, equal to zero, and the theoretical stock present in the system, greater than 0. Some studies by (Delen et al., have shown that this type of accuracy in retail stores can be as high as 70% at the SKU level. In other words, around 30% of the SKUs present in a store have a real stock different from the one present in the system.

In the second case, the problem of non-replenishment is due to the sales staff forgetting to replenish the sales area. In traditional stores, the sales staff carry out restocking once or twice a day, moving among the shelves and observing any gaps. In a store 4.0 model, a system generates an alarm whenever a product is not on the shelf but is in the store, generating a replenishment task. Extensive studies carried out by the RFID lab at the University of Parma (Bertolini et al. 2012) have shown that significant increases in turnover can be achieved, even in the order of a few percentage points, thanks to the reduction of OOS through RFID technology.

In addition to precise control of the stock, RFID technology applied to products makes the purchase process much more streamlined by eliminating queues at the checkout and therefore increasing the level of service since the customer no longer has to wait. Consider, for example, the experience of Decathlon, where the introduction of RF technology has made it possible to completely eliminate the queue at the checkout thanks to self-checkout systems installed in its stores. Before the introduction of this technology customers often avoided entering the store due to the high footfall. Self-check-out systems also make it possible to reduce the staff assigned to the checkout barrier, significantly impacting the cost structure of the stores since this item represents one of the predominant items of the Opex.

A further benefit to be obtained by applying RFID technology at the item level is that of being able to significantly reduce the waste associated with exceeding the shelf life of single items. In fact, each item acquires its own identity and can therefore be sold at a variable price depending on its expiry date. With the same reference, items close to their expiration date can be recognized at the checkout and discounted, unlike those with a longer expiry date which are sold at full price.

The Japanese Ministry of Economic Development METI (Meti 2020), launched an ambitious programme precisely with the aim of reducing waste and increasing operational efficiency in the check-out phase. Developing a low-cost RFID tag, in the order of Japanese yen or less, to be applied to all consumer products, and through this technology increasing the accuracy of inventory, reducing waste, automating the checkouts in more than 70,000 Japanese convenience stores, known there as *konbini*. These are small to medium-sized stores with a number of categories, and a high number of references for each category. The goal is to equip fresh products or products subject to expiration with RFID tags to keep inventory under control and manage the sales of individual items at variable prices according to the expiration dates. The programme is extremely ambitious given that it is expected to introduce 100 billion tags by 2025.

Worldwide distribution giant Walmart (Sweeberg 2022a) has also recently sent its suppliers a communication requesting the application of an RFID tag at source to certain categories of products in order to manage them in a supply chain. This tag must be encoded with an SGTIN, in compliance with GS1 standards. Initially, the application will be for home goods products, hardware and automotive products received directly from suppliers, which must be equipped with RFID technology by September 2022 in order to be received at a DC or a Walmart Supercenter. The declared goal, however, is to subsequently expand the mandate to other categories and other products.

Walmart's strategy is to increase inventory accuracy, and thereby ensure the availability of product to customers and increase customer satisfaction. Especially in a scenario where, according to the BOPIS model, online sales are growing to double or triple digits year on year, it is essential that the inventory accuracy is close to 100% in order to confidently put SKUs with few stocks units on the market, and improve online order fulfilment and decrease order cancellation rates. And this becomes feasible thanks to accurate inventories which can also be carried out every day or several times a day in a very short time, considering that the productivity which can be achieved is in the order of 30,000-50,000 items/hour (depending on the storage density; the greater the density, the greater the productivity since the operator does not have to travel to collect the readings, hence shorter travel times), with an accuracy close to 100%, and in any case higher than that of a barcode inventory.

Another example of an unmanned store based on RFID technology, the Internet of Things and other Industry 4.0 technologies (Big Data, Cloud Computing, Simulationsabove all), this time in consumer products, is that of the Spanish startup Ghop (Sweeberg 2022b).

Ghop has been deploying an unmanned smart convenience store the size of a small shipping container, which can be installed in neighbourhoods or at busy commuter sites where larger, fixed convenience stores are unavailable. The retailer is offering a variety of snacks and beverages for individuals on the go, with no employees onsite to accomplish the sales. Customer simply use an app to enter the store, then select and pay for items within a matter of minutes or seconds. The unmanned functionality is provided by RFID technology in the form of tags affixed to products, readers in the store, and a software platform.

The RFID system provides loss-prevention, inventory-management and mobile-payment solutions. Since goods are stocked at the store, Ghop employees apply an RFID Flag Tag to each item. By scanning every tag as it is encoded, the system updates the inventory data regarding what goods are onsite and available for purchase. Those products are then put on display on store shelves or in coolers.

Once shoppers have selected everything they intend to buy, they can place the items in an RFID reader terminal that is insulated so that it will not pick up stray readings from products not being purchased. The reader antenna reads the tag IDs, then a screen displays these products and prompts the customer to confirm whether these are the items being purchased. Once customers approve their purchase, the cost is deducted from their credit card account.

And once this is done, customers can simply walk out of the store. At the exit, an RFID reader, serves as an RFID loss-prevention system by reading all tags and detecting if any items have not been purchased. If any unpurchased items are detected, the system will display a notice on a screen requesting that the customer go back inside and pay for them. Periodically, an employee conducts a stock count using a dedicated handheld UHF RFID reader, then restocks all the depleted items.

The idea behind Ghop is to bring convenience shopping to places retail previously could not reach, without the expense of sales personnel and the need for permanent facilities. For the solution to work properly, the company needed transactions to be easy, without requiring queuing for payments. Customers could simply select an item, place it under a scanner, approve the sale, and walk out. By automating the store, the retailer makes its products available at any time of the day or night. The 15-m^2 (165-square-foot) modular structure was first launched at a pilot site in July 2021. Many of the targeted sites are at gas stations. Ghop expects to see its stores deployed in places such as college campuses and airports, and it plans to have expanded outside of Spain by 2023.

Self Service Retail

Another type of Retail 4.0 Model based on technologies of the fourth industrial revolution and in particular of the Internet of Things, Big Data and Cloud Computing, is that of the smart cabinets.

By smart cabinet we mean advanced display cases, able to keep current stock under control in real time, to calculate what the consumer withdraws thanks to the use of 4.0 technologies, and to charge for the consumption. Without the need for any action on the part of the final consumer. The data flows in real time on the Cloud where the updated status of stocks and consumption is visible, and artificial technologies applied to the Big Data generated allow optimization of the processes linked to all the functions involved in Porter's value chain. For example, it is possible to decide which products to promote, and where, according to consumption or expiration dates in a single display. For instance, there may be two pieces of the same reference with different expiration dates in real time, the Smart Cabinet can then be programmed to promote the item with a shorter expiry date. At the same time, promotional messages can be shown on the display's screen, or push notifications can be sent to a consumer's app, informing him or her that a favourite product is available at an advantageous price in a nearby display case.

The consumer typically accesses the display case via an app after registering, or even directly with a credit card, with which he or she is recognized and is enabled to then make a purchase. At this point, the consumer can take out the products he or she wishes to buy, and the smart cabinet can recognize what has been purchased and charge the consumer for it. Typically, smart cabinets are Smart Fridges, intelligent refrigerators managed by RFID technology, or Smart Vans. An example of the former type is the Fresco Break https://www.frescobreak.it/. This Italian startup has installed around twenty smart fridges to date, in strategic locations such as offices and high-density transit points, with which it markets a wide range of high-end ready-to-use products that can represent a substitute for a good healthy meal.

Each fridge is an RFID device capable of constantly inventorying the products and transmitting data on consumption. There is also a Cloud system on which a Business Intelligence dashboard runs, with which the company monitors the entire supply chain process, from sales to replenishment, to production and distribution. Marketing initiatives can also be managed centrally and enabled in cascade on the individual refrigerators in individual zones, both at the reference and single item levels.

Also the case of the Californian startup Robomart https://robomart.co/ can be included among the Smart Cabinets based on Industry 4.0 technology, although in this case, the application of Industry 4.0 technologies is even more powerful and more integrated since not only RFID and the Internet of Things, Cloud Computing and Big Data, but also Artificial Intelligence, Computer Vision and Autonomous Driving technologies come into play. In fact, Robomart wants to push beyond the Smart Cabinet concept and bring customers basic necessities thanks to self-driving vehicles, making the concept of home shopping even more futuristic.

Among the various technology options available to track products individually, Robomart selected RFID for its reliability and accuracy. According to some Morgan Stanley research, Less than 20% of groceries are sold online in the U.S., because 84% of consumers do not trust anyone else picking their fresh groceries, especially when it comes to perishables like fruits and vegetables. Robomart is intended as a solution to that problem, It gives consumers the ability to pick their own groceries, right at their doorstep, adding that no delivery fee is added to the cost of the products.

For the moment, the US startup Robomart has launched an invitation-only Beta test in the city of West Hollywood, California, through which local residents can download the app and request an invitation code.

The app allows consumers to "call" a special van stocked with basic necessities and items for the daily routine such as paper handkerchiefs, medicines, snacks and drinks. The vehicle arrives at the place where the customer is in less than 10 min and allows them to buy the desired items wherever they are, without the need to go to a shop or wait around during a delivery window. In times of Covid-19, a useful idea also to avoid crowds in supermarkets and the passage of products through the hands of cashiers and other customers.

In the test taking place in California, a Robomart vehicle which delivers pharmaceutical products is already available. The Pharmacy Robomart has over 500 packs of 50 everyday products, including over-the-counter medications, first aid items, toothpaste, personal care items, as well as kitchen and household goods. A second vehicle, called Grocery Robomart, will be making its début in the coming weeks, offering consumers access to fresh foods and dairy products. Users select the goods they want to buy from the Robomart app, and the vehicle brings them to their location. Unlike Glovo, Deliveroo, and other such delivery companies, Robomart is designed for minimal expenses and for buying just one item. In addition, the founders plan to use self-driving vehicles for deliveries in the near future.

Robomart is based on RFID technology. There is an RFID tag affixed to each product and a reader built into the vehicle. The unique ID number of each tag is linked to a particular product in the system's cloud-based software. When the item is removed from the vehicle, the system will detect that the product's tag is no longer being read. The app will then list a full breakdown of the products removed and provide a receipt for the purchase. Based on the total cost, Robomart charges their saved debit or credit card accordingly.

The software related to sales provides stores with analytics regarding sales and the consumption patterns of their customers. Retailers can stock any products they choose to in the Robomart.

For retailers, Robomart offers the ability to expand their footprint without up-front capital expenditures. National supermarket chain Stop & Shop is the first retailer with plans to partner with Robomart to deploy a driverless, RFID-enabled mobile store that will offer a selection of goods to shoppers wherever they are located.

For now, Robomart uses vehicles with a driver, but soon it should only use self-driving vehicles. The ultimate plan is to make deliveries via autonomous vehicles when all the necessary regulatory and technological hurdles have been overcome.

Automatic Home Replenishment

Finally, it is interesting to observe that some patents and methods deposited by major retail multinationals suggest a further evolution of the concept of e-commerce towards a new model of Retail 4.0, in which the products materialize directly in the consumer's home, without the need to reorder, and this is thanks to the technologies of Industry 4.0, which will completely change the way we shop. We will call it hereafter Automatic Home Replenishment (AHR).

Conventional shopping requires the end customer to physically go to the store to purchase the needed product; E-commerce is based on the manual reordering of products on a website, and requires access to an online store, either through a desktop website or through a smartphone app, where the customer places the order; AHR methods are automated replenishment service methods.

The Vision is that, thanks to RFID technology enabled products, and RFID enabled smartphones and appliances, the AHR system gets consumption visibility straight to the shelf in the consumer's home and automatically refills that shelf when it is empty or nearly so, without the need of any action by the end customer.

The advantages of AHR methods are self-evident. Unlike traditional and e-commerce shopping methods, consumers neither need to personally go and walk along the aisles of their grocery store to pick goods, nor do they have to realize their need for a product, visit a e-commerce website/app to place their order for that product, and wait for it to be delivered over the following days.

Thanks to RFID technology, RFID based AHR systems will automatically and efficiently:

i)   register our routine daily consumption of products at home;
ii)   charge us for product consumption;
iii)   automatically replenish products according to our own replenishment programme;
iv)   optimize our product consumption based on target parameters, such as costs or nutrition facts;

Other companies have also tried to develop automated replenishment services. In 2015, Amazon introduced Dash buttons, Wi-Fi–connected devices programmed to order specific products, while other companies like Samsung or Whirlpool have developed prototype appliances like refrigerators which use barcode technology or artificial vision to warn of products close to the end of their life or to reorder predetermined products. Compared to previous methods, AHR exploits RFID technology and RFID enabled appliances and rubbish bins and/or smartphones to enable automated replenishment services for domestic products.

All products are equipped with RFID tags. The application of the RFID tag takes place either at the point of manufacture or at a distribution centre, when the products are received from the AHR suppliers or when a customer order is fulfilled.

These tags are read through RFID read points—aggregation and shipping, receiving, RFID rubbish bins and inventory apps installed on consumer smartphones.

Automated readings (at least tag ID, read point, timestamp, event type) detect product tagging, shipment and delivery to the end consumer as well as domestic consumption. This information is registered through the internet in the AHR cloud system. The system uses this information to feed artificial intelligence and simulation algorithms to:

- update end-user inventory levels for items every time a delivery is either shipped or received and a product is thrown away;
- invoice each single item to the customer when it is thrown away (each individual unit is invoiced at the shortest timestamp between due date and an RFID reading from a bin/ app);
- automatically reorder the product from the retailer when it reaches a certain reorder level set by the end consumer or optimized by the system;
- trigger alarms for products close to their end of life and help the consumer to localize them at home;
- propose optimized replenishment programmes to end users based on previous consumption and target parameters (costs, nutrition facts like calories, carbohydrates/proteins/fat consumption).

The AHR cloud system not only serves to order products, but also has an inventory function. The cloud system keeps a record of each product discarded in the household's or office's waste disposal device(s), as well as each product ordered. The AHR cloud system also receives a notification from the ordering platform for each product effectively delivered to the household or office after an order. Knowing this, the cloud system can provide the user with an accurate and real-time inventory of the products available in his or her household or office. The user can also be provided with an overview of the products he or she threw away into the waste disposal device, which products will be ordered, when they will be ordered, and when/where/how they will be delivered. All this information is accessible in the household or office's cloud system account, for example via a remote device.

In parallel, the AHR retailer can also keep track of products discarded into the household's or office's waste disposal device(s) on the one hand and of the products ordered and delivered to the household or office on the other hand. This provides an accurate and real-time inventory of the products available in the user's household or office and of the products discarded into the waste disposal device, which will be ordered again soon. Based on this information, the retailer can easily and reliably anticipate the household's or office's future orders and optimize their stock management.

Currently, in the absence of information concerning the use/consumption of a product in a household or office after it has been bought, online retailers operate based on the assumption that a product is needed by the user at the moment he or she orders it. As a result, typical stock management methods of online retailers are optimized for delivering a product to the user as quickly as possible as soon as an order is processed. So, for each product, there are actually two successive and utterly uncoordinated stock levels before and after the order by the user: a stock of products in the storage facility of the online retailer and/or of individual suppliers offering products on the retailer and a stock of products in the user's household or office.

However, the point when a product is actually needed by the user is not the moment the product is ordered, but the moment it is effectively used/consumed. So, knowing exactly when one or several households or offices use/consume and discard a product, a new stock management method can be provided for coordinating the stock of products in the storage facility with the stock of products in the different households or offices to decrease the total overall stock of products in the storage facility and in the households or offices, thus improving the efficiency of the overall supply chain.

For example, knowing that a stock of products in households or offices is above a certain threshold, e.g. in the event that many households or offices still have several identical unused products in stock, can allow an AHR online retailer to temporarily reduce their own stock for this product in their storage facility and thus save storage costs.

Conversely, knowing that several households or offices will soon consume the last units of a certain product in their stock can trigger a reorder from the storage facility's own upstream supplier in order to anticipate the user's future orders and make sure that the required number of products is available in the storage facility's stock just in time when needed.

Advantageously, the relative stock of a certain product in a first stock in a storage facility and in a second stock inside a household or office can be varied according to the respective constraints on the first and/or second stock. For example, the storage or processing capacity inside a storage facility is typically under great strain and pushed to its limits in festive periods because of the high volume of orders for food and presents. In order to temporarily increase their storage or processing capacity in these periods, the online retailers managing the storage facility may encourage users to order products they will probably need during holidays ahead of time and to stock them in their household or office, e.g. by offering individualized exclusive offers or discounts for these products depending on each household's or office's current stock and anticipated use/consumption. This allows an online retailer to consider the user's household or office as an extension of their own storage facilities. In another scenario, online retailers could remunerate users for storing certain products in their household or office rather than in their own storage facilities by making the above mentioned individualized exclusive offers or discounts, which saves the online retailer both storage space and money.

The stock in the user's household or office can also be managed by the AHR retailer on a consignment basis: one consumed, one reordered. The product may not be billed upon order or delivery to the household or office, but upon use/consumption of the product in the household or office or after a certain time period previously agreed upon (i.e. expiration date of the product), the payment being guaranteed by registering the user's credit card details online.

1.3.8 Third-Party Logistics (3PL)

Definition of Third-Party Logistics

In the distribution part of the supply chain, third-party logistics providers operate—known as 3PL.

3PL involve third parties with respect to the two main interlocutors at the opposite ends of the distribution chain: typically on the one hand, the manufacturer/focal company which produces the product; on the other, the retailer who deals with selling it to the end customer.

These third parties are supportive, inasmuch as they deal with managing the distribution in whole or in part, providing focal companies/manufacturers and retailers with the infrastructure and logistics services necessary to deliver the product to the final consumer. The role of 3PL is summarized in Fig. 41.

A 3PL typically offers the following services:

- Storage: by means of a proprietary network of warehouses and distribution centres, in which to keep the product in stock.
- Capacity for sorting and consolidation of flows for order fulfilment: through transit points and distribution centres (sorting or consolidation centres). We talk about fulfilment centres or distribution centres which carry out the following processes:

  - Receiving
  - Storage/Retrieval
  - Picking & Sorting
  - Packing & Marking
  - Shipping
  - Other particular re-elaborations, known as *Value Added Services* (VAS):

    Repacking: typical of the FMCG sector, in which, before being destined for a large-scale distribution chain, a product may be re-elaborated for a particular promotion (e.g. product taping for a promo—pack opening, taping, pricing, cartoning), or the product needs to be placed in cardboard displays to be located around the store in strategic positions (e.g. at the checkout, beneath a headboard, etc.).

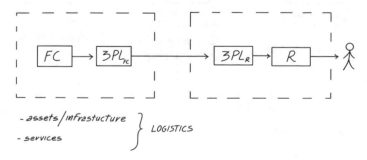

Fig. 41 The role of 3PLs in the SC

In the fashion & apparel sector: quality control of returns, reconditioning, washing and ironing and bagging, unbranding and labelling, pricing.

- Transport: through a fleet of means of transport in the broadest sense, from tractor units and semi-trailers, swap bodies and containers, up to proprietary railway wagons, ships or cargo planes.
- Asset pooling: rental and integrated management of transport packaging (pallets and rolls, bins, etc.) and secondary returnable packaging (fruit and vegetable boxes). We shall return to this aspect below.

The rationale behind this type of partnership is that investments in infrastructure and logistics assets require huge resources, in terms of both Capex and Opex. Thanks to a medium/long-term partnership, the focal company or retailer can continue to invest their resources in the core business (for the focal company, for example, in the development of new products, the development of production plants; for the retailer, the opening of new stores), while the logistics operator's core business is specifically logistical and it therefore directs its Capex towards infrastructures (DC, warehouses, TP) and other assets (physical and technological) of a logistical nature.

In addition, a logistics operator is also able to optimize its Opex thanks to economies of scale (linked to the volumes of the same product) and of scope (different products for different customers), which it manages to generate by managing several customers at the same time. These concepts are summarized in Fig. 42.

The Levels of a Logistics Partnership

There are different levels of integration in the transition from in-house management of logistics activities and services to that of full outsourcing. The level of integration and partnership makes it possible to classify logistics service providers. On a scale from a minimum level of integration to a maximum level of integration and partnership (see Fig. 43), we therefore have:

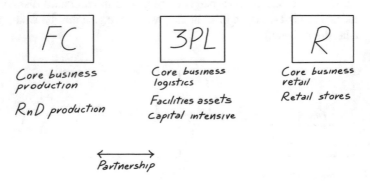

Fig. 42 The rationale behind logistics partnerships

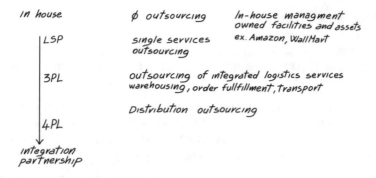

Fig. 43   The 4 levels of logistics outsourcing: from in house to 4PLs

- In-house management: We speak of in-house management when the SC player manages the distribution processes internally, using proprietary assets and logistics equipment. This is the case of companies such as Walmart or Amazon, which consider distribution logistics a strategic process. One that represents a real competitive lever on which to base their business strategy, and consequently they decide to invest in resources on a core business, strengthening and maintaining it internally, or at least in part (the case of Amazon which relies on both Amazon Logistics and other players such as US Postal to reach suburban areas, but also FedEx and UPS for some cities).
- Logistics Service Provider (LSP): we speak of an LSP when the SC player simply purchases a service from the 3PL (e.g. a transport service from an origin to a destination, a pallet storage service at a warehouse), but maintains management of it, whether for distribution, storage or transport activities.
- Third Party Logistics (3PL) proper. The SC player, manufacturer or retailer, relies completely on the logistics operator, with whom it shares pickup and delivery plans, and it is then the logistics operator that organizes and manages the logistics services for the customer in an integrated manner, dealing with transport, storage, order preparation, consolidation and sorting activities, and value added services, to deliver the requested product to the end customer in the manner and status requested and at the required time.
- Fourth Party Logistics (4PL): There is an additional level of integration, so-called 4PL, in the event that all of the distribution part is managed directly in outsourcing by the logistics operator, and therefore not only the operational part typical of a 3PL, but also the management side such as order collection and the planning of deliveries, or the management of stock levels at various warehouses.

Examples of 3PLs or 4PLs in the Italian food sector are the *Number 1 Logistics Group* (logistics operator of such leading dry brands as Barilla, Lavazza, and many others) *STEF* (which operates on the contrary in the fresh goods sector, In Italy among others for Danone, Muller, and Mondelez formerly Kraft), *Italtrans* (logistics operator for fresh and dry goods for Auchan and Esselunga).

The Choice of the Type of Partnership

The choice of the level of logistics outsourcing is made by the manufacturer or retailer from a qualitative/quantitative perspective, considering both quantitative and strategic aspects:

- The quantitative part is in fact a *Make-or-Buy Analysis* based on a comparison of the net cash flows generated by Capex and Opex in the *Make* case (in-house management), and in the *Buy* case (outsourced management). The costs to be considered for a Make-or-Buy Analysis are typically:

  - Capex: cost of facilities (warehouses, distribution centres, transit points) and storage infrastructures, cost of handling systems, cost of means of transport (tractor units, semi-trailers, railway wagons, ships, aircraft), cost of IT infrastructures (hardware, software). In the *Buy* scenario there may be a one shot cost for IT integration between the manufacture/retailer and the 3PL.
  - Opex: cost of management, organization, and operating costs of warehousing and transport activities (personnel, energy, consumables). Cost of software licenses and after-sales service. In the *Buy* case, there is only one Opex item equal to
    $$Handling\ cost = \frac{€}{handling\ unit} * \frac{handling\ units}{year}$$

- Qualitative Analysis. In a *multi-attribute logic*, this answers the following questions:

  - How much core business is represented by logistics in the process falling within the competence of the SC (i.e. manufacturer/retailer)?
  - How much flexibility is required? A 3PL is more flexible as regards volumes than a *Make* case. Think, for example, of the case of *transportation*, or of pallet spaces in a warehouse. Faced with a contraction in volumes, from a *Buy* point of view it is sufficient to simply reduce the number of tranports supplied by the 3PL or the number of pallet storage shelves; in the case of *Make* there may be an unused asset park or warehouse, whose depreciation cost negatively affects the total logistics cost of the product. Therefore:

    Make case: when handling volumes decrease $\Delta V < 0 \rightarrow \Delta C_u \gg 0$
    Buy case: within a certain threshold, $\Delta C_u = 0, C_u = cost$

    In addition, in case of in house logistics, a high variability of volumes can lead towards new investments (as handling volumes increase $\Delta V \gg 0$), If volumes then decrease, the need to contain unit logistics costs returns.
  - How reliable does the service need to be? It is true that distribution may not necessarily be the core activity of an FC, but it is equally true that a decline in the logistics service could be critical for an FC's ability to deliver a product (effectiveness) or in terms of cost (efficiency). Reliability, as will be seen in detail in Chapter 5 on customer service, is measured with reference to so-called *transactional service indicators* such as:

Lead Time and frequency
Punctuality of deliveries
Frequency of deliveries
Accuracy of deliveries
Flexibility

Precisely with reference to punctuality and accuracy, a KPI called *OTIF* is used in practice. This is an acronym that stands for *On Time In Full*, defined as in (5), referring to a generic time interval ΔT. The operator $\wedge$ is the *AND* of the Boolean functions (two concurrent events):

$$OTIF = \frac{n° of\ punctual\ \wedge\ accurate\ deliveries}{n°\ deliveries\ completed}_{\Delta T} \qquad (5)$$

The contracts signed between a client and a 3PL are medium-long term, especially if the 3PL needs to make considerable investments to guarantee the service standards requested by the client. E.g., guaranteeing delivery in an uncovered area requires investments in a new DC in that area, an investment which must be amortized over a medium-long period of time and must therefore be covered by a service contract of comparable duration.

At the same time, however, the contract must include termination clauses, e.g. specific bonus/malus clauses for fees based on quantitative and objective service KPIs. For example, a target KPI value (e.g. OTIF) and a deductible within which the consideration remains constant is established. Apart from the deductible, compensation mechanisms (service credits) are triggered, with variations in the consideration increasing (in the case of higher performance than agreed) or decreasing (in the case of deterioration), ending with proper termination clauses which include termination of the contract for just cause in the event of non-compliance or a worsening of service levels as well as predefined limit thresholds for prolonged periods of time. The service performance measurement system in a 3PL contract is shown in Fig. 44.

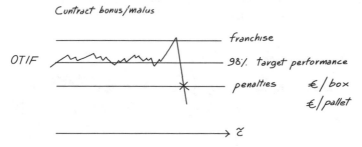

Fig. 44 Service performance measurement through OTIF in a 3PL contract

1.3.9 Asset Pooler

Asset Poolers deserve a particular in-depth analysis in the panorama of logistics operators. These are not strictly logistics service providers, but providers of rental and management services for transport packaging and returnable secondary packaging (i.e. so-called *Logistics Assets*).

In the case of the FMCG sector, the use of asset poolers is quite frequent for different categories of packaging and procedures, such as:

- Transport packaging,

  - Pallets: 1200 × 800, suitable for replenishing hypermarkets and superstores. These can be brought into the sales area with equipment such as hand pallet trucks.
  - Rolls: 80 × 80, used for the supply of small stores (supermarkets), convenient for the direct replenishment of a sales area with narrow aisles and without the need for hand pallet trucks to move them.
  - Isotainers for fresh and frozen products.

- Secondary packaging:

  - Bins and mini-bins for fruit and vegetables (Reusable Plastic Containers RPC)
  - Crates for fruit and vegetables or meat (reusable crates are also widely used in the automotive sector).

All of these packs, given that they cost tens or hundreds of Euro, are not disposable but returnable, and there is therefore a need for backward logistics for the recovery, reconditioning, and reuse of the asset.

The network structure is described and represented in Fig. 45.

The manufacturer receives the asset from the asset pooler and uses it to move and ship the product. The product (complete with packaging) is shipped to the retailer's DC. Once the package has been received at the retailer's DC together with the goods, the product is sorted for each store. The packaging is brought to the sales area together with the product. At this point, the asset must be picked up at the store, and taken

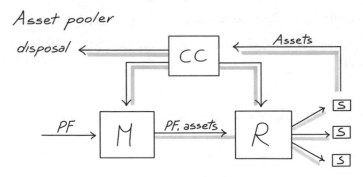

Fig. 45 Asset pooler, the closed loop supply chain of assets

to a consolidation centre (*CC Asset Pooler*) where the asset is subjected to special processes (e.g., checking, repair, sanitization/cleaning, removal of logistics labels) to then be put back into circulation and redistributed to the manufacturer and/or retailer.

As in the case of logistics services, the rationale behind a 3PL able to provide an asset pooling service is the fact that the third party has its core business in the assets sector and therefore invests resources both for the purchasing of the assets and in their management, from picking-up, reconditioning, and making the required quantity available directly at the point of use.

In this way, the SC player can focus on its core business, and need not invest Capex resources in the assets (a pallet used to cost around € 15–20, after the pandemic rised up to € 30; a roll can cost € 100–150, while an isotainer costs thousands of Euro)—in repair and sanitization lines, or in storage and Opex plants in non-core activities such as packaging management, transport, handling, repair and conditioning of assets (see Table 8 for details of terminating costs in the case of outsourcing). The asset poolers' core activity is the management of packaging, therefore, by exploiting economies of scale, they are theoretically able to offer a better service at a lower cost.

On the other hand, outsourcing does include a source cost, represented by the Capex for IT integration (one shot) and the Opex of the rental fee for each €/pallet or €/roll of packaging handled.

Different levels of assets outsourcing are possible, as shown in Fig. 46

Table 8 Capex and Opex terminating in the case of asset management outsourcing

Capex	Opex
Asset cost (initial investment and reinstatement)	
Logistic infrastructures for asset reconditioning	Staff (for asset management activities)
Technological infrastructures for asset management	Operating costs (transport, asset reconditioning, repair, disposal) Activity management costs

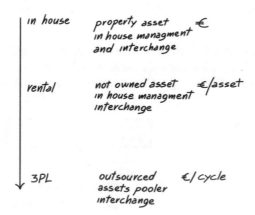

Fig. 46 Asset pooling—possible outsourcing scenarios

- In-house: We talk about in-house management to identify a situation in which manufacturer and/or retailer are owners of the assets, and manage handling and repairs internally.
- Rental: This is a situation in which manufacturers and retailers do not own the assets but rent them from a third party. However, reverse logistics are the responsibility of manufacturers and retailers, e.g., picking up from a store and returning to a DC. Any repairs are carried out by the asset pooler to which the assets to be repaired or disposed of have been sent.
- 3PL asset management is completely outsourced. The asset pooler owns the assets, and deals with the management of reverse logistics, conditioning, and eventual disposal.

Interchange of Assets

An important issue at the manufacturer/retailer interface is that of the interchange of assets. In point of fact, SC players must be able to exchange among themselves not necessarily the same asset (typically a pallet, therefore we still refer to pallets in the further discussion) but also equivalent assets (pallets), and not necessarily at the same instant but also at different times. The term used is *equal interchange*—the interchange can take place in two ways (see Fig. 47):

- Immediately: i.e.., upon delivering the palletized goods, the transporter receives and collects an equal number of assets of the same type. This is certainly the most straightforward way, since the supplier collects an equal number of equivalent assets for the number of assets delivered. However, equivalent assets are not always available, so the need for a deferred interchange can arise.
- Deferred, i.e., with a asset voucher: this occurs due to the unavailability of interchangeable assets (absent or damaged) and the inability of the transporter to pick up (saturated load trailers), at which point a voucher is generated for a deferred pickup. In order for the deferred pickup to take place, the quantitative and or qualitative unavailability at the point of delivery must be temporary and not endemic.

In both cases, however, the standardization of assets plays a pivotal role, in order to exchange equivalent assets not only from the point of view of size but also of performance. For example, the EPAL (*European Pallet Association*) standard ensures

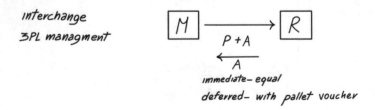

Fig. 47   Interchange of assets

not only that the pallets are 800 × 1200 mm in size, but also that the quality and manufacturing specifications and therefore the performance of the wood between two EPAL pallets are equivalent, and they can therefore be exchanged.

In the case of asset pooling, the management of the interchange is the responsibility of the 3PL which can organize it immediately, or defer it, according to its needs.

If the asset pooler then uses a closed circuit of proprietary assets, the management of the interchange does not arise, since the assets are easily identifiable, usually in colour (e.g. In Italy CHEP uses blue pallets, CPR green pallets, CTL red pallets) and also recoverable.

Examples of asset poolers are:

- CHEP, which supplies blue pallets and black crates.
- CPR System, a consortium which brings together the main fruit and vegetable producers (i.e. Apofruit and Apoconerpo), and distribution brands such as COOP, Conad, and PAM Panorama. It uses pallets, bins, mini-bins, crates with folding sides, and green crates.
- Number 1, which uses standard EPAL pallets and arranges interchanges by means of pallet vouchers.

RFID for Asset Management

RFID asset tracking solutions can help track assets as they move from one partner to the other, not only for real-time information on shipped goods, but also to speed up processes, reduce labour, increase accuracy, and locate and reduce issue related to losses and theft. Indeed, by using a combination of RFID tags, RFID readers, fixed and mobile devices, as well as tracking software, these assets and the goods transported can be accurately and real time located.

For example, in a SC composed by a manufacturer, retailer and the asset pooler, as that depicted in Fig. 45, an RFID tag with the electronic product code EPC is affixed on pallets designated to a particular FMCG manufacturer at the 3PL warehouse. The EPC contains information the pallet ID, according to the GRAI standard detailed below. As the pallet exit the dock door of the asset pooler, RFID readers capture and store the relative information in a data warehouse DW, making it available to both parties. Once arrived at the manufacturer production plant/DC, the pallets are loaded with the goods to be shipped to the FMCG retailer (to the DC, to the TP in case of cross docking, or directly to the retailers' stores for direct shipments) and scanned through RFID readers. Once the pallet is off-loaded, it is scanned again and information can be shared among shareholders. The asset then returns to the manufacturer or goes back to the asset pooler. In case the pallet returns to the asset pooler, the tag is scanned one last time to verify the authenticity of the pallet and note any damage before it's repaired. Since the system knows the journey and which parties were involved along the journey, all data can be shared to gain visibility of the entire process (Bottani and Bertolini 2009; Green et al. 2017).

Therefore, implementing RFID/EPC technology for asset management could help retrieve valuable information to reduce losses and manage cost involved, improving also the rotation index and optimising capital associated with logistic assets. RFID implementation could also definitely lead to labour saving as it potentially reduce manual operation involved in manual checking and documentation (Rizzi et al. 2011).

According to GS1 standards, The standard for RFID assets identification that enables real time or deferred interchange is the GS1 Global Returnable Asset Identifier (GRAI), which is part of GS1 standards for identification. It is a numeric code $N1,\ldots$,up to N29.

The GRAI is composed of:

- Extension digit N1—always 0
- the GS1 Company Prefix of the company assigning the asset identifier $(N2,\ldots,N9)$
- the asset type (N10–N12). The asset type is assigned to uniquely identify, together with the GS1 Company Prefix, a particular kind of asset.
- The check digit—N13
- the serial—$N14,\ldots$, up to N29 that identifies the single asset. Although consecutive numbering is recommended, the structure is left to the discretion of the assigning company.

Asset Pooler—Make-or-Buy

Also in the case of asset poolers, the choice between in-house or outsourced management takes place in a make-or-buy logic by comparing the Capex and Opex of the two solutions. As previously mentioned:

- In the case of *Make*, the main items to consider are:

  - Capex:

    cost of assets (initial investment plus annual replenishment due to lost or damaged assets),
    cost of repair, reconditioning and handling infrastructures

  - Opex:

    management costs of the assets park (organization of collection, restoration, distribution),
    costs of pickup, transport and handling,
    conditioning and repair costs.
    disposal costs.

- In the *Buy* case, there is the Capex cost for IT integration, while there is only one Opex item:

  - Opex: rental cost $(cost\ per\ handling\ = \frac{\text{€}}{handling\ unit} * \frac{handling\ units}{year})$.

In addition to the strictly quantitative part, linked to the analysis of make-or-buy cash flows, it is necessary to add a qualitative part, related to the strategic and tactical implications that the choice to outsource a key service can entail.

Tactical aspects typically concern the flexibility of the service since the quantities requested can vary. Outsourced management is certainly more flexible to changes in demand since increases or decreases in demand for pallets or assets simply translate into a greater or lesser demand for assets and are absorbed by the asset pooler who merely makes them available (or not). Conversely, in the case of in-house management, an increase in volumes requires investments in new assets, while a reduction in volumes involves unused assets which clearly have a negative impact on handling costs. Therefore, if strong fluctuations in volumes are expected, it may be reasonable to move towards a more flexible, outsourced management; on the contrary, in the face of stable volumes, it is possible to think of opting for in-house management.

Strategic aspects may involve considerations on a service's reliability. Again, an asset pooler's deterioration in service could be critical in terms of effectiveness or efficiency. For example, non-deliveries of crates, bins or pallets which could block a palletization process, or deliveries of products that do not comply with quality standards (e.g. deliveries of pallets of the wrong size) which could result in the impossibility of a pallet being properly stored in the warehouse, or non-standard assets that can break and give rise to damage and loss of the product in addition to the costs of repacking.

In this case too, the contracts must verify the termination for cause clauses, such as the specific bonus/malus clauses for fees based on quantitative and objective service KPIs (as seen above), and similarly, termination clauses that protect those who request the service in case of non-fulfilment or worsening of service levels in addition to predefined limit thresholds for prolonged periods of time, with the possibility of exercising termination of the contract.

Asset Pooling—A Make-or-Buy Exercise

A company must decide between the in-house or outsourced management of its pallet fleet in light of the following numbers, referring to a service 7 days a week, 365 days a year:

Data	Values	Units of measurement
Volumes (7/365)	1,500	Pallet/day
Turnaround Time	2	Weeks
Purchase cost	20	€/pallet
In-house management cost	3.5	€/handling
Cost of 3PL	4.0	€/handling
Capex depreciation	$i = 0; n = 5$	Years
Reintegration (restoration)	15%	1/year
Tax rate	33%	

We need to calculate the annual operating cost for in-house and outsourced management, and establish which is more advantageous.

A feasibility study is carried out in the following table.

MAKE-OR-BUY ANALYSIS					
• N° of assets necessary					
$n°\ asset =$	$volumes * turnaround$				
	$1,500\frac{pallet}{dd} * 14dd = 21,000\,pallet$				
Initial investment	$21,000\,pallet * 20\,\frac{€}{pallet} = 420,000\,€$				
Depreciation in constant instalments at 0 rate	$\frac{420,000\,€}{5\,yy} = 84,000\,\frac{€}{yy}$				
Reintegration	$15\% * 21,000\,pallet = 3,150\,\frac{pallet}{yy}$				
	$3,150\,\frac{pallet}{yy} * 20\,\frac{€}{pallet} = 63,000\,\frac{€}{yy}$				
Depreciation	$\frac{63,000\,\frac{€}{yy}}{5\,yy} = 12,600\,\frac{\frac{€}{yy}}{yy}$				
• In-house management cost—MAKE					
Handling costs	$\frac{handling}{yy} * \frac{€}{handling} = 1,500\,\frac{handling}{gg} * 365\,\frac{dd}{yy} * 3.5\,\frac{€}{handling} = 1,916\,\frac{k€}{yy}$				
Costs of managing 3PL—BUY					
Handling costs	$\frac{handling}{yy} * \frac{€}{handling} = 1,500\,\frac{handling}{gg} * 365\,\frac{dd}{yy} * 4.0\,\frac{€}{handling} = 2,190\,\frac{k€}{yy}$				
• Tax credit					
T	$Taxes = grossprofit * taxrate$				
– MAKE					
Capex	$Y = 0$	$Y = 1$	$Y = 2$	$Y = 3$	$Y = 4$	$Y = 5$
Investments [€]	420,000	63,000	63,000	63,000	63,000	63,000
Opex	$Y = 0$	$Y = 1$	$Y = 2$	$Y = 3$	$Y = 4$	$Y = 5$
Management [k€]	1,916	1,916	1,916	1,916	1,916	1,916
EBITDA [k€]	−1,916	−1,916	−1,916	−1,916	−1,916	−1,916
Amortization [€]	84,000	96,600	109,200	121,800	134,400	63,000
Year I [k€]	84	84	84	84	84	–
Year II [k€]		12.6	12.6	12.6	12.6	12.6
Year III [k€]			12.6	12.6	12.6	12.6
Year IV [k€]				12.6	12.6	12.6
Year V [k€]					12.6	12.6
						12.6

(continued)

(continued)

MAKE-OR-BUY ANALYSIS

$GP =$ $EBTDA -$ $Amm.$ [k€]	−2,000		−2,012	−2,025	−2,038	−2,050	−1,979
Tax credit T [k€]	660		664	668	672	676	653
$NP = GP + T$ [k€]	−1.340		−1.348	−1.357	−1.365	−1.374	−1.326
$NCF =$ $NP + A -$ $I0_0$[k€]	−1.676		−1.315	−1.310	−1.306	−1.302	−1.326
NCF_{MAKE} [€]	$\sum_i FCN_i = 8,237,355$						

− BUY							
Capex		$i = 0$	$i = 1$	$i = 2$	$i = 3$	$i = 4$	$i = 5$
Investments [€]		−	−	−	−	−	−
Opex		$i = 0$	$i = 1$	$i = 2$	$i = 3$	$i = 4$	$i = 5$
Management [k€]		2,190	2,190	2,190	2,190	2,190	2,190
EBITDA [k€]		−2,190	−2,190	−2,190	−2,190	−2,190	−2,190
Ammortization [€]		−	−	−	−	−	−
$GP = EBTDA - Amm.$ [k€]		−2,190	−2,190	−2,190	−2,190	−2,190	−2,190
Tax credit T [k€]		−722	−722	−722	−722	−722	−722
$Net\ Profit = Gross\ Profit + T$ [k€]		−1,467	−1,467	−1,467	−1,467	−1,467	−1,467
$NCF = Net Profit + A - I_0$[k€]		−1,467	−1,467	−1,467	−1,467	−1,467	−1,467
NCF_{MAKE} [€]		$\sum_i FCN_i = 8,803,800$					

Simplified Approach

It is possible to further deepen the study with a simplified approach, i.e., considering only the Capex and Opex (both for the MAKE solution, only the Opex for the BUY solution, respectively), without considering the ammortization and tax rate benefits. In this case, the analysis is as follows.

$$\text{MAKE} \sum CAPEX + OPEX$$

$$\text{BUY} \sum OPEX$$

MAKE-OR-BUY ANALYSIS

– MAKE						
Capex	$Y = 0$	$Y = 1$	$Y = 2$	$Y = 3$	$Y = 4$	$Y = 5$
Investments [€]	420,000	63,000	63,000	63,000	63,000	63,000
Opex	$Y = 0$	$Y = 1$	$Y = 2$	$Y = 3$	$Y = 4$	$Y = 5$
Management [k€]	1,916	1,916	1,916	1,916	1,916	1,916
$TOTAL =$ $CAPEX + OPEX -$ [€]	2,336,250	1,979,250	1,979,250	1,979,250	1,979,250	1,979,250
FCN_{MAKE} [€]	$\sum_{i} FCN_i = 12, 232, 500$					

– BUY						
CapEx	$Y = 0$	$Y = 1$	$Y = 2$	$Y = 3$	$Y = 4$	$Y = 5$
Investments [€]	–	–	–	–	–	–
OpEx	$Y = 0$	$Y = 1$	$Y = 2$	$Y = 3$	$Y = 4$	$Y = 5$
Management [k€]	2,190	2,190	2,190	2,190	2,190	2,190
$TOTALE =$ $CAPEX + OPEX -$ [€]	2,190	2,190	2,190	2,190	2,190	2,190
FCN_{MAKE} [€]	$\sum_{i} FCN_i = 13, 140, 000$					

1.3.10 Marketplace

Another category of SC players who interact with Focal Company, Retailer and 3PL, is that of the Marketplace. Its development is mainly due to the dizzying growth of online transactions in recent years, linked to e-commerce.

Born from services (Booking, Flixbus, Enjoy, Airbnb, Just Eat), marketplaces are gradually extending to the world of products (Farfetch in the case of clothing, Amazon itself playing the role of a marketplace for most of the products it sells, Cortilia for fruit and vegetables), becoming, to all intents and purposes, players in the supply chain, given that they contribute to bringing a product and a service to a final consumer.

A marketplace is a body that provides an online platform, which in turn itself becomes a focal company, within which producers or distributors meet the consumer's demand.

A marketplace manages the relationship with the supplier—in terms of products offered for sale, and with the final consumer—in terms of order collection. Once the product is put up for sale on the marketplace and the order is collected from the platform and returned to the manufacturer, the manufacturer takes care of the order fulfilment and delivery cycle, while the marketplace retains a percentage of the transaction (premium or fee). The distribution structure scheme is represented in Fig. 48.

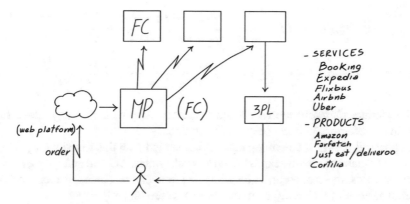

Fig. 48 Functioning of marketplaces

1.4   The Organization of the Distribution Channel

This chapter describes how the players seen above fit into the distribution channel. The structure of possible distribution channels is presented with reference to the fast-moving consumer goods sector, whose peculiarities can also be extended to other product sectors.

1.4.1   Introduction

Before tackling the discussion, however, it may be useful to introduce some details about the cost of stocks, the cost of transport, and the cost of service, since many distribution choices are intimately linked to their impact on the two components of the costs of stocks, in this case **PC** (*physical costs*) and MMC (*market mediation costs*).

The Costs of Stocks

The costs of stocks are divided into two components.

Physical Cost

A first function of holding stocks is to allow economies of scale in procurement and take advantage of quantity discounts, but also to absorb changes in demand during lead time and lead time variations with respect to anticipated values.

We talk about the Stock Rotation index IR and its inverse, the days inventories, which are defined according to the following ratios.

$$rotation\ index\ IR = \frac{demand\ in\ period\ t\left[\frac{item}{t}\right]}{average\ inventories\ in\ period\ t[item]}$$

$$days\ of\ inventories = \frac{1}{IR} = \frac{average\ inventories\ in\ period\ t[item]}{demand\ in\ period\ t\left[\frac{item}{t}\right]} \qquad (6)$$

The former therefore represents the frequency with which average stock turns over in the reference period t (day, week, or month).

A physical cost of maintaining stock is associated with this functionality. This is none other than the cost of capital fixed in cycle stocks SC and safety stocks SSI. In other words, the cost incurred by allocating financial resources to keep a product in stock instead of allocating them to other uses (see Fig. 49).

The physical cost of stock is calculated as in the equation below, where $\underline{G}$ is the average stock $\underline{G} = \frac{CS}{2} + SS$, p is the price (but we can also use industrial cost C_I equal to the sum of the costs of production, management, distribution, and sale), τ is the capital valuation rate (which may be bank interest i in the case of a bank loan, or ROE in the case of equity).

$$\underline{C}_s = \underline{G} * p * \tau \quad \underline{C}_s = \underline{G} * C_I * \tau$$

Market Mediation Costs

An equally important function of stocks is to allow a product to be supplied in advance of the moment when its actual demand will occur. This is the case of seasonal or recurring products demand for which is concentrated in extremely short periods of time, or products which need lengthy supply and production times, as in the case of traditional fashion goods. As a consequence of this advance, however, a mismatch may be generated between the demand anticipated by forecast F at time τ_0 and the real demand, $D_{(\tau_0 + \Delta\tau)}$, which will only occur after a while, $\Delta\tau$.

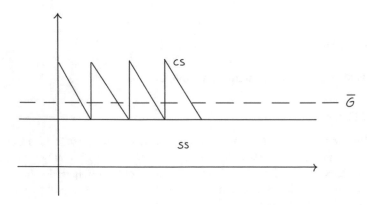

Fig. 49 Stock, cycle stock and safety stock

Fig. 50 The origin of market mediation costs: a mismatch between demand and forecast

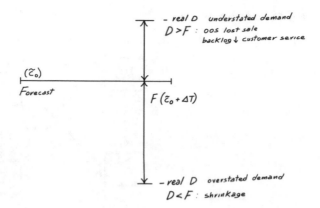

In other words, two situations can occur, which both correspond to market mediation costs, as represented in the following Fig. 50:

1. $D > F$: demand at time $t_o + \Delta T$ greater than forecast at time t_o. These determine the costs of backlogs or out-of-stock (OOS) events. Out of stocks are real lost sales, while in case of backlogs, the customer is willing to wait, and orders will be delayed and fulfilled once the product is available. Therefore, in the case of backlogs, the cost incurred is due to the poor customer service perceived by the customer. Out of stocks are *opportunity costs*, which can be expressed through the following relationship,

$$C_{OOS} = (D - F) * (p - C_{ind})$$

   where C_{ind} represents the industrial cost of the product, and p the selling price. Although it is true that there are lost sales, the costs for the production and distribution of the unsold product may not been incurred.

2. $D < F$: demand at time $t_o + \Lambda T$ shorter than forecast at time t_o. This results in a *shrinkage cost*, due to the fact that since the product has no market it must therefore be sold off at a lower price (e.g. recurring products) or disposed of, for example, in the case of fresh products which have remained beyond their expiration date.

The shrinkage cost can be expressed through the following relationship, depending on whether the condition is of *Disposal* or *Diverting*.

- Disposal: we speak of this if the product is disposed of due to obsolescence (technological, seasonal reasons, useful life).

$$C_{shrinkage} = (F - D) * C_{ind} + C_D$$

   where C_D represents the cost of disposa of the quantity (F − D).

Fig. 51 Shrinkage diverting

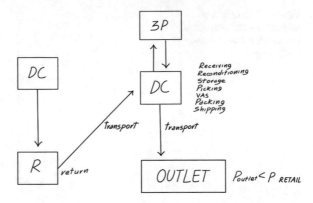

- Diverting: if the product needs to be placed on another channel (e.g. an Outlet), which brings with it $\Delta cost$ due to the handling for relocation (e.g., reconditioning, storage, picking, packing, shipping, etc.), and a $\Delta price$ due to the change of channel (e.g. outlet instead of retail) (see Fig. 51).

$$C_{shrinkage} = (F - D) * (\Delta_{price} + \Delta_{cost})$$

Even in the case of a physical cost, safety stock performs the task of absorbing the mismatch between demand and forecast, but in this case there are neither out-of-stock costs, nor shrinkage costs, since the product does not undergo either disposal or diverting (Fig. 51).

The Cost of OOS in FMCG

Out-of-stocks (OOS) is one of the major challenges for both retailers and FMCG manufacturers as they seek to maximize product availability and enhance the consumer shopping experience. Studies have shown that OOS levels at the upstream echelons in the supply chain are lower compared to OOS levels at the retail shelves. As cited by researchers, stock-out causes generally derive from; forecasting errors, ordering errors, shelf replenishment, distribution centre issues, manufacturer issues and other indirect issues (Bottani et al. 2009).

Incorrect forecasting and ordering practice is one of the major causes of OOS. Retailers need to forecast and order accurately, suppliers need to deliver the right quantities at the right time, distribution needs to ensure the product reaches the stores, and the stores themselves need to get it onto the shelves.

Products not being on the sales floor also contribute to stock-out. OOS situation affects sales of different products in different manners. Wrong forecasting (13%), ordering decisions (34%) are responsible for close to 50% of all OOS situations at the retail store together with shelf replenishment (25%) (Tellkamp 2006).

In the grocery environment, the fastest moving 25% of items account for 66% of lost sales, with promotional lines mostly affected, averaging an out of stock rate on advertised items of 15%. (PWC Consulting 2002)

For FMCG, out-of- stock situation in store is related to:

- Replenishment from back room (shelf stocking problems) In stores, OOS situation for products that goes out of stock for a few hours during the day, may be due to replenishment practices
- Poor inventory accuracy: overstated inventory due to theft, scanning errors and incorrect deliveries may lead to theso called "froze" OOS, ment to be a situation when on hand inventory is zero, while inventory in the legacy systems is more than the reorder point. In this situation the product can not be sold due to zero inventories, therefore the reordering can not be triggered
- Unexpected demand fluctuation (forecasting errors)
- Out-of-stock situation in the upstream of the supply chain
- Inaccurate deliveries to stores by DC.

As studies have shown, consumers are not indifferent to stock-out situations as they have gained more insight on the products they purchase. In stock-out cases, consumers can i) substitute one item for another, ii) switched brands (this entails a lost sale for the manufacturer—current and likely in future if the customer likes the new brand more—but not for the retailer), iii) delayed the purchase (lost sale neither for the manufacturer nor for the retailer), or iv) tend over to other retailers (lost sale for the retailer and not for the manufacturer). The tendency is a high degree of replaceability. The consumer is showing less loyalty to the brand and is ready to switch from one brand to the other, depending what is available on the shelf.

For retailers, a high level of product availability improves consumer's loyalty and definitely reduces losses in sales related to OOS situations. In order to manage inventory effectively to maximize product availability, RFID technology provides a solution for planning, forecasting, replenishment and transmitting information to enhance product visibility at all points of the value chain in real-time. RFID deployed at either item level or at case-level tagging can bring maximum benefit to track what is available in the backroom, on the sales floor, increase inventory accuracy, and prevent OOS (Hardgrave et al. 2006; Bottani et al. 2009).

Cost of Transport

The cost of transport is analysed in relation to road transport, by far the most common case when dealing with a distribution network.

The unit cost of transport C_U can be expressed through the relationship.

$$C_U = \frac{C_{trip}}{Q_{loaded}}$$

It is evident that to reduce unit cost it is necessary to increase the quantities loaded Q_{loaded}, tending to saturate the load capacity of the vehicle, in terms of either weight or volume. To understand whether the vehicle will be saturated by weight or by volume, it is necessary to compare the capacity of the vehicle with the capacity of the loading trail. In the case of articulated lorries, for example.

$$Q = 30.000 \text{ kg}; \ V = 2,5 \times 2,5 \times 13 \text{ m} = 82 \text{ m}^3$$

Accordingly, we talk of *heavy goods*, which saturate the vehicle by weight in a density higher than 370 kg/m^3, and *bulky goods*, which saturate the vehicle by volume for a density lower than 370 kg/m^3.

A second element which affects the unit cost are the economies of scale obtained as the load capacity of a vehicle increases.

The cost of a trip is given by the Capex quota represented by the depreciation of the vehicle, and by the Opex quota resulting from labour costs, tolls, fuel, maintenance, stamp duty and insurance, etc..

Some of these components are constant, others grow less than proportionally with the load capacity of the vehicle, therefore the unit cost decreases as the load capacity of the vehicle increases, and is therefore minimal (for full saturation, obviously) for articulated lorries, intermediate for a truck, and maximum for a van of limited capacity. On the other hand are the costs of the trip. Economies of scale related to load capacity are shown in Fig. 52.

Ultimately, to try to reduce the cost of transport, the available levers are essentially two:

(1) Try to saturate the load capacity of the vehicle
(2) Use means of a greater capacity in order to transport the maximum amount of goods possible in a single journey.

Service Level—Lead Time

The service level will be described more fully in Chap. 5. Here, for the purposes of the subsequent discussion and to understand the implications of the distribution system organization on the service level and vice versa, some basic concepts are introduced.

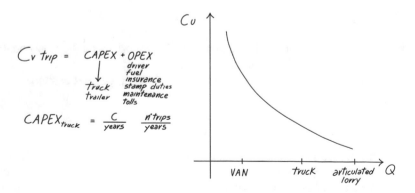

Fig. 52 The unit cost of transport according to the load capacity of the vehicle

As a starter, the service level is identified in a simplified manner with the lead time, which is the time that elapses between the issuing of an order by a buyer and receipt of the compliant goods from the vendor. As we shall see later, in reality, service depends on many other aspects, but for now we can go with this simplification.

That said, lead time LT basically depends on 3 factors.

- The degree of stock coverage: the greater the stock present at the vendor, the faster the vendor will be in fulfilling an order, not having to procure items upstream
- Processes at the vendor: the faster and synchronized the processes are, therefore without waiting between one process and another, e.g., to consolidate a large quantity of goods to carry out a process or a phase, the shorter the order fulfilment and transport times will be
- Proximity: the closer the vendor is to the buyer, the more the lead time is reduced, since transport times are shortened.

It is immediately evident that increasing the service level and therefore shortening the lead time tends to have a negative impact on other cost components: if the degree of stock coverage increases, the costs of stocks, both physical and potentially market mediation, increase, given that the likelihood of shrinkage grows; if we move close to the buyer, distances are reduced, but the structures and therefore the relative costs multiply; if the lead times are reduced with vehicles of lower capacity which become saturated more quickly, then costs are increased.

1.4.2 Traditional Normal Trade

Normal trade is the traditional channel, still particularly popular both in the FMCG sector in Italy, for example to serve small normal trade retailers located in cities and small towns, but also throughout the pharmaceutical sector, serving pharmacies where the sale of medicines takes place.

The organization of the traditional normal trade channel, the players involved, the scope of action, and the level of flows are shown in Fig. 53.

The FC has one or a few full-mix warehouses, usually located nationally (e.g. in Italy) or by geographic area (e.g. those of EMEA—Europe, Middle East and Africa). By "full-mix warehouse" we mean a place in which the entire range of products produced in different production plants is present, and which therefore, by having the whole mix available, is able to serve a vast territory and, in particular, the distributors. The full-mix warehouse is fed directly by factory production lines or end of production line depots. It is typically a distribution centre/warehouse, depending on the characteristics of the references. The longer the expiration period, the lower the rotation index IR and therefore the higher the stock compared to the flows, and vice versa; as the product's expiry date shortens it will be characterized by a "targeted flow", with the flow part predominating over the stock.

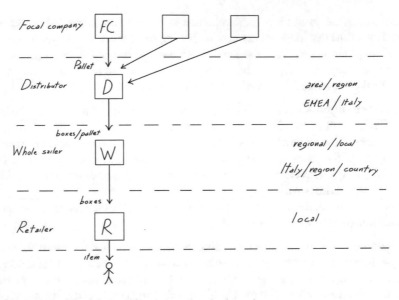

Fig. 53 Organization of the traditional normal trade channel

Distributors

A distributor also serves a regional market, typically an area (a nation, a region) and procures from manufacturers' full-mix warehouses, in order to satisfy the downstream demand made by wholesalers.

Since the downstream demand is high, the distributor procures supplies from a manufacturer, typically in whole pallets. In this way, it generates economies of scale both in procurement through large economical lots with which it can obtain favourable conditions from the manufacturer and quantity discounts, and in the transport. The distributor manages the stock at reorder intervals and order up-to level—typically low-rotation goods, or at a reorder point—typically for high-rotation goods, in line with economic concepts (i.e. economical lots), thus attempting to optimize stock maintenance costs with reordering and transport costs, further balancing opportunity costs and/or quantity discounts. Given the volumes, the FTL transport usually involves articulated lorries or trucks, but the volumes are such as to allow direct deliveries according to an FTL logic.

The inventory and transportation characteristics of a distributor are as follows:

- High stock (i.e. coverage of demand at least weekly if not monthly)
- High reorder frequency to the manufacturer/FC (weekly)
- FTL transport of high capacity articulated trucks
- LT: in days.

The distributor interfaces downstream with wholesalers.

Wholesalers

Wholesalers serve a national or local market, and procure their goods, as mentioned above, from distributors.

A wholesaler's order from a distributor is mixed, and may mean (i) full pallets for high-rotation or low value and/or long shelf life products (e.g. water, toilet paper, but also long shelf life dry foods like coffee or olive oil), and (ii) packages prepared using picking for low-rotation or high value and/or perishable products).

As we have seen for distributors, also for wholesalers the management of stocks takes place according to the same policies, therefore at a reordering interval and order up-to level for low-rotation products, or at a reordering level for high-rotation ones, with the logic of optimizing the costs of maintaining stocks, reordering and transport, and benefitting from opportunities related to quantity discounts.

As regards transport from the Distributor to the wholesaler, due to the reduced volumes linked to the limited size of the market to be covered compared to a distributor, these may be FTL if a sufficient quantity of goods is consolidated by increasing the lead time and therefore reducing the delivery frequency, or in the case of orders of reduced quantities and high frequencies, there are two options: (a) adopt low capacity transport, which lead to higher unitary transport costs; (b) have full load deliveries of high transport capacity articulated truck if fulfilled by a distributor in a multidrop perspective.

In this way, a high delivery frequency is possible (in the event that high volumes are consolidated for a single wholesaler), to delivery frequencies of even a few days, collecting the volumes of several wholesalers in a multidrop logic, as seen in Sect. 1.3.2

A multidrop delivery has a direct impact on the physical inventory costs for a wholesaler: In Table 9 is an example calculation of the change in the cost of fixed assets in the consolidation logic of an FTL or multidrop.

The multidrop logic is adopted above all for perishable or fresh products, in which the product market mediation costs linked to possible shrinkage if deliveries are delayed to consolidate transports, represents a higher cost than that for transport. On the other hand, for non-perishable products, the cost of keeping stock is strictly linked to the selling price, so that for high-value products the multidrop logic is expedient, while for low-value products it may be better to wait to consolidate volumes.

The stocks and transport at the wholesaler have the following characteristics:

- Intermediate stock (i.e. coverage of weekly demand)
- Relatively high reorder frequency from the distributor (1–2 times a week)
- FTL multidrop inbound transport
- LT: in days.

Downstream, the wholesaler interfaces directly with normal trade retailers.

Table 9 Example of distributor-wholesale distribution channel: the multidrop case

$$C_{imm} = Q_T * p * \tau_T = D * \frac{\Delta T}{2} * p * \tau_T$$

$C_{imm} = inventory\ holding\ costs$	
$\Delta T = Lead\ Time\ to\ saturate\ the\ load\ capacity$	
$Q_T = average\ demand\ in\ \Delta T$	$D = daily\ demand$
$p = unit\ price$	$\tau_T = annual\ interest\ T$
Case A (FTL)	Case B (multidrop)

Case A (FTL):

$D = 1\ pallet/day$

FTL transport

33 days $\longrightarrow$ 33 pallets

Q ↑ ... 33 ... 16,5 pallets ... 33 ... C

$CH = 16,5 \cdot P \cdot T$

Case B (multidrop):

$LT = 1\ day$ $PO = 1$

$O = 1$

$CH = 05 \times T \times \tau$

Retailer (Normal Trade Shop)

Finally, the last element of the distribution chain are the retailers, who operate at the proximity level and procure from wholesalers, usually by single at most by secondary packages or even in single items, according to reorder interval policies, with a high reorder frequency, e.g., every day, 3–5 times a week (typical example: pharmacies and low-rotation goods with an LT of hours or at most of 1 day).

Retailers have no inventory, if not minimal for high-rotation goods, that is, to meet daily demand or little more.

Retailers often buy their supplies by physically going to the wholesaler to make the purchase or, as in the case of pharmacies, they place the order and the wholesaler takes care of fulfilment, transport and delivery. Wholesaler transport takes place with vehicles of limited capacity (e.g. trucks or vans), in city centres or suburban areas, in a multidrop logic to contain transport costs, given the small quantities of the orders.

Sometimes, the wholesaler satisfies the retailers' demand using the *attempted sale* mechanism. This is, for example, the system for selling fresh milk at local stores, or any other fresh food products.

As seen previously, the wholesaler visits the retailer with a vehicle (e.g., a truck) containing stock. During the visit, the driver refills the shelf with a product sold directly by the store or placed on the shelf in an attempted sale. The next day, the returns are collected, and the difference sold is invoiced:

$$inventories(t-1)+delivered(t-1) = sold(t)$$
$$+inventories(t) + returns(t)$$

To sum up, the procurement and transportation characteristics of a retailer are as follows:

- Low inventory level (i.e., zero stock (Just In Time procurement) or coverage of daily demand)
- High reorder frequency (i.e., daily)
- Inbound Transportation from the wholesaler: multidrop with small capacity vehicles
- LT: hours or one day.

Critical Issues of the Normal Trade Model

The major critical issues of the distribution chain model of this type are down to the high number of players involved (at least three). This has a negative impact on the margins and duplicates certain logistics costs.

Margins

As regards the first aspect, that of the margins, the following considerations can be made. Having fixed the industrial cost of producing a product (basic cost for raw materials and for producing one unit of finished product), if each player in the distribution chain then adds a mark-up to the purchase price of the product, this translates into a higher price of sales with equal margins, or at parity of sales price, in a margin to be distributed between several players and therefore lower for each player.

In Fig. 54, we can see how the distributors (D), wholesalers (W) and retailers (R) divide the margins $M = P - C_p$, and that the only way to increase each margin without introducing other factors is to increase the selling price.

This is the case of the Italian pharmaceutical industry, where the selling price for the vast majority of drugs (those that are reimbursed by the national healthcare systems and can be sold only under a physician's prescription) is fixed in order

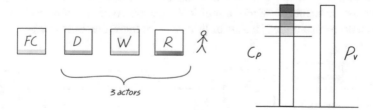

Fig. 54 Breakdown of margins in a normal trade channel

to guarantee each supply chain player (manufacturer, distributor, wholesaler and pharmacy) a fixed, predetermined, margin.

Duplication of costs

In addition, multiple players also lead to critical duplication of costs linked to various other aspects, first and foremost, the costs of stocks. The duplication of stocks between D, W, and R has an impact on both the physical costs of stocks, which are duplicated by each player according to its average physical costs $C_{physical} = \sum_{i=1}^{3}(\underline{G} * p * \tau)_i$, where $\underline{G}$ is the average inventory $\left[\frac{items}{year}\right]$ for each player, and on the market mediation cost, with the risk of shrinkage (expired, unsold, obsolete, disposal cost) duplicated at every level.

In addition, a whole series of costs for reordering, fulfilment, and product handling by the various players are duplicated. In fact, in detail, the following processes are present for each player:

- Focal Company

  - Receipt of client's order
  - order fulfilment
  - Shipping and transport
  - Active administrative management

- Distributor, wholesaler and retailer

  - Supplier order issuance
  - Receipt of customer orders
  - Inbound
  - Handling and Storage
  - Order fulfilment
  - Shipping and transport
  - Active and passive administrative management.

Transport deserves a special mention. In an SC like the one described, it takes place in the ways shown in the following figure (see Fig. 55).

In addition to there being three means of transport (so that the fixed quota of the transport cost is tripled), FTL (30-ton capacity) are optimized only for the manufacturer/distributor part, while for the downstream part D→ W and W→ R, they are LTT or must be managed within a multidrop perspective, or can be FTL but using vehicles with a capacity of 15 tons or less, and are therefore more costly.

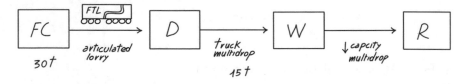

Fig. 55 Transport in traditional normal trade

The Bullwhip Effect

A further critical issue is linked to the depth of the SC which means that the upstream manufacturer, if the consumer demand data is not shared, sees an extremely distorted and delayed demand, due to a phenomenon known as the *Bullwhip Effect*, or sometimes, the Forrester Effect (Forrester 1961).

Typical symptoms of Bullwhip Effect sufferers are either too much stock (too many products in stock when demand is low) with consequent physical costs of stock or market mediation costs in the case of shrinkage; alternating with periods of high out-of-stocks, where there is no product to meet peak demand and therefore poor service levels due to backlogs, or market mediation costs due to OOS.

The Bullwhip Effect will be extensively dealt with in Chap. 4 in any case, considering a 4-level SC like the one represented, i.e., manufacturer M, distributor D, wholesaler W, and retailer R, each of them issuing orders O_i that are fulfilled with supply lead times LT_i, according to the scheme charted in Fig. 56.

The consumer's demand D is propagated to the manufacturer M, upstream through orders O, with a certain delay (given by the supply lead times LT_i) and amplified by the reordering mechanism of each SC player.

As we will see in Chap. 4, the Bullwhip Effect for level i + 1-th is defined by the ratio between the standard deviation of the orders of level immediately preceding Oi and the standard deviation of the real demand D of the final consumer.

$$BE_{i+1} = \frac{\sigma_{O_i}}{\sigma_D}$$

It will be shown that the BE typically increases upstream in the SC, and the further upstream the player is located, the higher the Bullwhip Effect is (see Fig. 57), according to the following relationship:

$$BE = \prod(N; LT)$$

Given that the BE grows exponentially from downstream to upstream, in a deep SC like that of the traditional NT, if demand is not shared among trading partners, in the face of small variations in demand, those upstream are subject to a significant BE, and therefore incur high physical and market mediation stock costs.

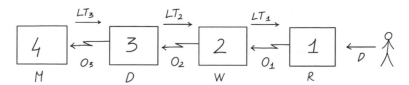

Fig. 56 Traditional demand transmission upstream in the supply chain

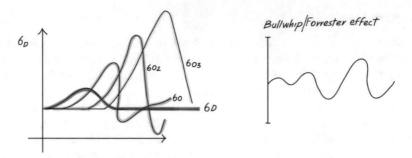

Fig. 57 The shape of the bullwhip effect

1.4.3 Normal Trade with a Logistics Operator

To try to overcome some of the critical issues listed above regarding traditional Normal Trade and in particular the length of the channel, some manufactures try replacing the traditional model to reach the NT retailer with one that includes a logistics operator, i.e. NT with a logistics operator. This is the model used not only in FMCG NT, but also in the fashion sector for both retail and wholesale outlets.

The logistics operator owns the infrastructure and covers the market through a network of regional depots and local transit points, which allow the manufacturer to directly reach the NT shops or retail and wholesale stores (proximity stores), without the need for distributors and wholesalers.

The general scheme is represented in Fig. 58. In this case, the manufacturer collects the order directly from the stores, typically through sales agents, and organizes the fulfilment either (i) through the stock present in a full-mix warehouse or (ii) in that present at a regional 3PL warehouse.

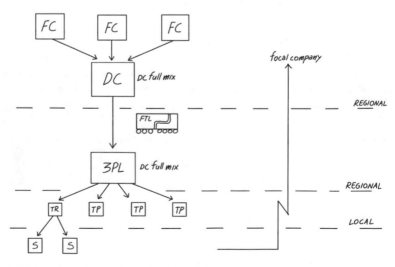

Fig. 58 Distribution scheme of the NT with logistics operator

The two configurations are represented in Fig. 59, on bottom (stock present in full-mix warehouses), and on the top (at regional 3PL depots), respectively.

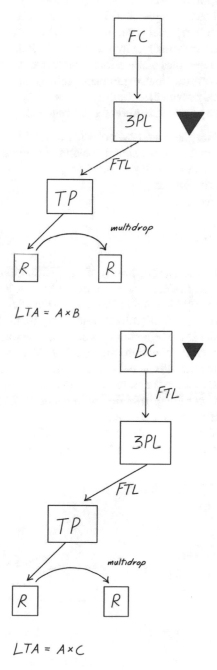

Fig. 59 Methods of order fulfilment and stock positioning in the case of an NT with 3PL

The stocks are only at a full-mix DC (i.e. factory depot), and partially at a regional 3PL warehouse (if this works as a DC), while at the transit points no stock is held.

The means of transport are FTL for the longest-travelled routes (i.e. from manufacturer full mix warehouse to regional 3PL and from 3PL DC to hubs), and LTT for short-distance ones at a local level (from transit points to store).

The fleet, as well as the infrastructure (i.e. DC and TP network), is owned by the 3PL which generates economies of scale for multiple manufacturers or retailers. The number of players involved is lower (i.e. manufacturer and retailer, while the 3PL gets paid for the service provided).

Ultimately, this type of structure allows a guarantee of:

- Greater margins, since there are fewer distributors and wholesalers involved and therefore there are fewer players among which to distribute it
- High delivery frequency
- Low LT and optimized transport
- Low stock level (general).

1.4.4   The Apparel Case

In the case of the fashion sector, upstream of the distribution chain there is an FC which owns the brand and produces and keeps in stock (at the level of its DC through a "shop box" approach) everything ordered and sold in the sales campaign to the single shop in terms of a model, variant (colour), and sizes (Stock Keeping Unit—SKU, understood to be the single combination of model/variant/size), and quantities.

The sales campaign and production usually takes place in advance of the season and therefore a customer's stock can be held at the manufacturer's full-mix DC, or at a logistics operator DC close to the store for fast replenishment, waiting for space to be freed up in the store for the distribution of the collection (2-up to 10 collections per season). The situation is, as depicted in Fig. 60.

The timing of a fashion season is depicted in Fig. 61. Fashion brands usually design their collection at least one year in advance, trying to envision and steer future trends

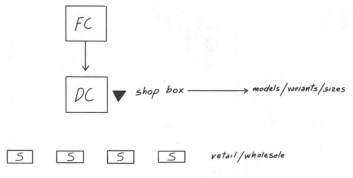

Fig. 60 Traditional distribution structure in fashion and apparel

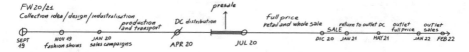

Fig. 61 Example of the timing of a fashion season

and styles. Once the collection has been industrialized, they then present it in fashion shows, and organize sales campaigns, where retailers and wholesalers can see and touch the samples and order them in model/variant/sizes. Based on sales, the supply chain is triggered. Materials are procured, production is scheduled, and distribution is organized to have the first garment available in store when the previous season is still ending.

Stores are present all over the world and as already fully described in the paragraph relating to them, they can be of two types:

- **Single-brand—retail**: brand-owned or franchised stores
- **Multi-brand—wholesale**: a completely independent third-party entrepreneur's store with regard to the management of the mix/quantity, and the display in the sales Area (when, what, and how to display).

Once again, the connection between DC and store can occur in two ways:

- Direct shipments—typically among large retail and wholesale stores: this is the case for early season supplies, or large stores that can stock the product in their backroom.
- Shipping through a logistics platform—typically among retail stores (owned or franchised): in this case the stock is moved in FTL from the focal company's DC to a 3PL DC, which works as a customer backroom for the stores, since it is located close to the stores, in order to ensure a daily or even two-day delivery frequency. This is the case for small retail shops that have no significant storage capacity. In this case, the 3PL practically operates out of the shop's backroom and its geographical proximity to the stores allows it to ensure the store timeframes of hours/days (see Fig. 62).

1.4.5 Organized Mass Distribution—Centralized Delivery

Centralized delivery is a distribution model with only two players (manufacturer and retailer), which has developed in recent years in the fast-moving consumer goods sector, following the gradual growth of Organized Mass Distribution OMD compared to normal trade.

The distribution channel for centralized delivery can be described in relation to Fig. 63.

Before centralized delivery, a manufacture's DC fulfilled orders from stores (hypermarkets/supermarkets) through direct delivery, the latter penalized by two types of inefficiency depending on the strategy pursued:

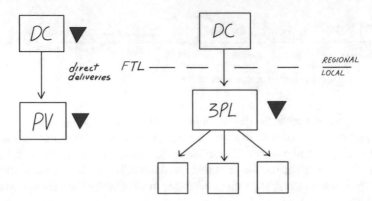

Fig. 62 Order fulfilment models of fashion and apparel stores: direct delivery and with logistics operator

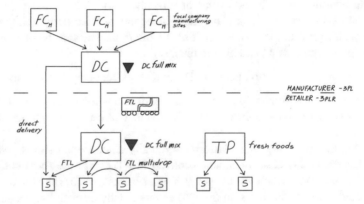

Fig. 63 OMD distribution structure with centralized delivery to retailer's DC

- The FTL strategy: inevitably subject to high LTs to consolidate a high quantity of orders and consequently high stocks in the store, equal to several days' demand.
- Low LT strategy and reduced stocks in the store: inevitably either low capacity or LTT trips and therefore high transport costs.

Consequently, the model has evolved, in an attempt to combine these two antithetical wants. To this end, a retailer's full-mix DC has been inserted, which acts as a 'decoupling' between the stores and the manufacturer's DC.

Centralized delivery is the mechanism also used for dry/non-perishable products in FMCG. The DC is replenished in FTL mode and approximately weekly (unit loads are whole pallets)—depending on the pattern of reordering. In this case, the delivery is typically $A \times B$ [next-day]. The stocks are both at the DCs of the manufacturer and the retailer, the supply usually takes place through the full-mix stock of the manufacturer's DC, but in some cases it may also take place from the manufacturer's factory depots.

The retailer's DC receives products from the manufacturer's DC in an FTL deliveries, and fulfils the orders of the stores by means of daily, or in any case high frequency, deliveries, always with multidrop FTLs for the chain's different stores. The stock level at the store can therefore be low, which allows the store to maximize its display area at the expense of the backroom area, provided that it only has to manage a minimum of stock on the shelf to meet consumer demand and maintain a minimum safety stock in the event of changes in demand (*e.g. peaks linked to weather conditions*) and/or lead time (inability to deliver due to strikes, catastrophic events, or unforeseen circumstances). Thanks to the high delivery frequency, the store can also receive product deliveries every day. To do so, the DC retailer uses the stocks present within it (days/week of demand).

Fresh products, on the other hand, are managed in a cross-docking logic, so we can refer to the relevant paragraph for the discussion of that. Minimum stock is held at the manufacturer's DC; the procurement takes place from the retailer's DC which is arranged as a transit point, to avoid having stocks of highly perishable products at the retailer's DC.

Deliveries which generate fully loaded trips to the stores (e.g. for the installation orders of promotions—an *installation order* being the first supply of quantities to cover the start-up of the promotion—or high-turnover products for large stores and hypermarkets) are still managed through direct FTL deliveries from the manufacturer's full-mix warehouse, or even directly from factory depots. This allows an optimization of transport, and at the same time avoids handling costs for the retailer's DC, as depicted in Fig. 64.

The result of the organization of this distribution channel is summarized below point by point:

- The number of players is reduced (Ideal situation no. 2; the 3PL of retailers and manufacturers may eventually come into play), and therefore greater margins or containment of sales prices are guaranteed.
- The number of infrastructures is also reduced (DC only and if necessary some TPs at a local level), while there is a large network of stores to serve the consumer locally, thereby providing a proximity service (brick-and-mortar retailers): ideally the consumer has no need to move (since it is the product that reaches the home via

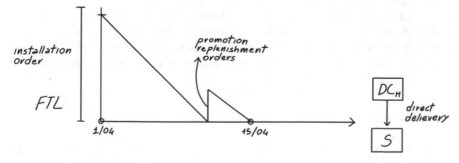

Fig. 64 Direct delivery of a promo installation order

Fig. 65 The effect a
two-echelon supply chain on
the bullwhip effect

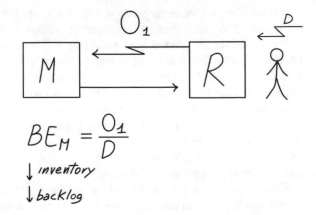

$$BE_M = \frac{O_1}{D}$$

↓ inventory

↓ backlog

supermarkets). The infrastructure may be owned (e.g., Walmart and Amazon own
DCs and fleets) or belong to a 3PL. Interestingly, if the 3PL is the same for
the manufacturer and the retailer, the logistics cost is further reduced since it is
unnecessary to move and transport goods, which become owned on the 3PL shelf:
these act both as a full-mix warehouse for the manufacturer and the retailer's DC.

- The transport involves FTLs with high-capacity vehicles (possibly multidrop),
  with the exception of small capacity transport to reach stores situated in city
  centres or out in the country.
- Both in case of direct deliveries and centralized deliveries, the chain is shallow (see
  Fig. 65), in a 2-echelon supply chain, the information on the variation in demand
  arrives earlier and is less distorted and therefore also the Bullwhip Effect, as we
  shall see later, is much reduced.
- As regards stocks, a distinction must be made between dry/non-perishable
  products and fresh products.

  For dry products, the stocks, even if smaller than in the normal trade case, are
  not optimized at the SC level since they are present at both the manufacturer and
  retailer levels, even if at the latter they are centralized in the DC—and therefore
  allow the retailer to benefit from *inventory pooling*; instead, stocks are optimized
  at the store level, since the store works with very few days' demand thanks to the
  high delivery frequency guaranteed by the upstream DC.

  As for fresh products, stocks are optimized at the SC level, since the retailer
  does not hold stocks but works in cross-docking. The stocks in the SC are present
  only at the manufacturer's DC level, while the retailer's DC is a transit point, and
  the stores hold very few days' demand thanks to frequent supplies ensured by the
  cross-cocking.

1.4.6   Online

Before the advent of online commerce, the only way to reach customers was to open
stores, so-called "brick-and-mortar", i.e. physical stores, where the consumer could
go in person and place his/her order/purchase. The situation is depicted in Fig. 66.

Fig. 66 The brick&mortar model: using a shop to let the consumer meet the product

The explosion of mass retail chains (which sometimes have thousands of stores for a single brand) is due precisely to the fact that large retailers in all sectors have tried to increase turnover T and EBITDA by increasing the number of shops, according to a principle of scale, whereby the turnover is proportional to the number of shops $T \propto n_S$, since the shop is the only way to reach the end consumer $n_S \propto n_{EC}$. In the same way, as the number of stores increases, the Capex and Opex of the store themselves increase.

Conversely, indirect costs, such as administration and finance, human resources, IT, and above all marketing, which is essential to promote a brand and a product, are almost independent of the number of stores; in some ways, even logistics and production can count on economies of scale which become extremely significant as the number of stores increases. Ultimately, therefore, the EBITDA increases as the number of stores increases (see Fig. 67).

The advent of the Internet and online commerce has broken the paradigm that in order to increase the number of consumers reached, it was necessary to increase the number of shops (Fig. 68).

In the case of online direct sales (e.g. Dell.com, Amazon.com, Yoox.com, but also in the Italian food sector, Olio Carli) the FC reaches consumers through a business

Fig. 67 Trend of revenues and costs as a function of the number of stores

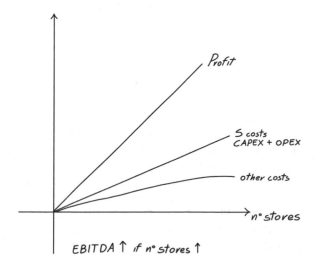

Fig. 68 The e-commerce model: using the web to let the consumer meet the product

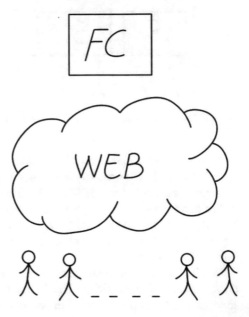

to consumer (B2C) portal which, via the internet, allows the product to be brought to the consumer. And meanwhile, the number of consumers reachable through the network increases exponentially, since every consumer, thanks to the fact that the scenarios envisaged by enlightened entrepreneurs such as "*A PC in every home*" and/or "*A smartphone for every pocket*" are becoming reality, and can be achieved from the FC through a web platform.

In the following paragraphs we will see how two e-commerce companies, namely, Dell.com and Amazon.com have structured their SC for direct sales to the final consumer and how they have used the SCM as a competitive lever with respect to their competitors.

Online—The Dell.com Case

Dell.com is one of the main players in the information technology market, in particular personal computers and servers. The company, founded by Michael Dell 40 years ago, now has a turnover of $60 billion and employs 100,000 employees worldwide. Dell's dizzying growth has been made possible thanks to the unique and innovative business model of direct sales—first by telephone and then online—and the structure of its SC, strategically designed to support this model.

In the IT market, the time factor makes a powerful distinction between a product that can be sold at full price and an obsolete product, due to both the potentially very low lifecycle and well-timed product innovations, in which the introduction of a new product or component can make existing products obsolete in a matter of days.

Before the advent of Dell.com, in a traditional supply chain made up of distributors, wholesalers and retail stores, like the model seen in Sect. 1.4.2, PCs were being produced *make-to-stock* (MTS), with a limited assortment. The average throughput time of a PC from production to sale through wholesalers and retailers was estimated at 2 months or more, with particularly high physical costs (holding costs due to high value products) as well as market mediation costs (due to the obsolescence costs).

On the contrary, in the Dell SC, the computers are assembled directly following an order, from a "Pull" perspective. The customer's purchase order is customized, and the breadth of the range is extremely high, allowing the customer to choose from many possible configurations. The customer order generates a production order, and no finished product warehouse exists. The PC requested is assembled to order and distributed directly to the end customer's home.

The basic strategy is to build and manage an SC capable of decreasing the consumer's lead time as much as possible, i.e. the time that passes between the procurement of parts, through the production of the PC and delivery to the final consumer, in order to ensure that consumers prefer to buy a customized product online and have it delivered to their home after a few days, rather than going to a store and buying a standard PC with a limited range of choices, but leaving the store carrying the PC. Simultaneously avoiding all the market mediation costs (mainly shrinkage from technological obsolescence), and the physical costs associated with stocks deriving from the stock in the distribution part, thanks to a Pull strategy, in which the product is made on demand and therefore there are no stocks in the distributive part. At the same time, in the absence of the economies of scale typical of a make-to-stock approach, but still achieving economies of supply, production, and distribution thanks to the coordination of the SC, necessary in order not to increase the industrial costs of the product and keep the price in line with those of the competition.

The Dell.com supply chain which allows to achieve this strategy can be schematized as in Fig. 69.

In the Dell supply chain, the *decoupling point* (the concept of decoupling point DP will be thoroughly addressed in Chap. 4, however, the DP is understood to be the point between the part that works on upstream forecasts and the one that works on orders downstream in Just In Time mode) is located at the procurement level. Consequently, only a part of the supply chain is managed from a Pull perspective, on demand, and this is the production and distribution part. Vice versa, the supply part is managed from a "Push" perspective, therefore on a forecast basis. In fact, the production and procurement times of raw materials would not be compatible with the desired delivery times to the final consumer, that is the lead time that the customer is willing to wait for, so that it is necessary to keep components in stock. However, components do tend to be standard for all PC models and therefore their demand is much more predictable than that of customized finished products; thanks to standardization, it is possible to take advantage of an inventory pooling effect.

To be able to reduce the throughput time of the SC and therefore the LT, the suppliers are few, and are located just a few kilometres from the production plants

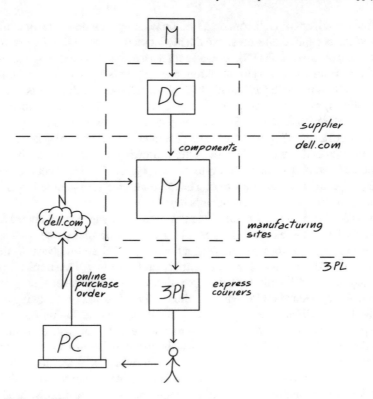

Fig. 69 Dell.com SC scheme

in order to reduce supply lead times. Indeed, one of the supplier selection criteria is the procurement LT.

In some cases, for certain key components, the stock is managed at the production plants with a view to VMI (Vendor Managed Inventory). The supplier therefore has proprietary stock at Dell manufacturing facilities and must maintain a minimum and maximum level for each code; the supplier has real-time visibility on the level of stock and component withdrawals, and invoices the component at the time of actual consumption. Clearly, such relationships necessitate long-term partnerships, and the supplier can count on significant sales volumes thanks to the VMI service offered (VMI means zero inventory management costs, zero fixed assets, financial leverage for Dell of at least 60 days on components—Dell receives payment from the customer while paying supplier invoices at 60 days from end of month of invoice date).

Therefore, once the standard components are available in stock, thanks to the huge investments made to maximize flexibility of the factories and assembly lines, the flow time of a personal computer from component to product available for shipment (the time to carry out the assembly processes, software customization which is the part that absorbs the most time, testing, packing and shipping) ranges from 48 to 72 h. The variable part of the time is related to the installation and testing of the software packages, which takes time. The package thus prepared containing the customized PC is then delivered to the logistics operator, typically an express courier, who ensures delivery within 2 days. Overall, therefore, from the closing of the customer order by clicking the "Buy" button on the Dell.com portal of a highly customized product tailored to the customer's needs, to the delivery of the PC to the customer's door, an average 3–5 days elapse.

In this way, and thanks to a structured supply chain managed from a pull perspective, Dell builds its competitive advantage. The direct sales model eliminates the intermediaries between Dell and the final consumer, ensuring Dell all the margins which in a traditional channel must be shared with distributors and retailers.

Furthermore, thanks to the pull management, Dell can enjoy elimination of the physical costs of maintaining stock of finished products and the market mediation costs related to non-sales due to out-of-stock or obsolescence costs associated with finished products compared to their competitors.

As for the other components of the total logistics cost, Dell aims to achieve the following optimizations.

As far as procurement is concerned, Dell contains procurement costs thanks to long-term partnerships with a few suppliers, selected on the basis of performance: (a) service in general and lead time in particular, and (b) product quality, which does not require controls but is delivered directly in-line to further reduce production lead times.

In relation to Operations, production costs are contained thanks to huge investments in flexible assembly lines, which make it possible to reduce assembly and customization costs even for minimum batches, as a minimum unitary—a single PC for a single consumer.

As for distribution, the physical costs of distributing the finished product are optimized thanks to the economics of scale of the logistics operators who are able to deliver at low cost thanks to the volumes they handle. In any case, Dell delivers free of charge in the case of standard deliveries, while it asks the consumer for an additional contribution for express deliveries.

Finally, in a single channel model of purely online sales, store costs are nil, and the conflict between the two channels does not need to be managed. Typically, in a multi-channel chain, online could become a competitor of a traditional channel if prices were significantly lower, which is not the case for Dell.com since the strategy adopted is solely online.

What was initially considered a niche player by its competitors quickly became a market leader. In fact, customers—those in both the consumer and business markets—became more and more confident in buying a product such as a personal computer or a server directly online, giving up the security of a physical place such as a shop to go to for support.

The development of an extremely sophisticated online support system and post-sales assistance with particularly high Service Level Agreements SLAs, able to intervene with a specialized technician directly at the consumer's home/office the day after a report, also means that the more demanding business customer or the retail consumer who is less confident in their IT skills, can confidently choose the Dell.com

Table 10 The Dell.com case and the competitive advantage of its SC

Dell	Competitor
High range of customized products	Limited range of models and standard versions
Price: in line with the competition	Price: in line with dell.com
Service: product delivered to the consumer's home within 3–5 days from receiving an order	Service: product available to try and buy in-store
After-sales assistance	In-store after-sales service
High margins thanks to a single-echelon SC and thus reduced number of players	Low margin due to the multi-echelon SC and thus a high number of SC players
No cost for physical store	High infrastructure costs
No physical cost of FP inventory	High FP physical costs due to inventories
No MMC costs associated with OOS or shrinkage	High shrinkage costs
Partnership with suppliers and economies in VMI procurement	Production and logistic economies of scale linked to MTS
Production flexibility and reduction in flow time	Reduced financial leverage
Logistic economies raised to the volumes of the 3PL	
High financial leverage	

option over a traditional store. Indeed, the level of service guaranteed is frequently far superior to that of a traditional shop.

In short, as summarized in Table 10, Dell's competitive levers are essentially:

1. Product: High-level customization of the product thanks to being tailored to the needs of the consumer;
2. Price: in line with the competition; but higher margins thanks to eliminating the number of intermediaries (online sales) and optimizing procurement, operations, and distributions costs.
3. Service:

   a. product delivered to the consumer's home; consumer confidence in purchasing a product online (pre-sales services through the web portal and call centres, after-sales service and return policies)
   b. Lead time of a few days, well-matched with the needs of the consumer
   c. High level first and second level support services, to gain trust of even uneducated customers, not able to handle after sales issues.

To do this, the SC is structured according to certain key points:

1. Eliminating intermediaries and reaching the end customer directly (higher margins).
2. A Pull supply chain, at least for production and distribution, which eliminates maintenance costs in stock (physical costs) and costs of obsolete and unsold goods (market mediation cost).
3. Long-term partnerships with suppliers (selected on the basis of quality and service),. High integration (e.g., sharing of sales forecasts and production plans;

VMI), (iii) participation of the supplier in product development and in-line control to speed up and standardize production engineering times and new product launches and reduce quality complaints.

4. Production flexibility and flow time reduced to a minimum; assembly and production of a customized PC in a few hours. Configuration of the manufacturing area in production/assembly cells: doubling productivity per m^2 of assembly area, and decreasing assembly time by 75%.

5. Distribution partnership with 3PLs for home delivery to final consumers all over the world and to cut distribution lead times to a few days.

One significant fact, reported in detail in the "Summary of the online distribution channel" section, is the increase in cash flow that an online sales system allows: Dell sees payments within 24 h, while on other channels the cash flow has an average turnover of 35 days

Further insights on dell.com supply chain can be found in (Strickland1999; Christopher 1992; Mars 2020).

On Line—The Amazon.com Case

Amazon is an online sales player born in the mid-1990s. Founded under the name of Cadabra.com by Jeff Bezos on July 5th, 1994, it began its activities in 1995 with the sale of books. Amazon soon expanded the range of products sold to include DVDs and music CDs, then added software, video games, electronic products, comics, clothing, furniture, food, toys, and much more besides.

Over the years, Amazon has evolved intensively using the online platform developed for the sale of books, and today is among the first online retailers worldwide (along with Ali Baba), second to Wal Mart only, with a turnover of 300 billion USD and a net profit of 11 billion USD, and employs 840,000 people worldwide.

Amazon's Roles in the SC

To be able to achieve these numbers, Amazon has grown according to different models. Born as a retailer, it then evolved into a focal company, 3PL, and marketplace. Today all four of these models coexist in the Amazon SC. We will deal with them in detail in the following paragraphs.

Retailer

For Amazon-owned products, fulfilled through the Amazon distribution structure, Amazon acts as a retailer. The distribution framework is charted in Fig. 70.

The Amazon DC (distribution centre) is called a *Fulfilment Centre*. The final consumer places an online order through the Amazon.com web platform. When the customer order is issued, the platform issues a *pick order* addressed to the DC structure responsible for the order. This process is typical for references managed in stock at the Amazon DC (high rotation references as category A products or Amazon Prime products, which are fulfilled within 48 h of ordering).

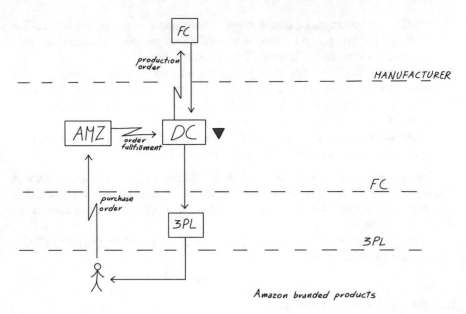

Fig. 70 Amazonl.com SC scheme—the retailer

For low rotation references (categories B or C), Amazon issues a purchase order (PO) to the Focal Company, on receipt of which it sends the material to the Amazon DC and from there it is sent to the 3PL to reach the final consumer.

Marketplace

About a quarter of the products that can be purchased on Amazon.com are third-party items sold by other players who market their products through Amazon.com.

In this case (Fig. 71), the platform advertises products that are not owned by Amazon. The products remain the property of the FC, including in terms of inventory, since the logistics flow is not handled by Amazon at its DC. Upon receipt of the customer order, Amazon transmits the order to the Manufacturer/FC. The FC, with its own DCs, prepares and fulfils the order and sends it by express courier to reach the final consumer.

In this case, Amazon operates as a *marketplace*: it retains a premium on the transaction and is not affected by physical product flows and stocks.

However, Amazon has full visibility of sale trends, and can exploit these data either to promote its own products, or to acquire potential high-growth companies.

Focal Company of AmazonBasics Products

i.e. Amazon-branded products, or products owned by Amazon and manufactured by Amazon through third-party manufacturers. In this case, a 3PM manufactures Amazon-branded products as a supplier or an Amazon investee/acquired company. In reality, this scheme replicates the retailer case, with the difference that the products

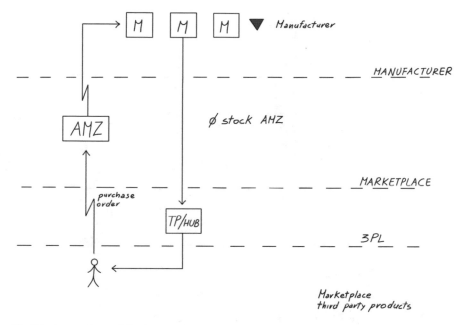

Fig. 71 Amazonl.com SC scheme—the marketplace

are Amazon-branded, and the stocks are owned by Amazon, typically at a DC level (Fig. 72).

Over time, a system of this kind has repeatedly raised competition problems, since many companies producing AmazonBasics products have been bought, or sometimes taken over by Amazon. Thanks to Amazon's knowledge of outstanding sales figures on the portal, Amazon can produce and propose a competitive AmazonBasic alternative at a bargain price and take over the competitor. For this reason, operations of this type have often been the object of attention by Antitrust authorities to monitor the real benefit for the final consumer of this type of operation.

Fulfilled by Amazon (FBA)

For third party products fulfilled by Amazon but owned by third parties and sold either through Amazon.com or another Internet channel, Amazon acts as a logistics service provider, through Amazon Logistics (see Fig. 73).

Since logistics is a competitive lever, Amazon believes itself so good at logistics and managing distribution centres, order fulfilment and transport processes that it has thought to offer them to the market, just as it has done with its Amazon Web Service (AWS).

The Amazon Logistics business model is based precisely on an ability to reach the customer in a very short time. The proprietary infrastructures are:

- Fulfilment Centres (where the references for FBA order fulfilment are stored)
- Fleets of planes, trucks and vans for home delivery logistics
- Logistics management by the customer (forward & return flow).

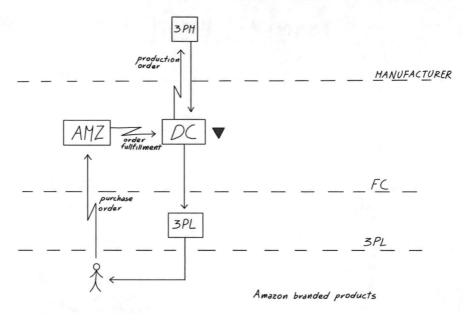

Fig. 72 Amazonl.com SC scheme—the focal company AmazonBasics

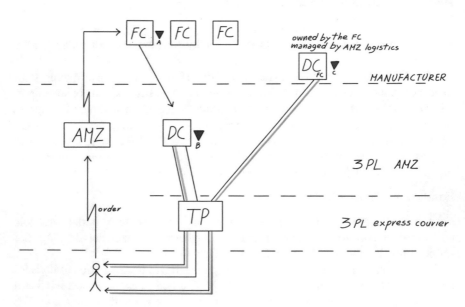

Fig. 73 Amazonl.com SC scheme—the 3PL fulfilled by Amazon

In fact, in this model, Amazon provides its own logistics infrastructure, its own staff and/or its own logistics know-how. The focal company's products can therefore be stored at an Amazon Fulfilment Centres or directly at the focal company's own DC. In the latter case, Amazon Logistics staff are outsourced and work at the focal company's DC.

Web Service Provider

Other particularly profitable Amazon business models concern aspects not strictly related to the SC, but to the provision of services, in particular web services. This resource is the Amazon Web Service—AWS, created to make the cloud power of the Amazon infrastructure available to third parties, and to compete with cloud services from such providers as Google Cloud or MS Azure. Today, AWS makes a significant part of Amazon's net profit and just like Amazon Logistics, Amazon believes its web service platform to be so efficient that it makes its own engine, AWS, available to allow those who wish to develop cloud platforms in order to use it.

The Competitive Strategy

To pursue this strategy and convince customers to buy on their portal and win their loyalty, thereby acquiring more and more market share, Amazon acts on the three competitive product/price/service levers in a very precise way.

- Product: the product sold by Amazon tends to be undifferentiated, and substantially the products available on Amazon.com are the same as those present in other supply chains. Thus it is clear that Jeff Bezos' company does not focus on product differentiation.
- Price: the price is also in line with that of traditional SCs. Often the price is lower than an SC based on a brick-and-mortar retail type, but it is still possible to find SCs where the same product is sold at lower prices. The price, being in line with other channels, does not represent a competitive lever for Amazon.
- Service: this is the main element on which the Value Proposition is based and therefore the competitive strategy pursued by the American company, and which distinguishes it from any other competitor's range of products, returns, lead time, and place of delivery.

The "value" of the service offered by Amazon is expressed in four strengths and three critical elements to be managed.

The breadth of the range: as we shall see better in the chapter dedicated to customer service, the breadth of a range is a service factor halfway between the marketing and sales support factors, and consists of the breadth of the assortment in terms of the number of references available. Amazon's figures on the breadth of its range are confidential, but are estimated to be in the order of several tens of millions of references, if not hundreds of millions. Impressive numbers if we consider that a large hypermarket has around 150,000 references in stock (three orders of magnitude less!). Amazon's goal is to constantly expand its offering to ensure a service that is unique in the world. On Amazon.com we really can find anything, and this has accustomed consumers to buying online only and directly from the Amazon site, without looking elsewhere, because they can find everything there. This phenomenon is well described by Amazon's motto: *"You name it, we have it"*. Virtually everything you can buy is on Amazon.

lead time: Another key factor in Amazon's success lies in the fact that the company has managed to compress delivery times to the point of being able to consign products to their destination within one working day. The Amazon Prime service—fee-paying, thanks to its convenience, has become an almost indispensable factor for customer choices, so much so that by 2017 more than 25% of the American adult population had already subscribed to Amazon Prime. Three years later, the percentage of American users subscribed to Amazon Prime had reached 44.9%, which meant that, in 2020, more than half of the American population was using the Prime service, considering that many accounts are shared by multiple users.

Place of delivery: the place of delivery represents the real e-commerce revolution. The customer can conveniently receive the product ordered online wherever he/she wants without having to go to a different place and therefore "saving" costs, time and the personal commitment of having to travel. Through a mix of online logistics (Amazon Logistics), express couriers (UPS, FedEx) and postal services (USPS), Amazon is able to arrive anywhere, from large cities to the most out-of-the-way places.

Returns policy: the ease with which products can be returned to Amazon and possibly replaced if damaged once they reach their destination has helped to increase consumer confidence in a company by which they feel assisted and considered wholeheartedly, with the result of making the purchasing process "easier", "safer", and therefore "recurring", generating repeat business for Amazon.com.

Nonetheless, three critical elements need to be borne in mind:

The intangibility of the internet: the product cannot be touched and tested in person by the consumer. For this reason, the first products to be sold online were books and consumer electronics, types of products for which it is not strictly necessary that they be examined personally. Initially, this critical issue led to the assumption that certain sectors, such as apparel and food, could never become involved in online commerce. Today, thanks to the returns service policies and the introduction of numerous features in e-commerce sites (definition of images and magnifying glasses, user reviews, filtering according to keywords and functions) it has become possible to overcome this barrier and e-commerce has thus been able to interest and include these sectors as well.

Inability to receive deliveries at home: to reach more and more consumers, even those who do not have the possibility of receiving deliveries at home or at their workplace, Amazon has set up various schemes. On the one hand, lockers, a series of cabinets in which to receive a delivery (see paragraph "Amazon Lockers"), but also the possibility of being able to receive deliveries at affiliated commercial establishments (bars, gas stations, tobacconists), which are always open and offer a pick-up service.

Delivery times: in order to comply with the delivery times guaranteed to the customer, which are now increasingly shorter, it has been necessary to carefully design and operate the supply chain and the distribution system. The general trend is to create a supply chain which, while seeking to keep costs low, is still able to guarantee lead times—the time between the moment in which the customer presses the "Buy It" button on the portal or on the Amazon app, and the actual delivery of the product to the doorstep—which fall within the order of 24–48 h

To guarantee certain delivery LTs while maintaining a price in line with the "brick-and-mortar" competitors who enjoy all the distribution and transport economies of scale, is a series of strategies and measures which Amazon has been applying for some time during its evolution.

First and foremost, the use of 3PLs initially and then the Amazon Logistics service later on. For its distribution logistics, Amazon has always made use of 3PLs—third-party logistics (such as the postal service or express couriers) to then build its own logistics operator, once the volumes were such as to make this option advantageous. This choice was mainly motivated by two factors which must be taken into account in designing a supply chain of any kind. The first is the saturation concept of vehicles from an FTL (full truck load) perspective. Using a third party, a win–win scenario is created where the 3PL is able to saturate the vehicles (considering that the costs for a generic transport from point A to point B are spread among several "senders") and consequently distribute the fixed distribution costs among several packages. As a result, Amazon enjoys lower transport prices thanks to the high volume of packages handled and the coverage of the territory offered by the organizations of the various subcontractors. Amazon's common practice is also to make use of the postal services already operational in the various countries where it is present. This system is named *"Postal Injection"*. The term "injection" refers to the physical act of moving a certain product, package, or lot from one level of the supply chain to another lower level, territorially closer to the final recipient. The term "postal" refers to such postal logistics carriers as USPS in the US, and in Europe, *Poste Italiane* in Italy, *La Poste* in France, *Correos* in Spain, etc., which are able to guarantee widespread coverage of the relevant territory. One of the main advantages deriving from the use of postal carriers is the possibility of making use of the breadth of their territorial coverage by means of their branches, sorting centres and intermediate warehouses, given that, wherever they are present, their logistics bodies are among the oldest ones still operating in the area thanks to a widespread network. In Italy, for example, *Poste Italiane* has at least one post office in each municipality, and it is the only logistics body to enjoy this peculiarity. A large e-commerce company can take advantage of this to bypass the supply chain and inject shipments for customers directly at the level of the last-mile distribution of the postal carrier (the last leg of the supply chain, the one that ends with the delivery of the package to the customer) (Youtube 2021).

A pared-down distribution structure, both in terms of physical infrastructure costs and stock maintenance costs. The key to Amazon's success are its Fulfilment Centres which, thanks to their strategic collocation throughout the territory, allow a reduction in cost of facilities, generating economies of scale due to the aggregation of logistics facilities and stocks in one or several large distribution centres. Amazon can count on more than 130 Fulfilment Centres (central warehouses with dimensions in the order of 100,000 m^2) located near the most important logistics hubs. These are the main assets of Amazon's distribution structure, which does not include wholesalers, distributors and proximity stores, nor shops (with the exception of Amazon Go and a few food courts).

This distribution channel configuration allows Amazon a double advantage. Not only does it reduce the costs related to infrastructure management, it also significantly reduces stock maintenance costs thanks to inventory pooling. By concentrating the stock in one or a few places it is possible to significantly reduce cycle stocks (CS) and safety stocks (SS) due to the fact that these are proportional to $\sqrt{N}$, where N is the number of peripheral warehouses. As a result, a considerable reduction in inventories is achieved by passing from N warehouses to a single centralized DC. Qualitatively speaking, the costs of duplicate stocks in a widespread distribution structure are avoided. Instead of duplicating the stocks in the various warehouses, one for each geographical area, they are aggregated in a single point which serves all the areas. The best result is obtained by balancing an increase in demand in one area with a decrease in that of another area.

Furthermore, only high-rotation products, characterized by more predictable demand and more stable lead times, are kept in stock in a push perspective at Fulfilment Centres, while low-rotation ones with less predictable demand and variable lead times are supplied upstream in a pull perspective following an order. To do this, Amazon makes massive use of AI (artificial intelligence) algorithms with which it decides what and how much to keep in stock at a Fulfilment Centre. With a range of tens of millions of references, it is impossible to keep all the stock available. AI algorithms try to predict consumer demand, deciding what to keep in stock, so that when an order is placed, the product is immediately available. E.g. in Rome it is unlikely that ice scraper brushes will be held in stock, since demand will be predictably low. Unlike in Castel San Giovanni (PC) where it is much easier for this reference to be present among the stock.

Automation of Internal Processes

In this and the following paragraphs, ample space will be given to a description of all those processes which take place inside an Amazon Fulfilment Centre, as the main hub of the entire distribution channel. In fact, the Amazon Fulfilment Centre processes represent a logistical best practice that deserves to be explored in detail.

The most significant idea on the real functioning of an Amazon DC is to visualize it *en bloc* as a single process made up of the aggregation/combination of various sub-processes and activities. Within the process, physical flows (products) and information flows are coordinated by a set of specific rules. The entire set of these rules is aimed at synchronizing the operations in a temporal sense on the basis of the time windows which synchronize the distribution by the logistics companies (third parties or Amazon Logistics). The main target is indeed to fulfil the order of the end customer in the shortest possible time and with the lowest possible use of resources.

It is therefore evident that two different entities, in this case Amazon and third party logistics, must operate in a coordinated and synchronized manner to guarantee what is defined as a "lean flow". This concept was partly borrowed from the "Lean Manufacturing" techniques developed between 1948 and 1975 by Sakichi Toyoda, engineer at the well-known Japanese car manufacturer Toyota, and was developed by providing for the elimination, within a process, of all activities without added value or with evidence of waste. The aim being, as far as possible, to achieve a significant reduction in dwell time, stocks and contain them within "rationally" minimal ranges from a quantitative point of view. This reduction having the purpose, on a practical level, to contain, if not indeed cancel the risk of obsolescence with the consequent related market mediation costs and achieve a double result on the flow of materials and, specifically, the greatest possible fluidity (minimization of flow time and thus lead time) along with elimination of delays (with respect to Takt times).

In the case of an Amazon DC, the waste is represented not only by materials or stocks, but above all waste in terms of dwell time. As mentioned, Amazon's competitive strategy is based on the fastest possible delivery to offer the best possible service to end customers. Waste is therefore any unproductive moment when the product is waiting to be processed, which slows down the flow and increases the lead time. Given the numbers that Amazon shifts (also in this case there are no official data available since they are strictly confidential, however on Black Friday and Cyber Monday 2019, 3 million orders were fulfilled in Italy alone, equal to 37 orders per second; in the USA during Cyber Monday 2019, over 800 items per second), even the saving of a few seconds in a single phase in the handling of a product in one process translates into savings of tens of thousands of Euro/year.

The six operations that take place in chronological sequence within an Amazon Fulfilment Centre are those described in Fig. 74, and the layout of the Fulfilment Centres itself reflects the sequence of processes and the flows of materials inside them.

At Amazon's Fulfilment Centres, operators are divided into teams, each assigned to a specific process (*receiving, stowing, picking, packing, SLAM, and shipping*), which is periodically changed. There are two reasons for this diversified periodic

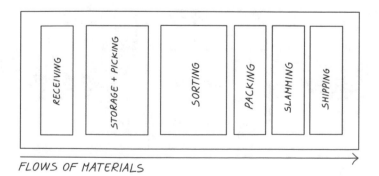

Fig. 74 Processes mirror the DC layout in a fulfilment centre at Amazon.com

assignment for each team; the first is connected to the ready availability of trained operators able, at any time, to be able to carry out all the operations that take place within the Fulfilment Centre and, the second, for problems and criticalities deriving from the repetition of tasks and work stress which the individual employees would be subject to.

In each process, the following are always involved: the operator who carries out the process; the WMS that governs the Fulfilment Centre processes; the problem-solver (a dedicated team for each process which deals with managing non-conformities/exceptions); and the Amazon.com web portal which is updated in real time.

The 6 main processes will be thoroughly detailed in the next sub paragraphs.

Fulfilment Centre Processes

Receiving

The receiving process is the first one which all the products entering a Fulfilment Centre undergo. These are therefore AmazonBasics production flows entered into the FC, flows of products ordered by Amazon (destined for a customer order or to replenish the stock), or flows to replenish products fulfilled by Amazon, in order to restore the range of offerings.

Like any receiving process, this too can be schematized according to the process represented in Fig. 75.

One exception is the presence among the controls of an ASIN code.

The Amazon catalogue contains literally tens of millions of different products, with different identification standards (*GTIN, UPC, ISBN for books, SKU for textiles, etc.*). Without a single product standard, managing this mass of sales items would be virtually impossible. Which is why Amazon has decided to adopt its own standard for product coding—the Amazon ASIN code. ASIN is an acronym that stands for *Amazon Standard Identification Number*, and consists of a ten-digit alphanumeric code. It is essentially valid for products, while for books it is replaced by or coincides

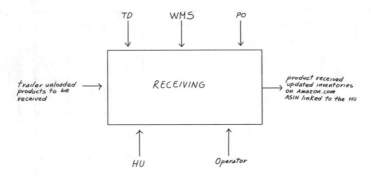

Fig. 75 The processes of the Amazon fulfilment centre: receiving

with the ISBN code. The Amazon ASIN is used to uniquely identify a reference offered for sale on the Amazon platform. In fact, it is used by sellers for a whole series of analyses and is also taken into consideration by Amazon's internal search engine for the organic positioning of the various products. In addition, a user can type an ASIN into the search box to immediately find the product it refers to. To all intents and purposes, when it comes to talking about a product, the fastest way to identify it without problems is to refer to its ASIN code.

Instead, the process flow chart is represented in Fig. 76.

The operator accesses the WMS via a terminal (fixed or mobile) and calls up the customer order to which the goods being inserted refer. He/she enters the Transport Document number; and if a DESADV is present, it is possible to directly retrieve the mix (ASIN) and quantity to be received for that TD, otherwise it is limited to a check between the physical item and the Purchase Order, while the matching with the TD will be done later in administration. In any case, the WMS retrieves the list with the references (ASIN) and the respective quantities that must be received for that TD and transfers them to the radio terminal (Fig. 77).

For each piece to be received, the operator checks the status, the presence of the ASIN code, and scans the ASIN barcode. The information arrives at the WMS, which checks whether the reference is among those to be received and proceeds by updating the stock on the Amazon.com portal. The update consists of increasing that specific code by one unit. From that moment on, the item is available to be ordered, with the

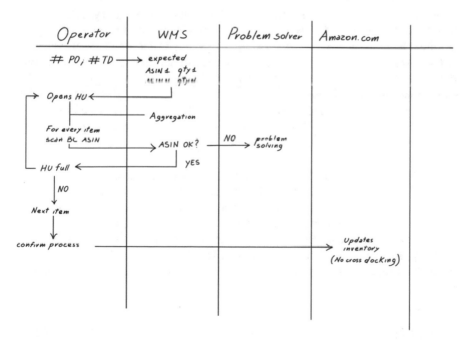

Fig. 76 Receiving process: flow chart of the process

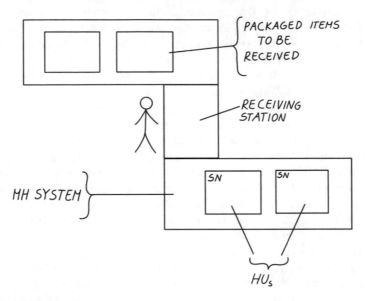

Fig. 77 Amazon manual receiving station

exception of orders received in cross-docking relating to references already ordered, which will therefore need to be sorted and will not increase the portal inventory. On the other hand, if an error is detected, the intervention of a problem-solver will be required which will activate a manual management of the exception (the latter activity should be cut to a minimum, since it involves a delay in the receiving operations within a time range of 2 h–2 days).

The operator works at a receiving station located between two roller conveyors: on one of these flow the packages to be received which the operator picks up every single selling unit and then scans its ASIN barcode using a portable terminal. On the other one run the destination bins (yellow reusable plastic containers RPC—which are part of the company assets) in which the references are placed. Each bin is equipped with an SN (serial number) which guarantees correspondence between the handling unit and the reference code. The association between SN and barcode is performed by the radio terminal. In this way, the content in terms of ASIN/Qty is available to the system for each single storage bin SN. When the destination bin is full, it is closed in the system and is taken to be stowed by an MH (Material Handling) system, while the operator opens a new one by a barcode scan, and the cycle is repeated until the TD is closed.

Stowing

Within Amazon Fulfilment Centres, storage operations can take place in two different ways: manual stowing and automated stowing.

Manual stowing

Manual stowing is used for all large-sized items that can be sorted (larger than 24 × 40 × 15 cm and weighing up to 27 kg), or large ones that cannot be sorted. Process flow chart is shown in Fig. 78.

This is a substantially manual random storage process of an "*operator-to-materials*" type. Each storage bin is identified by a unique SN. In each of these there may be completely different references, due to the totally random logic of choosing the bin in which to store an item.

Amazon's warehouses are divided into zones on the basis of the rotation index of references (Class-Based Storage A, B, C) so that the handling bins can be sorted in the different zones according to the frequency of handling of the products they contain; however within each zone the allocation is purely random.

The manual stowing process is represented by the flow chart of Fig. 79.

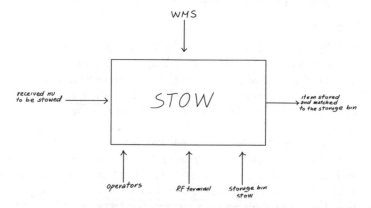

Fig. 78 The processes of the Amazon fulfilment centre. manual stowing

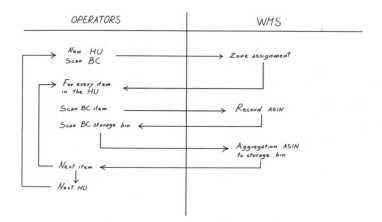

Fig. 79 Manual stowing process: flow chart of the process

The operator moves around the rack and allocates each item present in the handling unit (bin) to a storage bin, trying to saturate the available space. The storage bins are identified by alphanumeric coding which codes aisle/right-left/depth/height of the bin, e.g. 52d × 3204 identifies a bin located in corridor 52, on the right, place 32, height from the ground, 4. A double barcode—reference ASIN/storage bin—allows the WMS to record the process.

The random storage policy adopted by Amazon allows two main benefits: (i) on the one hand, the occupation of available space is optimized. Having to operate on millions of references, many with stocks of one or just a few units, it is not possible to dedicate a specific storage bin to each of them; (ii) the random storage policy allows a drastic reduction in picking errors. If the operator needs to pick a specific item from a bin which contains completely different items, he/she will have a much lower probability of making a mistake than if the objects in the bin were all similar (e.g. sweaters of the same model, colour but different sizes), and might have to search for one specific model, a specific variant/version, or a specific size.

Automated stowing

For small easily sortable items, 24 × 40 × 15 cm and up to 11 kg in weight, Amazon adopts automated storage in certain Fulfilment Centres. Thanks to the ease of handling due to the small size and weight, all of the processes can easily be automated. An automated *"materials-to-operator"* process is used. The process is carried out thanks to a fleet of self-guided storage/picking robots, called *Kiva*, named after the manufacturer acquired by Amazon (Youtube 2011, 2014) (see Fig. 80).

The Kivas are AGVs (Automated Guided Vehicles) in the form of electric mobile platforms, which translate and rotate on themselves, moving the storage bins in the process. The bins are configured as a sort of vertical rack with a square base (storage pods), beneath which a Kiva can move, with different shelves where products can be positioned. The Kiva robots weigh more than 100 kg and are able to lift a pod of more than 200 kg by rotating on themselves and operating a screw lifting system,

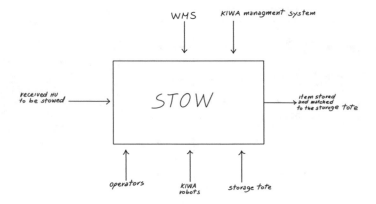

Fig. 80 The processes of the Amazon fulfilment centre: automated stowing

and then, following a QR code chessboard, to move and/or hand them directly to an operator who will physically pick up or stow the product. Once it is out of power, the Kiva goes autonomously to a charging station, to replace or recharge its battery (see Fig. 81).

Similarly to what happens for manual stowing, also in the automated system, to allocate references the barcode of each item is scanned, followed by that of the storage racks. Amazon recently also eliminated BC scan times and inversion errors by adopting *computer vision* systems. The operator no longer has to physically read the BC of the bin in which he/she will place the product, it is the computer vision system which recognizes it. At the same time, the vision system inhibits stowing in certain storage bins (e.g. already full or in which there is already the same reference to be allocated) by projecting a beam of coloured light onto them, and leaving the operator free to position the reference in any other bin (see Figs. 82 and 83).

Both in the case of BC reading and computer vision, the substantial difference compared to the manual system lies in the fact that thanks to the automation it is not the operator who has to move to reach the storage racks, but the pods that move from their position up to the operator to be filled with references and then returned to the storage location.

This eliminates all the variable times associated with the movement of the operator between the racks, effectively replacing teams of stowers and pickers with fleets of Kiva robots, which work 24/7/365 without making mistakes. Stowing/picking times are drastically reduced, and productivity is therefore increased, since all variable movement times have been eliminated.

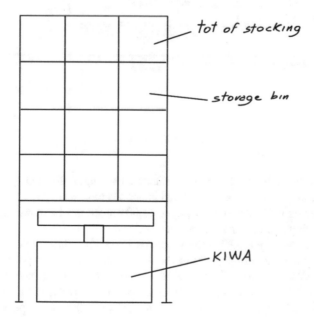

Fig. 81 Scheme of a Kiva robot in operation

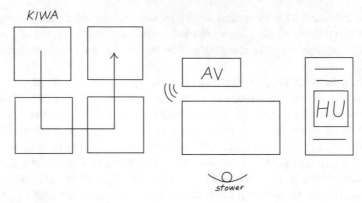

Fig. 82 Automated stowing station

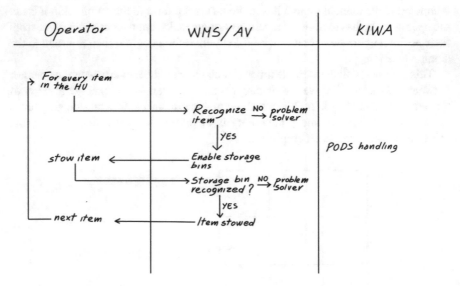

Fig. 83 Automated stowing process: process flow chart

It has already been noted that an infinitesimal saving multiplied by the very high number of times an operation is repeated generates huge savings.

At an Amazon Fulfilment Centre, up to 3,000 robots may be operating at the same time, for stowing and picking operations.

The racks moved by the Kiva robots can easily be placed next to one another, due to the fact that the anchorage point of the robots lies beneath the floor of the lowest bin. All the racks are also equipped with four support rods high enough to allow the free passage of the robot without these, or any other part, being collided with. In this way, real "aisles" are created beneath all the racks which make it possible to reach even shelves stacked in a pillar, more complicated to move due to their particular positioning.

The possibility of combining the racks allows considerable efficiency in exploiting the space required for storage (practically a double reduction, both in the number of aisles and the space they occupy). There is also a 50% reduction in the space occupied for the same amount of available room. In this case, the selectivity of the warehouse is not unitary, in the sense that not all the racks are directly accessible, since they are arranged in pillars. The less accessible racks, inside a pillar, can be freed by the robots themselves, controlled according to the schemes and logic programmed into the robot management software code, through the opening of an ad hoc aisle, to allow extraction of the rack from the storage pillar.

In DCs that use Kiva robots, the storage area is also divided into zones A, B and C according to the criterion of product rotation. The WMS algorithms control the logic both for the positioning of the racks in the designated areas, and to decide on preferential shelf-picking schemes which minimize movement and consequently also reduce travel times.

At the front of the storage area, an area known as a "Highway" is created (see Fig. 84), within which the Kivas move to bring the racks on which the stowing will take place towards the receiving stations.

The Kivas are forbidden from making curvilinear or Euclidean trajectories but are restricted to orthogonal movements which allow quantification of the metres to be travelled according to Manhattan Distances (see Fig. 85), at an average speed of 5–6 km/h, like that of a person walking.

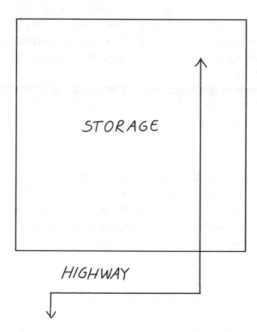

Fig. 84 Highway lanes area

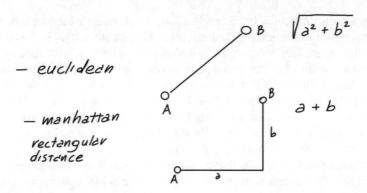

Fig. 85 Rectangular and Euclidean distances

Picking

The picking process can be subdivided, similarly to the stowing procedure, into two types depending on whether the Fulfilment Centre is using KIVA robots or not, hence Manual Picking (operator-to-materials) or Automatic Picking (materials-to-operator).

In both cases, the first step consists in turning customer orders into picking lists, and then from picking lists to actually taking the items from the racks.

Orders placed by customers on Amazon.com are grouped by time windows (known as waves), for example, hourly: orders issued between 12.00 and 12.59) so that, within each grouping, all orders issued in that period will be found, each with its relative reference. Once the orders have been aggregated, the references are divided into picking lists.

For Amazon, the average depth of the order, in terms of the average number of pieces per order, is very low and close to unity, so that the fundamental distinction is between two types of orders.

(a)   Orders with a single reference are grouped and aggregated into order picking missions. At the end of the picking they can be sent directly for packing, without going through sorting, since each piece picked corresponds to a customer order.
(b)   Orders that contain two references or more. Orders with quantities of two or more are grouped and aggregated into batch-picking missions. At the end of the picking they will need to be sorted to recompose the order before being sent for packing.

Manual picking

This is a picker-to-parts process performed manually by an operator, as represented in Fig. 86.

It is applicable to large products, which may be sortable or not, and for all Fulfilment Centres where Kiva robots are not in use.

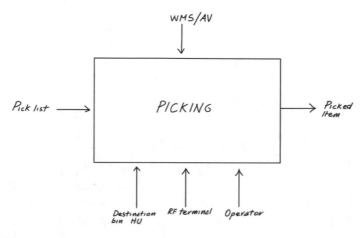

Fig. 86 The processes of the Amazon fulfilment centre: manual picking

The reference picking policy is that of zone picking (a picking mission contains references which belong to multiple orders—"batch picking"—and the operator moves within a specific area). Following the zoning of the warehouse, the picking lists are processed by the WMS according to the logic of proximity picking, to minimize the distances travelled by the picker, given the capacity C of the picking bin $\sum_{i=1}^{n} q_i < C$.

Once the picking mission—picking list (PKL) has been processed it is assigned to an operator. After opening the destination bin by scanning the BC SN, for each row of the picklist to be processed, the operator reads the picking bin code on his/her portable terminal, goes to the storage bin indicated by the WMS, scans the SN there and, having obtained confirmation that the bin is the right one, picks up the product and scans the ASIN code, while the WMS checks the correspondence between the product picked and that indicated by the warehouse management system.

This process is then repeated for all the items on the picking list and for all the picking lists in that particular wave. The process flow chart is detailed in Fig. 87.

Therefore, at the end of the picking process, all the N picking lists (PKL) relating to the reference time window will have been fulfilled. These consist of M bins HUs (Handling Units) of withdrawal, each identified by its own SN, of which the WMS knows the exact content and, for each reference, the order of belonging (see Fig. 88).

Picking bins containing orders of q = 1 can be sent directly for packing, while those with q > 1 must be sent to sorting for recomposition of the customer's order (see Fig. 89).

Automated picking

To automate the picking process, as in the case for stowing, the Kiva robots already seen for the stowing process are used (see Fig. 90).

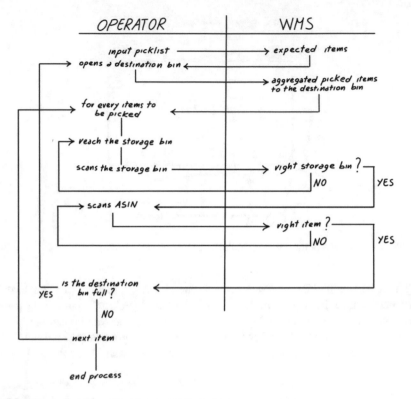

Fig. 87 Mmanual picking process: process flow chart

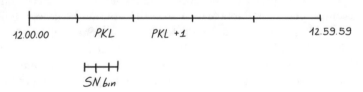

Fig. 88 Fulfilment of the wave picking lists

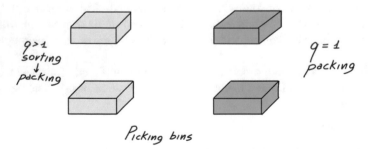

Fig. 89 Picking bin with single-reference (green) and multi-reference (yellow) orders

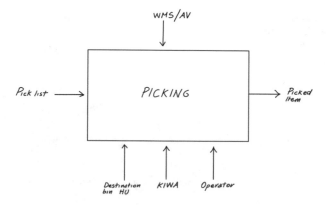

Fig. 90 The processes of the Amazon fulfilment centre: automated picking

Automated picking differs in two fundamental ways from manual picking.

1. Instead of being an *operator to material* process, it is of the parts-to-picker kind. The operator remains stationary, and it is the material handling system (Kiva robot) which carries the pods containing the order lines to be picked towards the picker. In this way, all the variable times of movement between the racks are saved, leaving only the fixed picking times. Picking productivity increases significantly, and also from an ergonomic point of view, the operator is saved from travelling kilometres between the warehouse racks.

2. The process can be both of the batch-picking type as seen in the case of manual picking, but also of the order-picking type, since it is the WMS which arranges the handling of the pods and presents the picker with the references necessary for order fulfilment. Also in this case, the fundamental distinction is between picking of references belonging to single article orders of quantity 1, whose bins can go directly for packing, and multi-article orders whose bins must go on to sorting to recompose the lines of the same order.

The use of Kiva robots makes it possible to move the mobile racks to the operator's station, who can then carry out the picking operations seen in the case of manual picking from the racks in front of him/her. It is the WMS which takes care of organizing the movements of the racks, so that the operator can find the right bins containing the order lines to be picked for each order. The operator simply opens the destination bin, and for each order line to be fulfilled, takes the product from the bin indicated in the racks in front of him/her. Also in this case a system of lights illuminates the bin from which to make the withdrawal to facilitate the operation and avoid errors. The operator then reads the ASIN code of the reference taken, and once confirmed by the WMS, adds it to the handling bin.

At a layout level, the situation is represented in Fig. 91.

In automated picking, thanks to the Kivas, productivity increases of up to 75% have been achieved, passing from a productivity of 100 pieces/h in the case of manual picking to a productivity of 400 pieces/h with automated picking. Note that in FMCG

Fig. 91 Layout of an
automated material picking
station versus operator

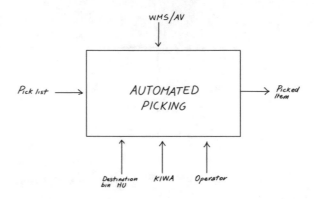

manual warehouses, carton level picking productivities can be double or triple (300 boxes/h), but the average number of pieces per product is far higher.

Sorting

Sorting is the process immediately following picking, to which all the pick lists processed in order picking are subjected, while as mentioned, the picking lists of items belonging to orders with a single piece can go directly for packing.

The items of an order must be put back together in the sorting, since products of the same order are divided between different picking lists and such a list can contain references from different orders. The purpose of this process is therefore to reconcile and group together the references belonging to the same customer order.

A black box of the process is shown in Fig. 92. In the case of automated sorting, a sorter system is used, while manual sorting is entrusted to operators who use mobile trolleys or other HUs, typically multi-bin racks.

Synchronization between picking and sorting is fundamental in order to respect the flow times and guarantee the service levels promised to the customer.

As in picking, there are "waves" of orders, based on time windows, in which they are frozen and grouped, to then be fulfilled all together. Therefore, the sorting of the same waves have to be synchronized (see Fig. 93).

The correct sizing of the waves is key to this synchronization: if there were only one window, the risk would be to see the references that make up an order assigned to pickers working at times that were very far apart, with an evident deterioration in service performance. At the same time, waves that are too short would not permit the typical benefits of batch-picking, since it would not be possible to aggregate a sufficient number of orders.

The fundamental distinction is between small items which can be automatically sorted, and non-automatically sortable items which require manual sorting.

Manual Sorting

This applies to items that cannot be automatically sorted due to shape or size constraints. In this case, either a mobile rack, consisting of n bins, or handling units

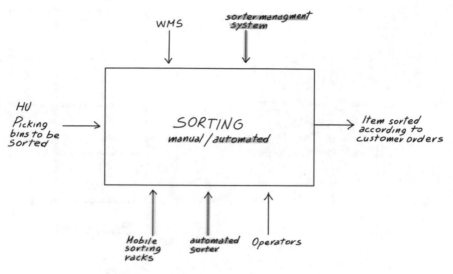

Fig. 92 The processes of the Amazon fulfilment centre: automatic and manual sorting

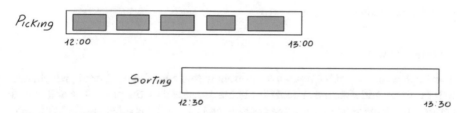

Fig. 93 Synchronization between picking and sorting waves

of appropriate dimensions, each of which will count the references of a customer order, are used (see Fig. 94).

The process can be described in relation to Fig. 94.

An operator takes the N HUs coming from the picking, which contains M customer orders, reads the BC of the bins and the BC of the sorting HUs (for example, in the case of small items, one rack) and matches them. For each reference, he/she reads the ASIN BC, and the WMS informs him/her in which bin of the mobile rack the article should be positioned (each bin being identified by a unique SN). The operator reads the BC of the bin and inserts the item in it. At the end of the process, each bin of the rack will contain the references of a single customer order, and the rack is then moved towards the packing tables. The operation can be performed by a single operator, or be distributed among several operators, who divide the sorting of the N picking handling units into K sorting handling units.

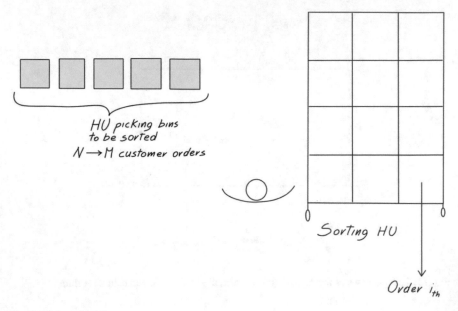

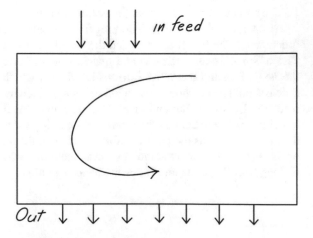

Fig. 94 Manual sorting

Automated Sorting

This applies to most of the references present within a Fulfilment Centre, which, due to their shape and size, are suitable for being processed by the plant. The automated sorter is the same seen for the cross cocking process, and it is schematized in Fig. 95.

The bins of multi-reference orders of the picking wave arrive at the infeed stations. For each of these, an operator reads the barcode of the respective SN (affixed to the bin itself). The reading of the ASIN BC is then carried out for each item in the bin

Fig. 95 Automated sorter

(this is the 4th reading so far to which the item is subjected during its path through the warehouse: *receiving, stowing, picking, sorting*). In this way, the WMS matches a particular item to a particular slot of the sorter (mapping 1:1). The most common and versatile types of sorter, as mentioned, are: (i) Conveyor sorter; the item is conveyed by a sliding pod towards the respective outlet; (ii) Tilt tray sorter; the item is placed on a "tray" which tilts to one side or the other, causing the item to fall towards the respective outlet. (iii) diverter belt—each sorter position is made of a diverting belt, which is actioned to divert the parcel to the desired outlet

In the sorting area (out) all operations are automated. It is the WMS which matches the orders to the outlet slots. The sorter is characterized by a very high number N of outlets: since each outlet corresponds to an order, the total represents the quantity of orders that can be fulfilled by the system concurrently and therefore defines the sorting potential. The sorter carousel moves the item at a certain speed and, when it reaches the appropriate outlet, automatically diverts the item to be sorted towards the outlet assigned to the customer order. Two different systems are possible, both equipped with light sensors, to materially facilitate the exit of the item from the sorter: (i) gravity chutes, which are the ones usually used; (ii) applications involving the use of roller conveyors, also using gravity.

The outlets can be doubled at different heights in order to optimize the use of space and consequently increase (double) the potential use of the sorter (see Fig. 96). When an order is ready, the green light relating to a specific outlet comes on, alerting a supervising operator who moves along the front to open the outlet. The operator moves towards the outlet to be opened, reads the BC of the bin previously matched to the outlet in which he inserted the items, and sends it to the packing tables using MH.

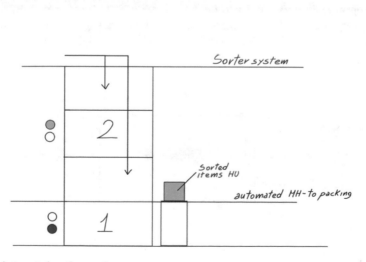

Fig. 96   Automated sorting: outlet

Finally, he/she opens the sorter outlet by pressing a button which informs the sorter that the particular outlet is now free and ready to be allocated to a new customer order.

Occasionally, at peak hours, several customer orders may be conveyed towards the outlet of a sorter, so a further manual sorting operation would be required downstream of the automated sorting.

Packing

This is the third last process in preparing a customer order, which makes the order suitable for shipment. Basically, it consists in the passage from the handling unit containing the references of the customer order, to a package suitable for shipping, unless the SLAM process is to be carried out. The process flow chart is depicted in Fig. 97.

The packing operation is difficult to automate and can only be done in-line, so it is carried out at a fixed station by an operator who takes care of all the activities, from the insertion of the items into the package, to the printing of the documents and the closing of the package. This is one of the most labour-intensive operations within the Fulfilment Centre, and also causes a bottleneck, in the sense that to increase productivity it is necessary to increase the number of workstations, and therefore the amount of labour and space employed.

Through MH systems, the operators receive the handling units to be packed directly at their workstations, which are equipped with video terminals, BC readers, printers and a whole series of "auxiliary" materials (see Fig. 98).

In input to the process there can be: (i) a bin of single-order items, coming directly from picking; (ii) a single order package, coming from the automated sorter; (iii) a mobile rack, coming from manual sorting, in which each shelf contains the references of a single customer order.

The optimization of times and methods is such that all the materials, from the packing carton to the filling materials and the final packaging belts, are specially tailored, prepared by the system itself according to the weights and volumes of the

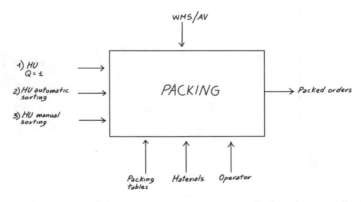

Fig. 97 The processes of the Amazon fulfilment centre: packing

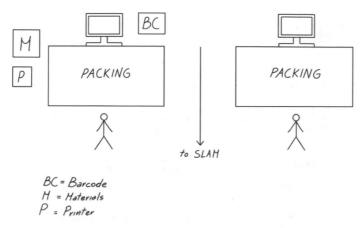

Fig. 98 Layout of a packing station

objects to be packed in the box. The WMS is able to process this information because it already has all the necessary data, from the volume and size of the item and weight to the quantity and mix of references, and therefore it can decide, e.g., on the size of the carton that best suits the package to be prepared. Although savings on a single package may be considered laughable and insignificant, with the extremely large volumes handled by Amazon they become crucially important. This attention to pre-determination of packaging materials allows an increase in productivity, even that of a single operator, who will not need to waste time and mental energy on the elements involved (how many and which materials to use, making mistakes and redoing), and who will already find everything at his/her disposal for operational use, without the likelihood of making mistakes.

In Fig. 99 is a swimlane BPMN mapping of the packing process, shown in reference to the case of bins containing orders with q = 1. Similar are those relating to orders coming from manual sorting (the process is repeated for each bin of the rack) and automatic sorting (where the package is already prepared in a carton ready for packing and labelled with an HU packaging sticker).

At the end of the packing process, a packaging handling unit sticker is applied to the newly prepared carton. This is a label which contains an alphanumeric SN linked 1:1 with the package and therefore with the customer order, essential for the *SLAM* process.

SLAM

The acronym SLAM stands for *Scan, Label, Apply, Manifest*. This process requires that the operations whose name constitutes the acronym SLAM are carried out in sequence.

- Scan: this is a further reading of the barcode, in this case referring to the handling unit packaging sticker (an SN label applied at the end of packing) in order to recall the customer order and all the information relating to it.

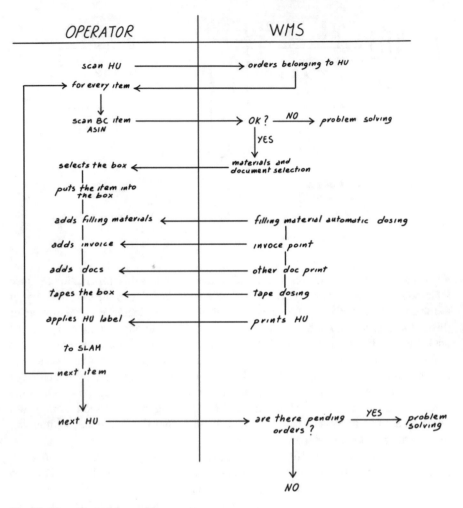

Fig. 99 Flow chart of the packing process

- Label: the printing of an additional label, the shipping one, relating to the courier used, equipped with a shipping SN and combined with a whole series of data such as the customer name, the shipping address, the weight of the package, the volume, the type (cardboard, envelope, special). A 3PL express courier is needed (UPS, FedEx, USPS or, more recently, Amazon itself) to manage the logistics processes (shipping, transport, last mile delivery, etc....)

    At the IT level, due to the integrated information flow between the two systems, there will always be a match between the 3PL shipping unit SN and the HU packaging sticker, which represents a key to the data held by Amazon.

- Apply: an operation that consists in the simple and material application of the label to the package. Extremely sophisticated automatic pneumatic vacuum applicators are used, capable of labelling packages of different sizes and types, without damaging their contents, and with extremely high productivity levels.
- Manifest: this operation involves the creation of a bordereau, that is a hierarchical tree, a structure whose root is a serial number from the TD (transport document) to which the various handling shipping units SNs generated during packing/labelling have been assigned. In turn, the TDs are in a 1:1 relationship with the carriers (e.g. truck [vehicle], carrier [who], route [from/to], time/date of departure). The assignment of packages to a transport document must therefore take into account not only the route and destination, but also a series of constraints, including the shipping window, the load capacity of the vehicle (saturation by weight or, more commonly, by volume...).

At the end of the SLAM process there are therefore a series of TDs (TD_1, ..., TDn) generated for one or more "waves" (picking-sorting-shipping, more than one if several are needed to give rise to a shipment).

The operation which allows sorting of the packages generated in packing according to the SLAM loading plans is automated sorting. The system is identical to the one seen previously. A barcode scanner at the entrance to the infeed identifies the SN of the package coming from packing and matches it to the tray of the sorter. Operating on the basis of the bordereau structure which the system knows from the 'Manifest' stage, the sorter automatically directs the packages between the N shipping docks (each corresponding to a vehicle, although it must always be considered that several TDs may be assigned to the same vehicle). The whole process is depicted in Fig. 100.

Shipping

Chronologically speaking, this is the last process carried out at the Fulfilment Centre. Shipping can be of two types:

- Automated, via a sorter, in which there is an automatic reading of the shipping label barcode (this operation, and the related flow of information, marks the transfer of ownership of the package from Amazon to the 3PL). In this case, the operator's activity is limited to the physical loading of the vehicle, which takes place via an extendable roller conveyor brought directly into the trailer;
- Manual, in this case it is the operator who reads the BC of the package and arranges the packages on the pallets, depending on the destination courier.

Also in this case (as already seen for other processes) the synchronization between all the Fulfilment Centre processes is maintained, and consequently there is an *absence of staging* between one and the other. This is a "lean flow" (*picking—sorting—packing—SLAM—shipping*) based on maximizing speed and a strict synchronization of the processes.

Basically, the focus of all improvement processes is that the customer order can flow between the various phases as quickly as possible and without interruptions.

Fig. 100 The sorting
process prior to shipment

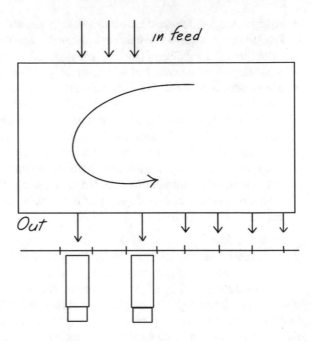

Consequently, it is necessary to act on two levers. On the one hand, maximum compression for each single process. Although an increase in speed is a necessary condition it is, in itself, insufficient. On the other hand, synchronization between processes is also fundamental. Both of these ingredients are essential to guarantee next-day prime delivery lead times. In a Fulfilment Centre what happens is the same as in a street with n traffic lights. To limit the travel time (order lead time), the speed of the vehicle between one traffic light and the next (process speed) is certainly important, however, the traffic lights must also be synchronized (process synchronization) to prevent the formation of tailbacks.

With shipping, and the loading of the package onto the means of transport, the sequence of processes of the Amazon Fulfilment Centre ends. From this point on, the package is taken over by the logistics operator, who delivers it to the final consumer according to the processes seen in the relevant paragraph (*transport, consolidation, transport, deconsolidation, transport, delivery*). Delivery can be $A \times B = local$ *route*, i.e. collection at the Fulfilment Centre on day A and delivery by 11:59 pm on day B, a or at the latest overflowing onto day C in which collection takes place on day A and delivery on day C, hence within 48 h.

Amazon: Evolutionary Scenarios

In the discussion of the previous paragraphs it emerged clearly that Amazon has earned recognition as a world leader, in particular for being a benchmark in logistics, distribution and more broadly, in supply chain management processes. This sector has been analysed and valorized in all its perspectives and strategies, to the extent that today it can be considered the real core business of the American giant.

The technical capacity of the organization, but above all of the continuous innovation which distinguishes Amazon, together with its foresight in long-term goals, represent the key factors that have enabled and consolidated an exponential success both in America and in the rest of the world. The preponderance and continuity of this markedly innovative footprint are highlighted with all their positive charge in the strong and constant expansionist growth of the market achieved and served by Amazon.

The Amazon principles: *"Customer Obsession"* and *"Invent and Simplify"* are precisely the basis of this growth: in supply chain management and distribution, innovating means finding new ways to improve speed and synchronization between processes in order to reduce lead time and give the best possible service to the customer, in order to increase the number of new customers and retain existing ones.

Over the last decade, in addition to diversifying and expanding the type of products and services offered by the Marketplace, the company has at the same time launched projects that can be defined as "futuristic" and certainly mark a new milestone on the path of innovative development in the retail, organized distribution, and grocery sectors.

A schematic description of these innovations will be provided below.

Amazon Go

We have seen how the trend, which has already been underway for some time and has now become an actual practice, is that of the progressive migration of business perspectives from the realm of physical retail to the online one in order to take advantage of the various opportunities offered by the direct sales channel, and which for some time now have found a full response and appreciation by consumers. However, in the reality of the market, we find examples of companies which, even if born on the net, have not neglected to add a brick-and-mortar retail channel alongside their online channels. Amazon is no exception, and in recent years it has implemented some initiatives which point precisely in this direction.

The *Amazon Go*! These are a new store concept (the pilot project was created in Seattle, the city where Amazon was born) in which consumers can enter, move around freely making their "physical" purchases, and leave without having to stop to pay at the checkout.

At the entrance, customers identify themselves by scanning a code from their smartphone using the Amazon Go app—based on a sophisticated technology called *"Just Walk Out"*—then once inside the store the desired products can be picked up and placed manually in trolleys or directly into their own shopping bag, and then the customers exit directly from the supermarket after making their purchases, without having to worry about paying at a physical cash desk (see Fig. 101).

What has happened is that the individual products have already been automatically detected and accounted for in the "virtual" cart of each individual consumer at the time of their removal from the shelf. A sophisticated machine learning system, consisting of hundreds of cameras and sensors, recognizes which products are introduced into

Fig. 101 The entrance barrier at an Amazon Go shop

the cart, also identifying a customer's change of mind if he/she later puts an item back on the shelf. The "Just Walk Out" technology described above takes note of the products removed and those put back in their place on the shelves, keeping track of them in a sort of "virtual trolley" matched to an individual. The contents of the cart and the total value of the products it contains are then confirmed at the store's exit and, as already mentioned above, charged directly and automatically to the credit card pre-registered on Amazon Go. For the moment, the artificial intelligence systems work correctly on packaged items such as FMCG, which are easily recognizable, while they are not valid such for sectors as textiles and clothing, since for the moment no computer vision system can distinguish the model or size of a folded garment. It is likely that the exponential progress of these technologies will make them applicable within a few years to other sectors where references are less easily distinguished (see Fig. 102).

The "stop at the checkout" stage is thus eliminated and replaced with a charge by Amazon on the account of the customer who has previously authorized the operation by using the relevant app.

This new shopping experience has advantages for both the business profile and that of the individual consumer.

On the one hand, customers see fruitless waiting times reduced considerably thanks to the fact that they no longer have to wait in line and stop at a cash desk; on the other, Amazon has the possibility to study directly, in a more precise and penetrating

Fig. 102 The artificial vision system supervising the sales floor in an Amazon Go shop

way, all the kinds of behaviour linked to a purchase, tracing the routes taken, the products looked at but not chosen, the customers' choices, any second thoughts, and to verify whether the elimination of the checkout increases the probability of a purchase, and if so to what extent.

The first store without cash desks and staff, inaugurated in January 2018 was called "Amazon Go" precisely to emphasize the speed of purchase, the lack of queues and therefore waiting times, the immediacy of certain passages, along with data recording thanks to a sophisticated automated system which uses both customers' smartphones as well as machine learning and computer vision techniques.

The "pilot project" launched by Amazon has produced satisfactory results, so much so that the Seattle giant has replicated the concept in about twenty other locations (8 in New York, 6 in Chicago, 4 in San Francisco).

Amazon Fresh

The discourse of *Amazon Fresh* is totally different. The foundations of this operation were laid in 2017, with the acquisition through an operation worth almost 14 billion dollars of Whole Foods Market, a US company based in Texas and operating in the production of natural and high-end organic foods. More than an acquisition, in this case we are talking about a sort of merger, however, the significant aspect to take into consideration is that this financial operation represented for Amazon a big step towards the world of "brick-and-mortar" by acquiring one of the largest and best organized supermarket chains in North America consisting of approximately 450 top-level physical stores.

This operation laid the foundations for the development of brand new "innovative" Amazon Fresh stores. Three years after the acquisition of Whole Foods Market, Amazon opened a new concept store in California, *Amazon Fresh Grocery*. This store develops over an area of more than 3,000 m^2, with a premium assortment (in line with Whole Foods Market which inspired it and whose team collaborated in the realization) and technological innovations in store, including contactless features.

Fig. 103   the Amazon dash cart at Amazon Fresh

Amazon Fresh has been designed to enable a shopping experience that links online and offline. The restaurant is equipped with both traditional checkout counters and a cashless solution, using the Amazon Dash Cart—an alternative solution to the technological infrastructure present at Amazon Go (see Fig. 103).

Amazon Dash Cart, which recognizes through AI and computer vision the products that are inserted into it, allows customers to skip the line at the checkout: payment is processed using the credit card registered on the Amazon account. The activation of a 'Dash Cart', through a QR Code and a specific Amazon app, allows customers to place the items to be purchased in it, which emits an acoustic signal to indicate that the product has been recognized and registered while a screen displays the count made thanks to a combination of algorithms and sensors. When shopping is complete, customers step out of the specific aisle and automatically complete their payment.

The structure is also equipped with *Ask Alexa* stations (see Fig. 104) where it is possible to take advantage of the new AI functions linked to Alexa, useful for guiding customers in the management of their shopping lists, or to answer any questions about the assortment and services.

When it comes to produce, Amazon Fresh offers a selection of fresh produce from fruit and vegetables to meat and fish, along with a range of baked goods and foods newly prepared every day by the store's culinary team, from freshly baked bread to made-to-order pizzas, rotisserie products, and hot sandwiches. There are both branded and private label products in the assortment, including Amazon and Whole Foods Market brands. The company also intends to introduce two new private label brands: *Fresh* and *Cursive*. In its choice of products, Amazon Fresh pays particular

Fig. 104   Ask Alexa at Amazon Fresh

attention to organic and healthy foods as well as local produce, and uses sustainable packaging wherever possible. Amazon is expected to expand this format with new stores in Chicago and Los Angeles.

Amazon Prime Fresh

The company's commitment to covering as many market segments as possible is also confirmed by the introduction of the *Amazon Prime Fresh* service on the American market and, for some years now also on the European market, which allows—for the moment only in metropolitan areas for reasons for economies of scale and service limits—the ordering of typical everyday products, such as fresh and frozen foods, housewares, etc. … Amazon Prime Fresh is a service reserved for customers registered on Amazon Prime who can subscribe to the Prime Fresh service and receive orders placed through a dedicated application in 1 h, or within a 2 h window.

In the United States, the products purchased and delivered with this new service cover almost all types of food sector goods: from bread to vegetables, from meat to pasta, from wines to condiments, from frozen foods to desserts. In Italy, as mentioned, the dry-goods service is available nationwide, while fresh products can only be ordered in Milan, Rome and Turin, and in some zones neighbouring large metropolitan areas. The freshness maintains its qualities and organoleptic characteristics unaltered thanks to the limited distances to be covered and the use for the transport phase of thermal bags that can maintain an optimal temperature.

Another very important aspect is the price. Currently, Fresh's prices are in line with the competition, but knowing Amazon's policies, it is thought that once the trial period has passed, costs will drop drastically and it will be precisely on this point that Amazon will beat the mass retail chains: aggressive prices, convenience of shopping, guaranteed punctual deliveries. All within a few minutes. Basically, a new way of shopping.

Amazon Prime Air

Amazon Prime Air is currently the most innovative home express delivery service, not yet available to the public, and based on the use of small flying drones, technically prepared for the air transport of objects.

This is a system which could speed up product deliveries exponentially, especially in low density areas. In fact, it has been estimated that a system such as Prime Air could reduce delivery times (from issuing the order to delivery to the consumer's home) to less than 30 min within 10 miles' distance for small packages. Basically, once an order has been received, the employees of the specific Prime Air distribution centre take care of preparing the package and loading it onto the drone (most likely this operation will eventually be done automatically), which will then take off, reach the destination, and once there, identify the best spot to land, delivering the package and, retracing the same route, return to the departure centre; all in a completely autonomous way. Amazon's goal is to revolutionize traditional delivery methods by supporting them with an additional, innovative service that is fast, efficient, safe, and respectful of all environmental issues. In this context specifically, the safety of the system has been confirmed as one of the company's priorities: the drones are equipped with numerous sensors which allow them to recognize, identify, and avoid obstacles, both in the air and on the ground. The drones developed so far by Amazon, although the company is continually working to improve their capacity and safety, are capable of flying for more than 20 km, carrying objects weighing just over five kilograms. Amazon is carrying out various tests with drones featuring different designs and delivery mechanisms, in order to understand how to prepare the most efficient service, which also depends on the specific environmental conditions of the journey and the place of delivery (see Fig. 105).

Various Prime Air project development centres are currently operational in the United States, Britain, Austria, France and Israel. As mentioned, Prime Air is still under development, and experiments are still ongoing to test the technology, also in order to comply with the laws and regulations of the various countries involved, both in the technical and legal fields as regards aircraft and "restricted areas".

Recently (September 2020), the FAA (Federal Administration Aviation) which is the American body that regulates the use of airspace, granted clearance to Amazon for the use of small drones for commercial purposes, thus recognizing it the status of an authorized air carrier. It is therefore clear that even this "dream" of Jeff Bezos will soon become a concrete reality and with the possibility of large-scale use, for a very near future in which "seeing Prime Air vehicles will be as normal as seeing mail trucks on the road."

Fig. 105 Amazon Prime Air home-delivery drones

Amazon Logistics

As already mentioned, Amazon makes logistics a competitive element: being able to fulfil an order in 24–48 h worldwide represents a strategic competitive factor which has allowed Amazon to become the leading online retailer in the world together with Ali Baba. Which is why the company has decided to make this distribution logistics capacity available also to third-party companies, rather along the lines of its IT services with Amazon Web Services.

Amazon Logistics has developed along two lines.

The FBA fulfilled by the Amazon model mentioned above: companies which wish to use Amazon as a logistics operator can ship their products to a Fulfilment Centre and Amazon will take care of keeping them in stock, putting them up for sale on the Marketplace, and once an order has been received, proceed to fulfil it, exactly as happens for an Amazon product.

Meanwhile, Amazon Logistics can also operate as a proper logistics operator, bringing its technologies, equipment, and staff to the Distribution Centre of the client company, and from there carrying out all the warehouse processes, from inbound to outbound, just like a Fulfilment Centre.

Thanks to the enormous liquidity generated each year and available for investments, starting from 2015, Amazon Logistics has equipped itself with a Prime Air branded cargo fleet with which it manages the distribution of its products in the United States, in competition with the express couriers, but which it continues to use in part. Currently, Amazon Air has its airport base at Cincinnati Northern Kentucky International Airport (KCVG) on a plot of approximately 180 ha (the equivalent of approximately 250 football fields), with plans to double it by a further 194 ha by 2027. In all major cities in the United States, with around 200 flights a day and about fifty active stopovers. The fleet currently consists of about seventy Boeing cargo planes, with the aim of reaching one hundred aircraft by 2022. A second hub in Europe has been operating since 2020 in Leipzig in Germany, with Amazon intending to extend its Prime Air express courier service throughout Europe.

In addition to this air fleet, Amazon Logistics is flanked by an impressive fleet of articulated trucks, branded 'Amazon Prime' for primary transport from air hubs to its Fulfilment Centres. The last-mile distribution is instead entrusted to independent patrons who are associated with Amazon Logistics through franchising. Each operator is therefore an independent entrepreneur, to whom Amazon Logistics entrusts technology and volumes (for a fee), while the operator invests in its own vehicles (100% electric) and staff, and carries out the home delivery service that the company assigns him/her every day.

Amazon Lockers

Amazon Lockers are a service which aims to provide a service to people who go to work and cannot therefore receive their purchases at home, or even have them shipped to the workplace. Thanks to Amazon Lockers it is therefore possible to receive purchases made on Amazon directly at a self-service collection point.

Not surprisingly, Amazon lockers are orange (in some cases blue) and are equipped with an electronic lock which can only be unlocked by the person in possession of the code which staff provided at the time of purchase from the platform. Which means simply selecting a locker as the shipping address at the time of ordering. To find a collection point, there is a search option by address, postcode or landmark; it is sufficient to choose a locker and save the address. Once the order has been placed, the customer is notified by e-mail that their package has arrived at the chosen locker and a code is provided which must then be entered to collect the item/s. In addition, at every Amazon Locker point, there is a computer to guide the customer through the unlocking procedure. The time available for collection of the package is 3 working days upon receipt of the e-mail. If at the end of the available time the package is still in a "not-withdrawn" state, it will be collected by Amazon and the order amount credited to the card used for payment. The use of this service involves no additional costs on top of the Prime subscription fee. For those who have not subscribed to Prime, the only cost incurred is that for shipping. However, those without a subscription pay the shipping costs in their entirety (see Fig. 106). When lockers are installed in shopping malls, the usually increase the mall traffic, therefore it is a win win proposition for both Amazon and the mall.

1.4.7 Returns

Returns are one of the key elements of Amazon's and other ecommerce retailers' service. Free, easy and hassle free returns is the pivotal element to create confidence in customers to buy. Nonetheless, returns are by far one of the largest challenges to

Fig. 106 Example of Amazon locker

e-commerce. Returns drain companies millions of dollars in unwanted inventory and extra labor, and at the same time they create billions of pounds of waste and entire "wall of shame" warehouses around the world. And as more consumer spending is shifting from brick-and-mortar stores to online, the size of the returns problems it's just exacerbating.

Amazon is trying to change all of that E-commerce giant is trying to be the easiest, lowest friction return experience for the consumer and thereby win customer loyalty and increase customer purchases. From free of charge home pickup returns, to in-person returns Amazon is redefining the return process. At the same time, as e-commerce grows smaller companies are finding ways to make money off returns.

In this paragraph we will tackle how Amazon on this process works, and what the company's doing to protect the environment and the bottom line.

High Costs of Returns

Inefficiency of the reverse logistics generated by returns is a huge waste of money for online retailers and the problem is just exacerbated by steady growth of online sales. Traditional brick-and-mortar stores may experience average return rates of 8 to 10%, but in ecommerce it is common to see 20–30% with peaks of 40% of all purchases returned. Before the pandemic, Forrester research (Kodali 2019) estimated the worth of returns in ecommerce in North America at 207 billion dollars/yy, and Amazon is about half of that, so slightly more than 100 billion: a huge expense. These are the root causes for ecommerce returns. 34% of returns are due to the wrong size, fit or color; 21% are due to item damage. 14% of customers returning products say the item wasn't as described, while 10% that simply didn't like it and 9% changed their mind. Only 5% claimed they returned the product because it didn't arrive in time or was late.

And then returns processmatters to customers. According to a market survey by Invesp (https://www.invespcro.com), 79% of consumers want free returns, and 67% check the returns page before making an online purchase. This has led retailers to offer free returns shipping, which is now offered by half of the online retailers. Therefore the key challenge is to reduce the unit logistics cost of the returns as much as possible, not to deprive the company's profit and losses. In order to keep the costs of returns as low as possible, he main advantages for Amazon are on the one end the economies of scale their return business generates, and on the other one, that they can accept that in the beginning they may do things at a loss, but then they figure out how to optimize processes and how to create efficiencies that will allow them to have the unit economics to work to their favor, and ultimately get those margins back.

Furthermore many customers abuse the free return policy. Under some circumstances, they simply borrow a product, use it at their convenience, and then return it for free. This is fairly common for suits, to be used for special evenings or parties, and are then returned, but also for TV sets. For instance during the Super Bowl, online TV sets sales oftens spike, but those items are simply returned a few days after. In other circumstances, and in particular for fashion, branded apparel, footwear and accessories, retailers suffered the return of counterfeited products, while the original ones are then resold on the black market.

In 2017, return frauds cost the industry 17 billions USD, therefore many retailers setup a score for customers based on how profitable and how good a customer is. If a profitable and good customer wants to return a product, retailers might continue to offer him/her free return service, while after the first or the second return without a good reason, they may charge him/her or even block his/her account in case of fraudulent misuse.

How Amazon Processes Returns

The reverse logistics journey starts when a customer decides to return his/her purchase. Amazon gives customers 30 days to return the product from the day he/she receives the delivery. Generally amazon gives the money back and even includes paying for shipping back. The consumer packs the items to be returned in the same box used for shipping, applies the return label that can be found in the package, and opens the return process on the web site. An Amazon driver shows up at the customer's doorstep, picks up the parcel and drives it back to a collection point that works as a transit point, where it is thus consolidated and shipped back to the fulfillment center for receiving.

That part of door to door pick up is by far the most expensive one, regardless it is carried out by UPS, USPS or Amazon Itself. Therefore, in order to relieve some pressure on these costs, Amazon started the in-person return service, in partnership with Kohl's, to allow items to be returned without a box at any of the 1.100 Kohl's stores for free. According to a survey carried out by Invesp, 62% of customers are more likely to buy online if they can return an item to a physical store. So, instead of going to 100 consumers' houses to pick up one box, Amazon has those 100 consumers all go to the same Kohl's. It is a win-win value proposition since Amazon needs collection points and Kohl's needs ways to attract people to the stores, and thus the traffic generated by Amazon consumers. Moreover those consumers have money in their pocket due to the return, so it's a great opportunity to generate sales for Kohl's.

In person returns without a box can also be carried out at one of 2.800 locker locations, that can be incidentally found also at Whole foods, and therefore increase sales of Amazon fresh food supermarkets.

Once the returned item arrives at the fulfillment center, the receiving process is completed, as described above, then, it begins another big challenge. Often products end up in a place of limbo, a place that some retailers call "the wall of shame". Under some circumstances, it can be a 6.000 m^2 area full of returned items, that may be worth million of dollars of inventories. Since for some retailers the cost of managing those items or inventories is by far higher than the cost of the items itself, they push it aside and the "wall of shame" grows until the products are eventually shrinked. In some DC, retailers like Rakuten have introduced autonomous robots, InVia robots, to manage the wall of shame at individual item level (up to 20kg/piece), and shrink the cost of managing returns.

Environmental Impact of Returns

As much as 5 million ton of waste gets thrown away every year in the US as a result of ecommerce returns that are not worth to be managed, handled and resold. And half of that is due to Amazon.

In 2019 it was reported that a single Amazon facility sent 293k products to the garbage dump in just nine months (Gibson 2019). That of course has a huge environmental issue. Amazon has spotted a light on that, also because consumers are becoming more and more sensitive to that environmental issue, and do not like doing business with ecologically unfriendly companies.

Therefore, even if destroying is the most economical option, Amazon found other options at its own expense.

One option is Fulfilled By Amazon donation. Launched in 2019, the program was aimed at turning unsold FBA products stored by Amazon's customers in Amazon FCs in the US and in the UK, into donations.

Another one is the use of the massive amount of data that Amazon has. Artificial Intelligence software drills those data to figure out consumers that may be interested in returned products at bargain prices. If there's an option, the returned product is reconditioned and proposed to the customer as an up-selling proposal at the next log in.

Another big issue is related to the packaging waste that Amazon is trying to reduce. Kohl's and Amazon pickup locations generally are using polybags and other kind of containers when they aggregate returns together to dramatically use less packaging.

Amazon has also replaced many cardboard boxes with more lightweight plastic mailers, reducing packaging waste by 16%. Given the huge numbers of Amazon packages shipped every year, it implies the elimination of the need for more than 305 million shipping boxes in 2017 in the US.

Furthermore, Jeff Bezos just pledged to make Amazon carbon neutral by 2040.

Returns Business Opportunities

There's also a growing market of companies that are trying to leverage Amazon's and other ecommerce returns and make money.

There are third party companies that buy returns in bulk and resell them at bargain price. Funny thing is that they sometimes go back to Amazon's marketplace. Other companies resell them in low cost markets, or get payed to pick up the returns and donate.

Another interesting example of companies doing business with returns is a company called Happy Returns (www.happyreturns.com). They currently have 700 return centers at malls and stores where customers can go and return their items from 30 popular online stores. Happy Returns gets paid by online retailers to aggregate returns, saving money on that first-mile pickup that would cost 20–30% of the total return logistics cost. The stores also pay Happy returns a fee, since the returns service increases the store traffic with consumers likely to purchase.

1.4.8   Online—Conclusions

Considering in detail the cases of Dell.com and Amazon.com and how these two 'online pure players' companies have been able to leverage the supply chain to build two companies which, in just a few decades, have become market leaders, we can summarize in conclusion the competitive strategies that have allowed these 'pure players' to establish themselves in the competitive scenario, and the strengths and weaknesses that characterize a purely online distribution supply chain compared to a traditional one

Online—Competitive Strategies

The difference in competitive strategy between a web retailer and a traditional brick-and-mortar retailer is summed up in Table 11, and can be recapitulated in the three competitive levers: (1) Product (2) Price (3) Service.

The set of product/price/service defines the target/market share that the SC will be able to capture, that is, it influences the ability to guide and retain the consumer towards a traditional channel or an online channel, as depicted in Fig. 107.

Table 11 Competitive strategies: internet distribution channel versus retailer

Leverage of strategy	Online retail	Brick-and-mortar retail
Product	Often the same, or at least with few functional differences	
Price	In line with the retail price or slightly lower thanks to savings on PCs or MMCs PC are optimized thanks to economies of scales related to volumes. Margins are higher thanks to the reduced numbers of players in the shallow SC, and thanks to the savings in MMCs and PCs	In line with the web price or slightly higher due to PC diseconomies (mainly store costs and duplicate inventory costs) and MMC (OOS costs, obsolete/expired goods). Margins are lower and have to be shared with a deep supply chain
Service	Reaching the consumer through the web very high range of products home delivery LT as low as possible and in any case acceptable to the consumer (e.g. Amazon Prime $A x B$ or $A x C$) Free returns service; after-sales assistance (for durable goods) additional services (e.g. cross selling/promotion; customer's opinion) extremely wide range of products	Reaching the consumer through physical stores Limited range Zero LT, the product is supplied in advance and available in the store to be touched and tried and purchased Advice from a store associate In-store returns Emotional in-store experience

Fig. 107 Traditional and on line supply chains

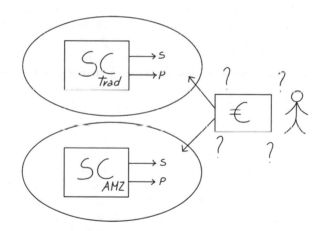

Online—Strengths and Weaknesses

In the Internet channel, compared to the traditional physical store channel, a series of benefits are obtained related to both physical costs (production, distribution and retail) and market mediation costs (out of stock costs, shrinkage/obsolescence/unsold products), which affect the cost of the product, and afford a player like Amazon.com or Dell.com a competitive advantage through greater profitability for the same price or lower prices with the same profitability.

This paragraph analyses these aspects in detail.

An online retailer, compared to a traditional player (*brick-and-mortar retailer*) can count on the following strong points:

- Zero costs for store infrastructures (*both Capex and Opex*). There are no store costs since orders are collected via a portal and then fulfilled by a DC (*proprietary product or logistics operator*), or directly by the supplier (marketplace). Amazon, for example, manages its entire business with a network of 130 Fulfilment Centres worldwide, and until recently without any proprietary stores: in reality, in recent times it has been experimenting with Amazon Go and Amazon Fresh in major cities around the world. Dell, on the other hand, produces make-to-order (MTO), and distributes in direct sales, without a store. To avoid competition between the online channel and the retail channel (multichannel retail) Dell does not sell on a retail channel but only online (single channel; online pure player).
- Transport is always optimized (FTL) with economies of scale both over long distances and in the last mile (multidrop). The flows are optimized either thanks to the volumes generated (e.g. Amazon) or by resorting to 3PLs and postal services which handle the consolidation, sorting and shipping in every small town.
- Optimized stocks in traditional brick and mortar retail, stocks are traditionally duplicated in width (stores) and in depth (among the N players that make up the SC at each level).

    In Amazon's case, the inventory is at the DC (only for high-rotation products, known as class A), which allows a reduction in both cycle inventory and safety inventory. In the next paragraph, we will analyse in detail this aspect known as *inventory pooling*, which enables a reduction in both cycle stocks and safety stocks by a factor $\sqrt{N}$, with N being the number of facilities. Low-rotation products (classes B or C) or for highly customized products, the logic is to avoid keeping them in stock. In the case of Amazon, there is pull management instead of push, with the consequent cancellation of the costs of keeping stock.

    In the case of Dell, an MTO strategy allows zero inventory costs and MMC related to obsolescence costs of finished products: in fact, in the case of MTO products, by definition, obsolete and expired products and shrinkages in general are completely zero, since the product is always made to order. Conversely, the components must be kept in stock, but their cycle and safety stock is much lower due to the principle of component commonality in different models/variants, also known as *product postponement* (managing the production in advance of "neutral" units to be customized to a customer's order).

- Additionally, in the case of textiles and clothing, by centralizing the stocks at a DC, the online channel allows a cancellation of the costs of replenishing sizes or more generally of reconstructing the model/variant assortment (backward and forward logistics costs, bw and fw respectively). In fact, in a network of outlets, stocks must continuously be moved upstream to rebuild the assortment thanks to "size gaps" (see Fig. 108). When a model variant begins to have "size gaps"

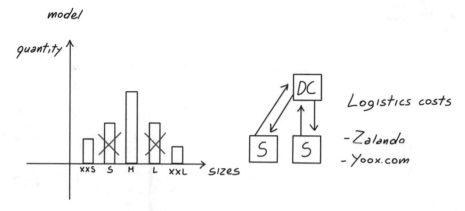

Fig. 108 The phenomenon of size gaps

at a store, typically for the intermediate sizes, it is returned to the DC, where the assortment of intermediate sizes is restored and sent back downstream, with significant logistical and handling costs (backward and forward), waste of time, and therefore also of sales opportunities.

- In addition to the cost savings discussed above, a traditional store is subject to MMC, which is much lower in an Internet channel, since the product is either made MTO or is fulfilled by a DC which holds smaller stocks; and consequently also the costs of obsolescence and expirations are lower.
- In addition, the important financial leverage of approximately 3 months remains on all online purchases compared to purchases in a traditional store. In fact, an online retailer collects the payments immediately and pays its suppliers within 3 months, while a traditional retailer produces in advance and pays its suppliers, and only later collects from the consumer and therefore often has negative financial levers.

Amazon Prime	
Liquidity	100,000,000 users (2018) → 4,000,000,000 € per year
Amazon and Dell	
Cash flow	Income: $\tau/0$
	Supplier payments: 60/90 days → 3 months of financial leverage

Conversely, a web-type distribution channel does suffer from certain critical issues:

- The lack of a store does not allow the consumer to "touch" and "try" a product. The emotional shopping experience therefore changes, and in certain sectors, such as fashion, this aspect can present a barrier to entering the market. In many other sectors, however, the product is standard (consumer electronics, books, etc.) and the consumer does not need to physically touch or try the product before buying

Fig. 109 Backward return
flows

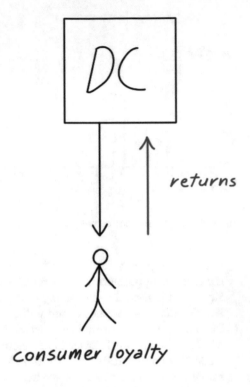

it. Even when it comes to clothing, the consumer increasingly disregards being
able to try on a product, or tries it on in a physical store and then buys it online.
In other cases, the consumer orders multiple sizes/models/variants online, tries
them on at home, keeps the one that fits and returns the products that do not fit.
The returns policy therefore becomes important. Traditional stores, as mentioned,
try to leverage precisely this physical characteristic to transform the stores from
mere retail outlets into points of contact and an emotional experience between the
consumer and the product/brand. In a modern-day store, a sale will not necessarily
take place, but the consumer will get to know the product and the brand, and then
complete the sale, possibly on another channel (e.g. online).

- Through the portal, the online player can adopt cross selling/cross promotion
  strategies either to steer or to increase the likelihood of success of the transaction. With cross selling, the portal tries to add other products to the one the
  customer is currently evaluating/buying and that are usually sold together or that
  match the current need. Products can be proposed at full price or at a discount.
  With cross promotion, instead, the online player can steer the demand to other
  brands/products, typically internal brands, where the online player has bigger
  margins or where availability is higher
- Returns management. Returns (backward flows) in the case of online sales can also
  constitute 50% of forward flows (see Fig. 109). Although this constitutes an important cost figure in the medium term (transport costs to 3PL, and returns manage-
  ment/handling, reconditioning, administrative and anti-fraud management), in the
  long term, the possibility of free returns granted to the final consumer which many

Fig. 110 SC lead time

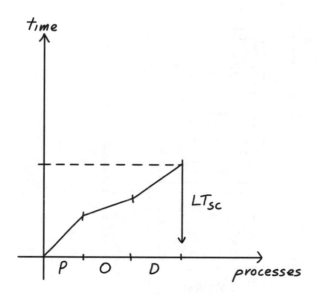

retailers on line adopt is a competitive lever of enormous impact, since it allows the seller to win the trust of the consumer, who knows that he/she can return the product if it is unsatisfactory, or can order more variants/sizes, try them on, and return those that are unwanted. All of which increases *customer loyalty and retention*, and therefore generates a competitive advantage (see Fig. 110).

- Lead Time greater than the zero LT of traditional retailers.

In a physical store, the consumer arrives and leaves physically with the product in hand, since the supply chain works in a Push mode. Instead, in the case of online purchases, an order fulfilment time is required, between the instant the customer completes the order on the online portal and the instant the order is delivered to his/her home, since the order triggers a part or the whole supply chain process, in a Pull mode. The longer the lead time, the worse the customer's shopping experience will be and therefore the lower the number of consumers available for online purchase compared to buying in a physical store, and vice versa.

In the case of a completely Pull process, the lead time will be equal to the sum $LT_{SC} = P + O + D$, with the time P required for the procurement process O, the time for the operations, and D the time for distribution. In a case in which only Operations and Distributions are Pull (*e.g. Dell*) it will be equal to $O + D$, while in the case in which only the distribution is Pull (*e.g. Amazon, Yoox.com, Zalando*) it will be equal to D (see Fig. 111).

For this reason, the logistics processes must be extremely fast to ensure a "lean flow" as much as possible, but equally, if the flow also involves production, suppliers must be nearby and fast, the components need to be in stock to start production quickly, and the production cycle must be fast and flexible, to process batches at the unit limit (*a single customer order*).

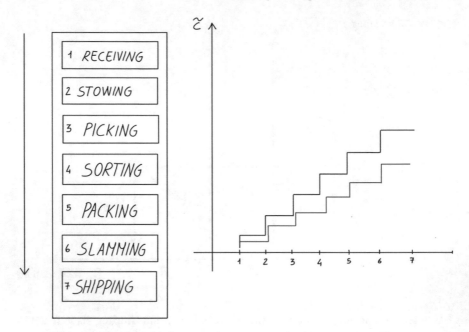

Fig. 111 Amazon BPR to fulfilment centre processes to reduce the SC lead time

These aspects will be explored in detail in the paragraph relating to strategic lead time management; in any case it can be premised that in order to try to make the flow ever faster, Business Process Re-engineering (BPR) techniques are applied. The goal is to analyse in detail each activity/phase of the P, O, D processes and eliminate/accelerate all operations that do not add value to the flow, but add time to the process.

The nature of re-engineering can take place in two ways: foreseeing the use of technological innovations, or defining new processes (a new way of doing things) (see Fig. 111).

Amazon, for example, has obsessively pushed this approach, to win consumer trust and guarantee "Prime" delivery within 24 h: the result has been obtained by continuously optimizing each of the 7 processes listed in Fig. 100, both in terms of time τ and quality of service. In this way, Amazon Prime customers are perfectly content to pay an annual fee in advance (around €40 a year) to guarantee Prime delivery within 24 h for millions of products. The competitive advantage is summarized in the scheme of the previous paragraph and includes being able to guarantee annual liquidity independent of orders.

1.4.9 Inventory Pooling

Inventory pooling is a phenomenon whereby cycle stocks $((\underline{D} * LT)$ in the case of management by economic batch or a reorder interval every $T + LT$ days, and regardless of the correlation of demand, and safety stocks $(f(\Delta T;\ \Delta LT)$ only in the case of non-demand correlated, are lowered by a factor of $\sqrt{N}$ by centralizing the stocks in a single point, compared to a "decentralized" structure in which the stocks are distributed over N points (DCs or stores).

Centralization may be physical, i.e. centralizing stocks in a single point, but also only virtual, for example by sharing stock data at the same level, and carrying out transhipments. We talk about transhipment e.g., in a chain of stores where the stock is shared virtually and, in the case of a product being missing at one store where it is requested and available at another, transport is organized at the same level as the SC between one store and another to fulfil the order.

A similar result is obtained by keeping in stock a neutral product or a more complex product (*product postponement*).

The reduction in stock constitutes one of the competitive advantages—in terms of the reduction in PCs (cost of inventories) and MMCs (cost of obsolescence/shrinkage) of a web sales structure (B in Fig. 112) which satisfies the demand of N consumers from a DC, compared to the retail structures of N physical stores which serve N consumers (A in Fig. 112).

Although the description presented, with reference to the diagram in the figure, constitutes a simplification, i.e. the same distribution of demand for all the stores, it can nevertheless be generalized, highlighting the fundamental hypotheses on which the pooling mechanism is based and from which the following numerical description has been developed.

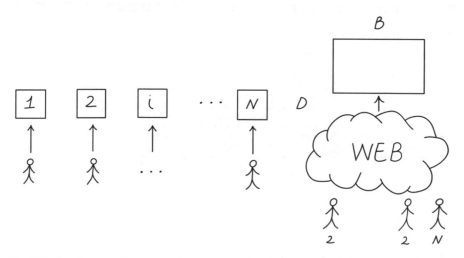

Fig. 112 Inventory pooling: comparison between the split scenario of physical stores (A) and centralized online sales

Both the d random variable relative to the demand seen by the single store, of mean m and standard deviation σ_d, which determines the random variable of aggregate demand D seen by the DC in the case of a centralized system, of mean D and standard deviation σ_D. In Case A, each store i verifies an average demand d and m_i standard deviation. σ_i. In Case B, the aggregate demand will be $D = \sum_{i=1}^{N} m_i$. Therefore, the relations in the following scheme apply.

The cov is the covariance which relates how the random variable varies d_i as the random variable varies d_j, calculated through the correlation coefficient r_{ij}.

Case A	Case B
$r.vd = (m_i; \sigma_i)$	$r.v.D\left(M; \sigma_D^2\right) = \sum_{i=1}^{N} d_i$
$mean = m_i$	$M = \sum_{i=1}^{N} m_i$
$standard\ deviation = \sigma_i$	$\sigma_D^2 = \sum_{i=1}^{N} \sigma_i^2 + 2 \sum_{i=1}^{N-1} \sum_{i+1}^{N} cov(i; j)$ $= \sum_{i=1}^{N} \sigma_i^2 + 2 \sum_{i=1}^{N-1} \sum_{i+1}^{N} r_{ij} \sigma_i \sigma_j$

The symmetrical matrix of the covariances (Table 12) allows a quick calculation of the σ_D^2 in Case B.

The hypothesis of the same distribution of demand for each store translates into $m_1 = m_2 = m_N = m_D$, and $\sigma_1 = \sigma_2 = \sigma_N = \sigma_D$.

Table 12 Covariance matrix

$cov(1; 1) = \sigma_1$	$cov(1; 2) =$ $cov(2; 1)$	...		$cov(1; N) =$ $cov(N; 1)$
$cov(2; 1) =$ $cov(1; 2)$	$cov(2; 2) = \sigma_2$			$cov(2; N) =$ $cov(N; 2)$
...	...	$cov(i; j) = r_{ij} \sigma_i \sigma_j$	...	...
$cov(N; 1) =$ $cov(1; N)$	$cov(N; 2) =$ $cov(2; N)$	...		$cov(N; N) = \sigma_N$

Inventory Pooling—Cycle Stocks

Analysing the cycle stocks CS in the two cases, in the case of an economical lot, calculated with the well-known formula $EOQ_i = \sqrt{\frac{2m_i C_{Ri}}{C_{TL} \Delta \tau}}$ in which C_R is the reorder cost, while C_{TL} is the total logistics cost incurred to supply the stock (e.g. industrial cost in the case of components) while τ is the valuation rate of the inventories, which could be the bank rate if the company uses bank capital, or the ROE in the case of equity.

From an analysis of the comparison of the results in (7) we obtain:

Case A	Case B
$CS_i \propto EOQ_i$	
$EOQ_i = \sqrt{\frac{2m_i C_{Ri}}{C_{TL} \Delta \tau}}; EOQ_i \propto \sqrt{m}$	
$CS_A = N * SC_i \propto N * EOQ \propto N * \sqrt{m}$	$CS_B \propto EOQ \propto \sqrt{N * m}$
$CS_A \propto N * \sqrt{m}$	$CS_B \propto \sqrt{N * m}$

$$CS_B = \frac{SC_A}{\sqrt{N}}, \ N \ number \ of \ stores \qquad (7)$$

As can be seen from the result (7), the cycle inventories in Case B (*centralized*) compared to Case A (*fractionated*) decrease by a factor of $\frac{1}{\sqrt{N}}$.

The same result is obtained in the event of a reordering interval policy, by reordering every T + LTA days, or by determining a reordering interval which optimizes stock maintenance costs with reordering costs.

Inventory Pooling—Safety Stocks

With similar reasoning, the safety stocks are calculated in the two cases, centralized (B) and fractionated (A), under the same assumptions. In general, safety stocks are proportional to a standard deviation in the random variable demand in supply lead time, which in turn can be expressed as a function of the average value and variance of the random variables "daily demand D" and "supply lead time LT", through the relationship (8).

$$SS = k\sqrt{\sigma_D^2 * LT + D^2 * \sigma_{LT}^2}, \ SS \propto \sqrt{\sigma_D^2} \qquad (8)$$

$$k = 1 \ for \ 84\% \ likelihood \ of \ avoiding \ OOS \ and$$
$$managing \ LT \ and \ demand \ variability \ through \ SS$$

$$k = 2 \ for\ 97.5\% \ likelihood\ of\ avoiding\ OOS\ and$$
$$managing\ LT\ and\ demand\ variability\ through\ SS$$

$$k = 3 \ for\ 99\% \ likelihood\ of\ avoiding\ OOS\ and$$
$$managing\ LT\ and\ demand\ variability\ through\ SS$$

Therefore, from (8) the proportionality factors in the two cases can be calculated as was done for cycle stocks.

Case A	Case B	
$SS_i \propto \sqrt{\sigma_i^2} \propto \sigma_i$	$SS_i \propto \sqrt{\sigma_D^2},\ \sigma_D^2 = N*\sigma_i^2 + 2*\frac{N(N-1)}{2}*r*\sigma \ con\ r_{ij} = r \ \forall i, j\text{(hp)}$	
$SS_{TOTA} \propto \sum_{i=1}^{N} SSI_i \propto N\sigma$	Absence of correlation $r = 0$ $\sigma_D^2 = N\sigma_i^2$ $SS_B \propto \sqrt{N*\sigma_i^2} \propto \sqrt{N}*\sigma_D$	Perfect correlation $r = 1$ $\sigma_D^2 = N^2\sigma_i^2 SS_B \propto \sqrt{N^2 * \sigma_i^2} \propto N * \sigma_D$

$$SS_B = \frac{SS_A}{\sqrt{N}}, \ N\,number\,of\,stores \qquad (9)$$

$$SS_B = SS_A \qquad (10)$$

From (9) and (10) it is clear that the fundamental hypothesis to see a reduction in safety stocks is an absence of correlation between the demands of the individual facilities (stores), so as to absorb the peaks in demand from one market with the decline in demand of another market. Furthermore, in the absence of a correlation, there is the same benefit factor $\sqrt{N}$ concerning the decrease in safety stocks and cycle stocks.

Virtual Inventory Pooling: Product Transhipment

As mentioned already, centralization can be physical, i.e. centralizing stocks in a single point, but also only virtual, for example by sharing stock data at the same level and operating transhipments (e.g. a chain of stores in which stock is shared and transport is organized at the same level as the SC between one store and another to fulfil the order in the event of a product shortage at a store where an item is requested but not available and another where the same item is available). This is the typical case of the BOSTS scheme (Buy On Line Ship To Store fulfilled by the store) or even BOSFS (Buy On Line Ship From Store) seen in the paragraph on omnichanneling.
In Table 13 is an analysis of the pooling possibilities for this type of configuration.

Table 13 Order fulfilment in the various types of online store

	Order fulfilled through specific store stock	Order fulfilled through any store stock	Order fulfilled through DC Stock
BOPIS	No pooling	X	X
BOSTS	X	Virtual pooling	Physical pooling
BOSFS	X	Virtual pooling	X
BSSFS	No pooling	X	x
BSSTD	x	x	Physical pooling
BOSTD	x	Virtual pooling	Physical pooling

1.4.10 Product Postponement

The same description seen for inventory pooling applies to so-called *product postponement*.

By product postponement we mean the deferred differentiation of the product, with the aim of keeping it "neutral" for as long as possible, in order to reduce stocks, and customize it as far downstream as possible, only once the demand for the customized product is reasonably certain.

This technique was first introduced by Benetton to manage variability in colour demand. The garment dyeing process was moved to the end of the production cycle. Neutral shirts were made and kept in stock, in order to dye them only once the most popular colours for the season were known. Today, we can see this applied to many contexts, from consumer electronics to pharmaceuticals (more complex product; the instruction booklets are printed in all languages or there are power plugs for all electrical standards) to textiles and food (in the form of deferred pricing).

Also in this case, the stock of an undifferentiated product is lower than the stock of a differentiated product by a factor of $\sqrt{N}$, where N is the number of variants/versions of the product, i.e. the variability of demand for a specific product variation is absorbed through customization of a "neutral" product (see Fig. 113).

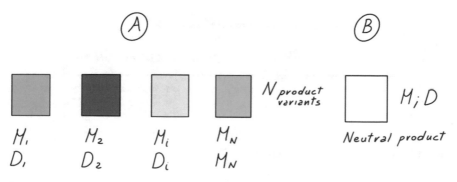

Fig. 113 Product postponement

$$CS_B = \frac{CS}{\sqrt{N}} \quad SS_B = \frac{SS_A}{\sqrt{N}}$$

For product postponement, the following concepts are used:

- **Deferred customization**: in this case the product customization phases are delayed (e.g. product pricing as the last phase to allocate it to the market at the time of demand; garment dyeing at the time of demand), with the benefit of maintaining product stocks at a component rather than FP level (in Fig. 114).
- **Standardization and commonality**: use of standard components for different models/variants (e.g. in a car, the dashboard is the same for all models) so that the same component can be used to satisfy the demand for products that may be not or even negatively correlated (an increase in the demand for one model corresponds to a peak in demand for the other).
- **Modularity:** ability to customize a product quickly through kits and/or modules (e.g. software installation in personal computers, colour modules for printers, but also the possibility to choose between several alternatives of standard components during assembly of a finished product) so that the product can be kept in its neutral form as long as possible, and then its functionalities are customized thanks to the proper assembly of the modules.
- **More complex products:** in this case the product is kept neutral/undifferentiated and supplied "customizable" through a supply of all possible variants (e.g. instruction booklets in all languages, different formats of power plugs).

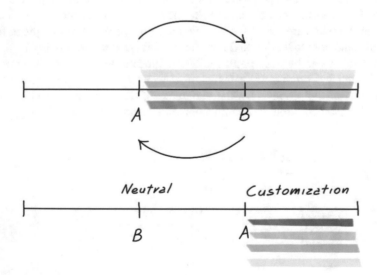

Fig. 114 Example of deferred customization

2 Annex A

2.1 Fast Moving Consumer Goods Industry

Consumer goods are products purchased for consumption by the average consumer. They are divided into three different categories: durable, nondurable goods, and services. Durable goods have a shelf life of three years or more while nondurable goods have a shelf life of less than one year. Due to the large variety of products available in this sector, products can be categorized as: convenience goods, shopping goods and speciality goods. This division is mainly due to different assumption as; purchase frequency, brand relevance, consumers' requirement and aversion to product innovation. The largest segment of consumer goods known as Fast Moving Consumer Goods (FMCG), as they are consumed immediately and have a short shelf life.

Fast Consumer Goods (FMCG) or Consumer Packaged Goods (CPG) is a general term to identify products purchased frequently, at relatively low price and sold in large quantities. These include non-durable household goods such as packaged foods, beverages, toiletries, candies, cosmetics, over-the-counter drugs, dry goods, and other consumables. As the name suggest, FMCGs usually have high consumer demand, high turnover and either high perishability (e.g., meat, dairy products, and baked goods) or low perishability (pasta, coffee, dry foods).

Bagnobianchi (2004), categorized the FMCG sector in; Food and Beverages; Home care products; Personal healthcare products; Home utensils; Spare-time items; Fashion accessories; Domestic animal foods and accessories and Consumer Electronics.

In the shopping rational of the consumer, if the product is not available now in store, they are almost replaceable with another available product. Since this sector is a consumer driven, goods have to be available on hand in the least possible time and in right quantity. For this reason, FMCG supply chains prioritize as intensive distribution system to render high product availability in retail stores. FMCG are characterized by the following traits:

- High turnover due to frequent, simple, fast and purchase.
- High substitutability and low brand loyalty as consumers switch to different brands among similar products.
- Convenience, as goods can be found in different stores.
- Highly advertised.

Beck (2002) stated that the fast-moving consumer goods (FMCG) industry includes *"those retailers and their suppliers who provide a range of goods sold primarily through supermarkets and hypermarkets. The core of their business is providing 'essentials' such as various fresh and processed foodstuffs, but they also stock a wide selection of other goods as well including health and beauty products, tobacco, alcohol, clothing, some electrical items, baby products and more general household items.*" Italian retailers such as Esselunga, Coop Conad, or world-wide retailers like Carrefour, Tesco, and Wal-Mart are often referred to as grocery retailers.

Data from OC&C Strategy Consultants (The FMCG Global 50' report) revealed that the world's largest FMCG companies entered the coronavirus crisis in better shape than they had been in for many years, with headline revenues growing 3.9% in 2019, rising from 3.4% in the previous year. As stated in the report (2021), Switzerland's Nestlé topped the list with solo-sales of $93.1 billion, followed by two US-giants: Procter & Gamble, both of whom commanded grocery sales of more than $67 billion.

There are different types of retailers and manufacturers in the FMCG industry. On the customer side, there are retailers with a strong emphasis in private label (Migros), channel retailers (Wal-Mart) with large number of branded products and global discounters (Lidl) which sell only a very limited number of SKUs.

On the supplier side, there are brand manufacturers which focus on the value-added of their products for customers, channel manufacturers which focus on both consumers and retailers and follow a pragmatic approach to private labels, and private label suppliers which are almost entirely dependent on their retail customers. Retailers may also be classified as "traditional format" retailers (e.g. supermarket chains), and "alternative format" retailers (e.g. warehouse clubs and mass merchandisers such as Wal-Mart). This classification does not take into account discounters. In contrast to mass merchandisers that offer up to several tens of thousands of SKUs, the product range of discounters is much more limited. (Tellkamp 2006).

3 Annex B

3.1 Case 1—Horse Meat in Buitoni Products: Nestlé Withdraws them From Italian Shelves

The decision concerned beef ravioli and tortellini: traces of equine DNA equal to 1%. The multinational reassured the public: "There are no food safety problems." The Carabinieri NAS (*Office against Adulteration of Foodstuffs*) inspected the company's registered office in Milan and its production plant in Moretta.

NEW YORK—Nestlé has withdrawn Buitoni beef ravioli and tortellini from Italian and Spanish shelves. A decision taken after traces of horse meat DNA equal to 1% were found in the products. Having informed the authorities of the results of the examinations, Nestlé reassured them: "There are no food safety problems." The products withdrawn will be replaced by others "which tests will confirm to be 100% beef" added Nestlé in a note, in which it specified that "all deliveries of finished products containing beef from one of our suppliers, the German company H. J. Schypke, have been suspended."

This morning the NAS inspected the registered office of the Nestlé company in Milan and its production plant in Moretta (Cuneo) to ascertain traceability, self-control procedures, and compliance with the obligations on the withdrawal of Buitoni's '*Ravioli di Brasato*' and '*Tortellini di Carne*' products. This was reported by the [Italian] Ministry of Health which added: "Italy introduced checks to combat EU fraud involving horsemeat last February 11, as soon as it heard the first news from the press, and therefore well before the approval of the European Commission's recommendation."

The spread of the scandal of horsemeat in burgers and lasagne prompted the European Union to take action and approve a series of tests on beef to verify its composition. A test which only Italy, the number one consumer of horsemeat in Europe, said it was against. Quite the contrary was the attitude of Germany which—the *Financial Times* reported—would follow a ten-point plan which went beyond what had been established in Brussels to verify the possible presence of other undeclared additives.

But Nestlé reiterated that it had strengthened its controls with new tests: "Ensuring the quality and safety of our products has always been a priority for us. We apologize to consumers and assure that the actions taken to address this problem will result in higher standards and enhanced traceability," Nestlé added in a note that its '*Lasagnes a la Bolognaise Gourmandes*', produced in France, would also be withdrawn from sale.

Meanwhile, the burger industry was suffering from the crisis: in the week that ended February 2nd, sales of frozen burgers in Britain, where the case had broken out and caused an uproar, plummeted by 40%, and two-thirds of the British—according to a Nielsen survey—said they were against buying frozen meat in the future.

3.2   Case 2—Toyota Again Recalls the Same Cars Because of Defective Airbags

The April 2013 recall would not have remedied all Corollas affected by the passenger seat airbag problem.

One of the maxi-recalls which sent a tremor through Toyota certainly concerns the repair of one million airbags in the USA. Over a year later, however, the Japanese car manufacturer has been forced to again recall many of the same cars that already passed through the workshop due to a defect in the front passenger airbag.

CHANGE OF STRATEGY—The previous Toyota recall over 1 million airbags did not repair all the cars in which the batch of defective airbags was installed, and for this reason Toyota Motor has decided to again recall the cars which at the previous check were found to be functioning properly. The reason, as stated in an official note from the company, is that the airbag supplier had finally provided all the serial numbers of the defective airbags to be replaced. Thus, of the 1.3 million cars recalled in the US alone, many had been inspected and sent home without a new airbag. Instead of sifting through the serial numbers, the Japanese giant has made it known that it will replace the component for all car models affected by the previous recall in which the new component had not been installed (see *How an Airbag is Made*), renewing the invitation also to those who have not yet presented themselves.

BLANKET REPLACEMENT—"For vehicles that have not benefited from component replacement since the 2013 recall, Toyota is changing the remedy from "inspect and replace the inflator if necessary," to "replace the inflator with a new one," we can read in an official note from Toyota Motor Corporation. This decision was made after discovering that many of the serial numbers Takata provided were incomplete and subsequent verification revealed that even some complete serial numbers of the defective airbags were among those that needed replacing. The recall for airbags produced by the Takata Corp. was triggered following verification of an unsuitable propellant used in an undefined number of cars. In the event of an accident, the gas generator could trigger a violent explosion or result in an ineffective inflation of the bag.

NOTHING TO DO WITH ITALIAN CARS—The models affected by the voluntary recall are Toyota Corolla, Corolla Matrix, Yaris, Vios and Tundra (from 2002 to 2004), Toyota Sequoia (from 2002 to 2004) and Lexus SC 430 coupé (from 2002 to 2004). The number of Toyota cars recalled for the defective passenger airbag thus rises to 2 million 790 thousand, of which 690 thousand are in Japan alone and 350 thousand are in the USA. The new recall, which should not affect cars sold in Europe, will make it possible to renew the invitation to go to the workshop for a free replacement of the passenger airbag, even for owners who have not yet done so following the first recall.

4   Annex C

4.1   Amazon Presents 350 Exclusive Brands, Including Private Labels

How many Amazon private labels are there? Many but not nearly enough. According to a Gartner report, announced exclusively by Retaildrive.com, the world's most popular marketplace had 119 private brands by the end of 2018. But again last year, the dotcom focused on a new strategic line, namely an alliance with suppliers.

For its US platform alone, the group has signed 223 partnerships for exclusive production and sale, a decidedly less expensive way to create private labels, given that these are goods whose potential the platform has already tested and do not involve particular development costs on the part of the distributor.

In fact, again according to Gartner, Amazon is increasingly aware of the need to differentiate itself through unique products, not available from the competition, products which can be delivered within two days at the most, marked by its own brand and the Amazon Prime brand, and heavily pushed on its site.

The agreement does not envisage high charges for its partners, who pay a commission of around 5% in exchange for very high sales volumes, against an amount which, for normal marketplace operators, starts from 5%, but can rise to as much as 20%.

The product areas covered by these alliances are extremely varied, but 'fast fashion', consumer electronics and information technology stand out.

Let us not forget that on February 11th, Amazon took over Eero, a manufacturer of Wi-Fi routers, signal repeaters, and mesh systems (*i.e. point-to-point*). Eero, founded in San Francisco in 2014, is the leading brand in its segment in the US and has very high ratings, on average of 5 stars, on the Amazon.com website.

Chapter 3
Representation of the Supply Chain Through the Business Functions

1 Introduction

A second way of representing the SC, in addition to the physical one described in Chap. 2 which contemplates its players and infrastructures, is through the corporate structure, i.e., the business functions involved in the SC process described in the first paragraphs of Chap. 1.

In other words, we are recognizing that the physical part of the SC described above, that process which brings products and services to the final consumer, is merely a tool. This tool is then managed by people working in different corporate departments and business functions, each of which manages a series of processes within its own remit. This happens within every single player that is part of the SC (n tier suppliers, manufacturer/focal company, 3PM, Distributors, 3PLs, wholesalers, retailers).

These people and business functions interact with one another both inside and outside the company, with their corresponding business function (e.g. procurement of the buyer with sales of the vendor), and the way they are organized and interact have a profound influence on the effectiveness and efficiency of the SC process as a whole.[1] For this reason, this chapter deals with which business functions are involved in the SC process and how they should be organized to maximize the efficiency and effectiveness of the entire process.

For the discussion, reference will be made to a diagram of the Porter's Value Chain (Porter 1985), shown in Fig. 1, which describes the core business functions and, transversely, the supporting ones.

---

[1] By *effectiveness* and *efficiency* we mean the ability to bring the requested product and service to the final consumer (effectiveness) while minimizing the resources used and therefore at a minimum total cost (efficiency).

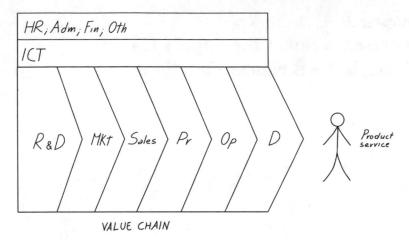

Fig. 1 Porter's value chain

1.1 Core Business Functions

According to Porter, the core departments are those directly involved in the SC process, and which directly contribute to bringing the product or service to the final consumer.

- R&D research and development
- Mkt: marketing
- Sales
- Procurement
- Operations
- Distribution/logistics.

In the following paragraphs we are going to analyse each of these in detail, highlighting those that are "traditionally" (in the sense of an approach that is not integrated with other business functions) the objectives of effectiveness and efficiency of each one, and how cross-functional conflicts may arise in the pursuit of these objectives.

1.2 R&D Research and Development

The research and development department in a company typically has the task of developing new products. This can mean either developing products from scratch, for example the development of a new drug by a pharmaceutical company, or improving existing products to respond to new or changed consumer demands, for example, the restyling of a car model by a manufacturer after a certain period of time has elapsed.

The role and importance of this department has gradually increased in recent decades, mainly due to the progressive shortening of product life cycles, which have gone from the years, typical of products of the '80s, to months or even weeks, up to today with the limit of the customized product for a single consumer.

The R&D department is central in so-called *technology-intensive* or *brand-intensive* SCs.

- Technology-intensive: these are sectors in which the technological content of the product, in the sense of its functional characteristics, is the first parameter that guides the consumer's choice.

  Think, for example, of the pharmaceutical sector, where the choice of a drug is dictated solely by its ability to cure a pathology. This is the reason why companies in the pharmaceutical sector are the ones which invest most in R&D—also due to the high mortality rate of research and development projects—annually ending up representing as much as 20% of this department's turnover. In fact, in this sector, research is becoming more and more expensive as it moves from a chemical to a genetic level. In this case, drugs are no longer chemicals, but genome modifiers. The R&D cost to develop a virus capable of carrying a molecule that can modify a gene and cure a disease within cells is huge—for scientists, laboratories, experimentation, and registration. The same goes for the petrochemical sector, where, for example, the buyer's choice of a catalyst derives solely from the chemical-physical properties of the component itself, and from its ability to cause a chemical reaction to take place effectively or efficiently.

  And the same can be said of some products in consumer electronics, in which we are always looking for increasingly high performance devices (e.g., telephones with folding displays, visors for augmented reality, etc.).

  Even in some specialist food sectors we can speak of technology-intensive products. For example, those functional foods defined as "a food that has a beneficial effect on one or more functions in the body, beyond its nutritional effects, in a way that is important for improving the state of health and wellbeing." E.g. foods for celiacs (individuals with celiac disease now count for 1% of the population), and the ability of a food to avoid generating allergic reactions combined with its organoleptic properties are the parameters of consumer choice.

- Brand-intensive: these are those sectors, typically fashion, in which the value of the product is linked to the experience and status that the product is able to convey.

  In this case, the R&D Department overlaps the Styling Department, which completely renews the production and range of "catalogue" products twice a year, something that does not happen in any other industry. First twice a year, then with four collections a year, now with the phenomenon of "fast fashion" every 2–3 weeks, a product is completely redesigned, and the bill of materials is completely redone by the Styling Department, to convey a product to the consumer, but above all an image, a certain status linked to that product.

Those involved in R&D within an SC cannot operate independently but must work in agreement with all the other players in the value chain, since the choices made in this area have substantial repercussions on other business functions, particularly on Procurement and Operations but also on Marketing & Sales.

The choices in R&D also have repercussions on the industrialization of the product, both in terms of the procurement of raw materials and on the components and operations. E.g., raw materials and components may not be available on the market at compatible prices and within timeframes compatible with the required production standards; again, even if the raw materials are available, it might well be impossible to industrialize an effective and efficient production process that allows the production of the volumes requested in the times required for the costs required. These are the reasons why, for example, in the fashion sector, the ratio between the garments conceived by the Styling Department and made as models, and those that are actually industrialized to become samples can even be 10 to 1. In other words, out of 10 of the prototypes from the Styling Department, only 1 will actually be industrialized and produced, precisely due to the impossibility of making it to meet the required quality, timeframe, and costs.

Also with regard to distribution and logistics, the impact of R&D choices can be significant. Think of the logistical implications of the vaccine developed by Pfizer which requires a cold chain of $-80\,°C$, with a vaccine survival outside the temperature threshold of only a few hours. There is no other product that is handled, stored and transported at those temperatures, so it became necessary to literally invent a new distribution model from scratch, based on special dry ice isotainers capable of ensuring the maintenance of such an extreme cold chain.

From a marketing & sales point of view, the impact of R&D can be extremely important, both on the saleability of the product and on the image which the Marketing Department builds. As for sales, it is never certain that a product developed by R&D can actually be sold. Research and development might develop a product with limited commercial impact, a product that does not sell, therefore. The contrary also applies. A failed R&D project may result in a product with very important commercial repercussions. In this regard, Annex 4.1 explains the case of 3M's Post-it Notes, which was actually an unsuccessful project to produce a tenacious glue, in which some researchers managed to recognize the potential value as a notepad.

As regards the impact on marketing, the products developed by R&D may not correspond to the image of quality, reliability, luxury or convenience that the Marketing Department seeks to promote for the company, and therefore not match the image that Marketing is promoting. Also Annex 4.2 *But does Volkswagen actually need a new Phaeton?* is an example of a conflict between the objectives of an R&D Department and a Marketing Department: Volkswagen's attempt to promote a luxury flagship, the Phaeton. The project was unsuccessful not due to the car's lack of technical features nor the quality of the materials, but, according to marketing experts, the fact that Volkswagen's image was that of a "popular" car (as the company's own name says, literally a "people's car"), in stark contrast with the image of its new luxury flagship.

1.3 Marketing and Sales

The second business function in Porter's Value Chain is the Marketing and Sales department. In structured companies these two business functions are distinct, just as the tasks and roles are distinct.

1.3.1 Marketing

The Marketing Department has two fundamental tasks.

First of all, marketing must understand the needs of the final consumer in terms of product, service, and price. This is a sort of "eye" and "ear" aimed at the consumer, from which the Marketing Department tries to capture the need in terms of product and service and the value which the consumer attributes to this combination. Once these needs are perceived, it is the responsibility of the Marketing Department to transfer them within the company to other corporate departments, in order to create that product and that service at that price.

On the other hand, the corporate Marketing Department is also responsible for creating the need for the product and service in the final consumer, conveying to consumers the message that the system (SC) is able to create that product with that service and at that price. It is a sort of "voice" to shout to the outside world what the supply chain can guarantee in terms of product/service/price. The megaphone available to the Marketing Department are the levers of the marketing mix, the so-called "4Ps", *Product, Price, Placement and Promotion.*

- Product: this is the set of functional material performances (related to the product in the strict sense and its functional characteristics) and intangible collateral aspects (related, for example, to the brand/status).
- Price: the one at which the product is sold, to be seen in close correlation with the *placement*. Clearly, the same product can be sold at completely different prices depending on the commercial channel—e.g., the same component sold as a part to a manufacturer or as a spare part to a retailer for after-sales service; the same finished product sold at completely different prices in different regions, in relation to the average spending capacity and the commercial and communication efforts which the brand can represent in each region.
- Placement: from a marketing perspective this is the commercial placement, the channel in which the product is marketed, and which frequently coincides with the service provided to the final consumer. In the case of food, we have seen how it is possible to have a traditional *normal trade channel*, or how it can be sold through *mass retail chains*, but also online, or through the *hospitality industry channel*. In the case of an item of clothing, there may be a retail channel, a wholesale channel, an outlet channel, and an ecommerce channel.
- Promotion: this is the promotional strategy which marketing intends to follow to spread the message of the product, service and price characteristics (radio, TV, door-to-door, word-of-mouth)... With this lever, marketing also promotes

so-called "brand awareness", and product information. Commercial initiatives based on price, sponsorships, and public relations also fall into this category.

At the head of Marketing sits the Chief Marketing Officer, CMO, who may be the head of the Marketing Department, and who coordinates the activities which the various managers develop for the different product lines and for different countries. E.g., a company like Barilla will have a marketing manager for the pasta division and another for the sauces division, who can operate at a single country or corporate level.

From a functional interrelation perspective, it should be emphasized that in addition to creating a need, Marketing also creates in the final consumer the level of product/service/price expectation that the other business functions involved in the SC process will need to meet. Accordingly, if Marketing promotes a high-tech or high-end product (food or fashion), high reliability, high after-sales service, particularly short lead times and/or widespread distribution, and a particularly competitive price, then the other departments involved, such as R&D, Procurement, Production, and Logistics, must then find the components and raw materials to create and distribute a product/service with those characteristics at that price. And they may not always be able to do so.

Consequently, Marketing can make choices which conflict with other departments, or with the guidelines that the Board of Directors have defined as a competitive strategy. This can potentially happen with all 4 levers of the marketing mix.

- Product: We can think of a case in which a product is promoted that research and development is unable to produce (e.g. a high-end food product which R&D is unable to produce, or Production is unable to industrialize).
- Price: or the case in which Marketing promotes a particularly low-cost product that does not meet the costs of Procurement, Production and Distribution.
- Placement: Commercial placement may be at odds with other SC strategies. E.g., Marketing marries a widespread commercial placement strategy on the normal trade or hospitality industry channel, which entails logistical costs that cannot be managed with the company's available distribution structure.
- Promotion: a push towards promotion, while the strategy of the other departments involved in the SC process is that of a daily low-cost product in order to avoid the distortion of demand introduced by promotions (*Bullwhip Effect*).

It goes without saying that a close coordination between the Marketing Department, the Board of Directors (BoD) and other corporate business functions is essential, so that marketing is promoting a trinomial—product/service/price—consistent with what the SC can actually achieve or which the BoD is betting on in terms of competitive strategy. This concept, known as *strategic consistency or strategic fit*, will be extensively covered in the following paragraphs.

1.3.2 Sales

If Marketing transfers to the market the message of what the SC is able to bring the final consumer, in terms of product/service/price, and generate the relevant demand, this demand must then be transformed into actual sales for the company, and thus sales orders. The Sales Department takes care of these. The Sales Department is managed by the Chief Sales Officer, CSO, and its main task is the transformation of demand into orders, and therefore sales of products and services at a certain price.

In a traditional approach, the Sales Director is evaluated on the basis of the sales and therefore on the basis of the turnover generated. Certainly, the more sales there are, the better for the company, and the more the sales manager sells, the better he or she is at the job—at least in theory.

In some cases, however, behind the sales there may be greater or lesser hidden costs for the SC which could negatively affect margins. Some of these examples are typical of fast moving consumer goods, and are described in the following paragraphs:

- *Listing fees*: To bring the product onto the shelf of an FMCG retailer, and thus ensure a significant sales volume, there may be entrance fees, also rather steep ones. These are costs which the manufacturer must bear simply to be able to enter and remain in retail chains and publicize a reference on a particular shop sign. Listing fees can range from 500 to 1,500 Euro/store/reference. Taking into account the number of references and the number of stores, the impacts on costs can be significant.
- *End of year discounts*: It is common practice for retailers to demand retroactive discounts (or "bonuses") at the end of a period, or year end, based on the sales volumes achieved. But also in general, regardless of the achievement of volumes, FMCG retailers may apply discounts that have not been agreed by a Sales Director and can have a negative impact on margins (once a certain amount of sales is reached on that reference, a discount of xx% is applied to the purchase price).
- *Discounts for compliance with payment terms*: Even before the entry into force of Italy's "Liberalization Decree", Legislative Decree 231/2002 set the maximum payment terms for perishable foods at 60 days from delivery. However, no authority supervised compliance with this term (*as the Antitrust Authority is now required to do, pursuant to Art. 62 of Legislative Decree no. 1 of 24.1.12*). In the absence of controls, some chains demand a discount not only to pay for the goods within the agreed terms, but also to comply with the legal terms.
- *Unilateral discounts not agreed upon*: Discounts applied unilaterally to deal with particular contingencies, e.g., a drop in consumption due to recession or particular occurrences. In this case, mass retailers can apply discounts to the sale price of the products which are then passed on to the purchase price of the product. In the years following the 2008 crisis, there have been cases in which suppliers have received letters from some mass retail chains, informing them of the application of these unilateral discount policies—to protect the purchasing power of the final consumer—which the supplier would have had to adjust retroactively with a 20% discount on one week's delivery. In the high inflation period of early 2022 that followed the Covid 19 pandemic, retailers did not recognized the price rise of

raw materials and energy that suppliers faced. They refused to recognize price increases to suppliers and to transfer them to end customers, and forced suppliers shrink their margins. Take it or leave it.

- *Recourse to promotions, also tactical*: Mass retailers sell more and more through promotions, trying to attract customers to the store thanks to the presence of discounted products. In other cases, if the product is on promotion, it sells, if it is not on promotion, it does not sell. In some product categories and for some references, also 90% or more of sales are reached during promotions, since the consumer does not recognize the value of the brand and simply buys what is currently being promoted. Moreover, mass retail chains often tend to buy more of a product than is actually put on sale during a promotion, creating stock to draw on once the promotion is over, or from which to draw from the stores or affiliates where the product is not on promotion (e.g., a product on promotion at Pam [supermarkets] but not at Panorama, or at Auchan but not at Conad, which are affiliates of the same group, however).

All of these phenomena can negatively affect margins of the vendor, so a good Sales Director should not only be evaluated on sales but on margins, and therefore also considering all the extra commercial costs which reduce margins (quantity discounts, listing fees, unauthorized discounts, promotional sales which erode parallel and/or future sales) used to generate that turnover. For this reason, the Sales Director is often entrusted with a budget for these initiatives (*promotions, year-end discounts, listing fees, etc.*)

But a really good Sales function should be able not only to sell a lot with a good margin, but to produce a good sales forecast and then stick to it. An overestimated or underestimated sales forecast leads to extra costs for the SC as a whole that may be difficult to shoulder. An overestimated forecast will lead to surplus stock and remainders—therefore physical inventory costs—which in turn will give rise to expired or unsold goods and therefore market mediation costs for the SC. By the same token, even an underestimated forecast will result in additional costs related to out-of-stock costs (opportunity costs of lost sales) or backlog costs—when the customer is willing to wait—due to non-optimized procurement costs, the production and distribution necessary to create small batches of products, which have not been budgeted.

Ultimately, not only the amount of the sales is important, but also the reliability of the forecast.

In a traditional approach, the salary of the Sales Director is based on a fixed part and a variable part, which for simplicity's sake is almost always calculated on the basis of the sales volumes reached with respect to the pre-established budget. However, circumstances may arise in which also the objective of achieving the sales budget at all costs for the achievement of an MBO (Management by objectives) bonus (and therefore a company bonus), could lead to conflicts with other business functions and contraindications on the effectiveness and efficiency of the SC process.

- *Unauthorized promotions*: the use of unauthorized promotions to increase sales can give rise to Bullwhip Effect phenomena deriving from the variation in demand, with repercussions on the whole of the SC (*Procurement, Production, Logistics*) which must, on the one hand, organize itself to cope with an unexpected spike in demand due to the promotion, and on the other must also interpret this signal as a promotion and not as an increase in demand itself, in order not to generate excessive stocks when the promotion comes to an end. From this point of view, it is much better for other corporate business functions to manage a stable demand policy linked to a low price every day (See Fig. 2).

- End-of-period budget: in order to reach the end-of-period sales budget, a Sales Director can fuel the end-of-period demand (e.g., with a particular discount), a demand that therefore rises above the average at the end of the period, and then drops suddenly or vanishes entirely in the following period, given that downstream there is still stock deriving precisely from this policy (See Fig. 3).

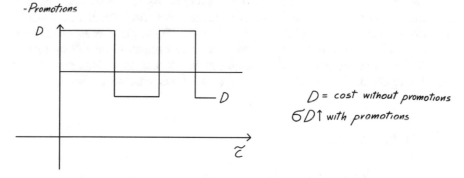

Fig. 2 Trend of demand as a function of promotions (peaks) versus "everyday low price" strategy (stable)

Fig. 3 Trend of demand following unauthorized promotions or end-of-period sales and tactical purchases by the buyer

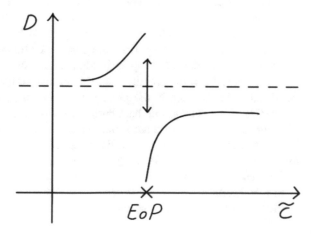

All of this, as well as the unauthorized promotions, give rise to a Bullwhip Effect and therefore extra costs, if compared to a situation with stable demand.

- Finally, the Sales Department, in order to reach its sales budget, may be led to overpromise, both in terms of product characteristics (*reliability, functionality, etc.*) and in terms of service (*sales volumes/minimum lots, delivery times, place of delivery, etc.*) which cause extra costs for the other departments of the SC (e.g. transport not optimized to cope with an unscheduled and urgent LTT transport; unscheduled production to cope with an unplanned and urgent production batch below the minimum order quantity (MOQ); extra raw material costs to cope with an unplanned and urgent production batch, and so on).

In an evolved approach, therefore, sales need to be assessed both on the basis of the turnover generated and on the basis of the margins. Additionally, a parameter that should enter the sales MBO should be the reliability of the forecast made, to be measured in terms of differences between what was estimated in the budget phase and what was actually achieved. It is important that deviations are both positive and negative, to avoid the Sales Director being too cautious in the budgeting phase.

1.4   Procurement

The first business function that directly enters the SC process of sourcing raw materials, transforming them into finished products, and distributing the finished product to the final consumer, is the Procurement Department. In charge of this department is the Chief Procurement Officer (CPO).

The purpose of this department and of the entire procurement process is to select suppliers and negotiate contracts for the supply of goods (raw materials, components, MRO) but also services (transport, warehousing, cooperative services to support logistics, etc.) that are instrumental in the realization of the SC process, and then to make daily purchases to make goods and services for the various processes available.

Procurement relates directly to the vendor's Sales Department, on which it has a great impact, since its choices affect the quota of business which the vendor will be able to obtain. Therefore, first and foremost, the ethical aspect of those who cover this role in the company is fundamental, especially in retail companies—think for example, of the Procurement Department of distribution companies which can decide what to keep in the price list and in which stores—and in any case in companies which manage purchase budgets worth millions of Euro. The Procurement Director and his/her subordinates may be prone to attempts at corruption, and it is therefore necessary that they are of flawless ethics and trustworthiness, and that they are motivated to negotiate in the interest not only of the company, but above all of the final customer, who is the final recipient of the finished product.

In a traditional approach, the Procurement Department is judged within a company on the basis of its effectiveness and efficiency, and therefore according to its ability to supply products and services in the required quantities and timeframes (effectiveness), and the resources used (costs incurred for sourcing, typically the costs of goods purchased). In the case of a mass retailer, for example, effectiveness is the ability to make products and services available for the functioning of the distribution chain and the filling of store shelves, while efficiency is typically in the cost of purchasing products and services.

In this approach, the Procurement Director therefore has every interest in trying to lower the cost of procurement, especially in those sectors where the cost of raw materials also represents 70–80% of the COGS (Cost of Goods Sold—if 100 is the cost of the finished product unit, the cost of raw materials can weigh up to 80) which tends to seem a correct enough ballpark figure, but in an advanced approach this does not necessarily correspond to the minimum overall cost of the supply chain. In other words, it must be stressed that reasoning solely by functional efficiency objectives does not necessarily lead to overall excellence in terms of system efficiency. This consideration certainly applies to the Purchasing Department but applies in general to all the business functions in Porter's Value Chain.

Precisely in relation to the Procurement Department, we can consider the impact that an approach based exclusively on the minimization of the purchase price can have on the SC process as a whole.

Quantity discounts could encourage a buyer to purchase much more than necessary, transferring to the Logistics Department downstream the cost of storing and stocking the purchased product, or the costs of obsolete and expired items. With this in mind, the concept of an *economical lot* is introduced, which seeks to balance the opportunities associated with quantity discounts against the physical and market mediation costs associated with stocks.

In addition, particularly low prices could lead to a decline in performance in terms of service (e.g. lengthening of lead times) or product quality, with consequent costs borne by the production department due to delays in the start-up of batches, production waste, and so on.

A buyer typically tries to understand the finer details of a vendor's industrial costs, going to explode direct costs (Procurement, Production and Distribution process, i.e. how much the vendor spends on raw materials, components, production labour, logistics systems, etc.) as well as indirect costs, and to negotiate/minimize the possible markup of the product by the vendor.

All with the aim of minimizing the purchase price, given by: $p = C_{Dir} + C_{Ind} + MarkUp$ (see Fig. 4).

If the negotiation is pushed beyond the minimum markup, the vendor may well abandon the negotiation, but in other cases, attracted by the turnover and the fact of being a supplier of a large group, he/she could even close the agreement with profitability that is low or close to zero, attracted by an agreement on volumes. Always remember that profit is given by a product between unit margin and volumes, so that a buyer's approach could be to compress the vendor's margin to lower the purchase price, in the face of a more or less explicit promise of an increase in volumes.

Ultimately, however, if a negotiation is not fair and transparent—e.g., the promised volumes are not guaranteed, or the margins are too low to allow the vendor appropriate

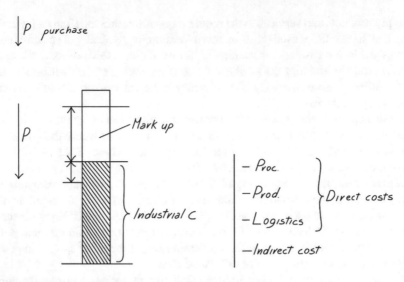

Fig. 4 Analysis of the structure of the unit cost of a product

remuneration—the vendor will try to recover the margins by reducing direct costs and therefore the quality of the product and service. And this can have profound cost implications on the links further down the value chain. They can, for example, increase production waste, affect the reliability of a finished product due to unreliable components, produce delays in production or logistics due to a worsening of the service level, or result in product recalls resulting from poor quality components and raw materials, etc.

Therefore, a negotiation on the purchase price is absolutely correct, but if pushed to the extreme it may have negative impacts on the SC process as a whole, since the procurement benefits deriving from the reduction in the purchase cost (purchase cost delta) are lower than the extra costs in the SC downstream, deriving from a deterioration in product quality or a drop in service level.

The win–win approach for vendors and buyers in the medium to long term is a transparent one, in which the Procurement Department first of all informs itself of the product/service/price specifications required downstream for the different categories of components and raw materials. On the basis of these specifications, the department analyses and transparently negotiates with suppliers the direct and indirect costs associated with the requested product and service (specific direct costs correspond to specific product and service performances). Furthermore, in a virtuous approach, the Procurement Department guarantees vendors the correct margin and therefore the right amount of profit, and may also contract and negotiate certain volumes through medium-term agreements (e.g., framework orders that commit the buyer to a calendar year or to 18 months). Faced with an approach of this type, the selected vendor (for example by means of a tender) undertakes to guarantee certain product/service/price performances, for the benefit of all the potentially impacted SC processes downstream.

1.4.1 The Total Transaction Cost

Before analysing the sourcing and procurement process and related activities, let us open a brief parenthesis on the definition and components of the transaction cost of purchasing. The Cost of the Transaction (CTR) is the set of costs that must be incurred to make an exchange.

$$CTR = C_{Sourcing} + C_{Procurement} + C_{Quality} + C_{integration}$$

It is considered to have 3 internal components (Richardson 1993), and a fourth component resulting from integration. The first three components are:

- Sourcing costs: these are the start-up and switching costs: they represent the fixed part of the CTR and include the costs of research and the developing of suppliers (*sourcing*).
- Procurement costs: these constitute the part of the variable CTR linked to the coordination of exchanges which results in activities such as ordering, scheduling, delivery, and so on (procurement). Also falling within this part are:
  - the purchase price of the component, raw or other material;
  - the direct costs of issuing and administration of an order, including non-conformities (the latter component could be counted among the quality costs).
- Quality costs: these constitute the remaining part of the variable CTR typically linked to:
  - costs of non-production and/or lost sales due to poor supply quality,
  - reprogramming and backlog costs resulting from unreliable deliveries and non-deliveries;

the Integration component is detailed below:

- Integration costs: alongside the direct costs, some "hidden" costs deriving from the lack of integration between customer and supplier need to be considered. These are therefore costs which could have been avoided if the customer and supplier had exchanged information transparently, accurately and in real time, and if they had reasoned from a SC perspective as if they were a single player, rather than in a conflictual manner. These include:
  - possible costs of process duplication: low supplier quality levels which require checks on 100% of deliveries, with the duplication of processes.
  - any costs of processes that could be carried out by the supplier: for example, in the case of suppliers available for vendor management inventory, the cost of inventories and the cost of procurement can be avoided.
  - inventory costs that are both physical (linked to high and duplicated stocks on both sides) and market mediation (due to shrinkage and OOS). These costs are often the result of a buyer's lack of sharing of demand and inventory data, which generate a Bullwhip Effect for the supplier.

Multiple studies have confirmed that in organizations the costs related to the entire purchasing transaction represent on average at least 75% of the costs attributable to the product (Fiocca et al. 2003).

In an advanced approach, therefore, a buyer is evaluated on the basis of the CTR value which must be as low as possible, in this way he/she will be constrained to try to minimize the total cost of the transaction, rather than the single cost item.

1.4.2 Sourcing and Procurement

A first fundamental distinction needs to be made between the terms *sourcing* and *procurement*. They are often used interchangeably, but in reality they are not the same, involving two different stages of the procurement process (Sollish and Semanik 2011).

Both aspects are evident immediately from the definition of the procurement process given above. A first aspect identified with the term sourcing, is more linked to the strategic plan, takes place on a one-off basis, and concerns the qualification of suppliers, their selection through tenders (RFPs—Requests For Proposal), the definition of purchase contracts and SLAs (Service Level Agreements).

A second, more operational aspect in which buyers operate, on the other hand, identified with the term procurement, is linked to day-to-day activities, and therefore to purchases, contacts and management of the relations with a supplier (e.g. requests for individual quotes), issuing of the purchase order up to the authorization for payment of invoices (see Fig. 5).

Identification of Needs and Definition of the Strategy

According to Sollish and Semanik, sourcing begins as a first step with the identification of requirements and strategy. The head of the Purchasing Department must therefore align him/herself with the CEO and the heads of the other business functions to understand the strategic objectives of the SC to be pursued, an aspect which will be extensively explored in the rest of this chapter. In fact, with purchases, we can aim at different objectives that can often be antithetical to one another, so it is advisable to identify those on which to focus from the outset. For example, we can favour low

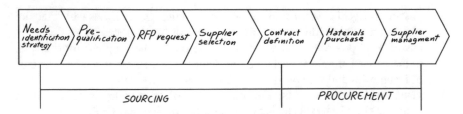

Fig. 5 Activities related to sourcing and procurement

purchase prices, or the quality and reliability of a product, but also the service, for example. In terms of supplier lead time, flexibility, and delivery frequency; then it is possible to focus on the supplier's ability to develop new products and innovate. In the case of textiles and clothing, for example, a Chinese façon may be able to produce at low cost but has no innovative capacity, and has extremely long lead times and little flexibility, in the sense that the production planning, once scheduled within a given time window, cannot be changed either in duration or in terms of the time slot; conversely, a local supplier is more flexible, quick to deliver, has a craftsmanship that can be a source of innovation, but unquestionably will cost more.

Another aspect concerns the definition of the sourcing strategy. The purchasing strategies that can be followed in the procurement of products and services are essentially three (Richardson 1993):

- Sole sourcing,
- Multiple sourcing.
- Parallel sourcing.

In a nutshell, the first consists of a single supplier for each purchase component, the second represents the opposite extreme and sees multiple suppliers for the same component, the third is a hybrid of the previous two, in which a limited number of suppliers, at least two, is used for each supply.

In the literature, a clear preference is not expressed in favour of one or other alternative, there are authors who favour one strategy and authors who encourage another: for example, the first is strongly recommended by Deming as the only method of quality assurance, believing that the investment necessary to effectively control it is sustainable only for a buyer–supplier relationship, while the prevailing tendency of other authors is to support the idea that multiple sourcing is the only method to guarantee low costs and high supply performance thanks to the competition generated and the possibility of assigning business quotas that are proportional to the price/service. There is therefore no better strategy at an absolute level since each one has its pros and cons, and the decision on which one to use depends on various factors such as the product category and the relative supply risk, the size of the supplier market, and the business sector in which the company operates.

In order to better understand the characteristics of each, the three strategies are initially considered separately and then a final comparative analysis is carried out.

Sole Sourcing

In the course of this examination, *sole sourcing* is considered equivalent to *single sourcing* even if, for an impartial discussion, we should specify that there is a substantial difference between the two terms.

In fact, sole sourcing implies the availability of a single supplier for the requested product, while in single sourcing it is the company that chooses the only supplier from which to procure, from within a set of possible alternatives.

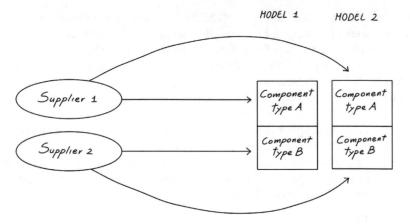

Fig. 6 Sole sourcing strategy (Richardson 1993)

Single sourcing establishes exclusive long-term relationships with a few suppliers, typically one for each reference procured, based on mutual trust. The scheme is the one represented in Fig. 6

This purchasing strategy seems to come into conflict with most of the literature concerning the strategic management of suppliers, which show that investing in few suppliers increases their bargaining power as a result of high switching costs and that this is also reflected in price increases. Furthermore, the choice of single sourcing exposes the company to supply risks (single sourcing is by definition a far from resilient solution), and to opportunistic behaviour by suppliers, and also prevents it from having information on the cost and performance of the same. However, there are also some positive aspects, mainly related to the standardization of the references procured, quality assurance, and a reduction in transaction costs thanks to well established relationships and greater process efficiency.

In the case of single sourcing, low transaction costs can be achieved thanks to minimal start-up and transition costs, and quality costs can also be reduced; at the same time, the purchase price could well be suboptimal.

Multiple Sourcing

Multiple sourcing, schematized in Fig. 7, represents the diametrically opposite choice to single sourcing. Porter (1985) defined it as the most effective way to guarantee low costs and high performance from suppliers.

This strategy consists in turning to several suppliers for the same reference to be procured in order to develop maximum competition between the available suppliers. In this way, the supplier is encouraged to increase its performance also thanks to the promises of assigning greater business quotas, and so the company consequently decreases the costs dependent on competitiveness. However, the same cannot be said for start-up, transition, and negotiation costs: in fact, an initial investment and a

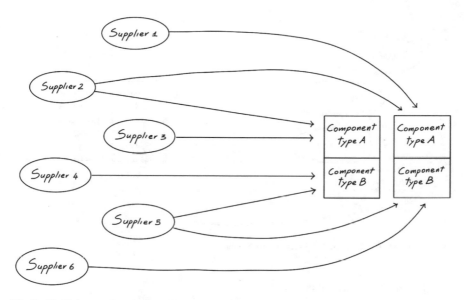

Fig. 7 Multiple sourcing strategy (Richardson 1993)

transition investment must be made every time a supplier is changed since it is more difficult overall to coordinate with multiple subjects.

Parallel Sourcing

Parallel sourcing represents an intermediate strategy between the two previous ones, capable (at least theoretically) of reaping the benefits of both. Therefore, several suppliers are employed for each component, but they are kept separate and dedicated to different product lines.

The process behind parallel sourcing is represented in Fig. 8.

On the one hand, this allows buyers to influence supplier performance through the threat of changing, and on the other, it keeps transaction costs low. In particular, it has lower negotiation costs than both single sourcing and multiple sourcing respectively, thanks to the lower performance of suppliers and the use of one supplier at a time for each reference procured. Furthermore, like multiple sourcing, it has the lowest competitive costs.

Therefore, while reiterating the impossibility of determining an absolute best choice, it can be said that parallel sourcing shows how the commitment to a long-term relationship with suppliers through specific investments combined with a credible threat to change in the case of inadequate performance, form the basis for a fruitful relationship with suppliers.

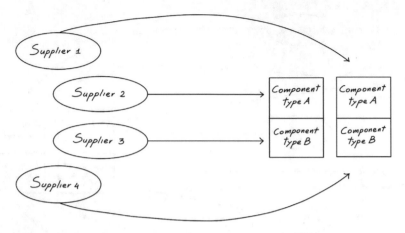

Fig. 8 Parallel sourcing strategy (Richardson 1993)

Pre-Qualification—Supplier Audits

Once the strategic objectives have been defined, potential suppliers who meet the requirements are identified through market research.

These are subjected by the Purchasing Department to a qualification audit. This audit consists of a structured process to verify whether the supplier possesses the necessary requisites to be able to establish stable and lasting business relationships and does not apply to one-off supplies of non-critical materials (a typical example being that of stationery materials).

During the audit, some preliminary background checks are carried out, by contacting the supplier and requesting the Chamber of Commerce registration, an up-to-date DURC [social security contribution certificate], the latest financial statements (P&L and balance sheets) or a certificate of the administrators' criminal record, the Organizational Model 231 against offences envisaged by Legislative Decree 231/2001, or any other documentation deemed relevant.

It is also typical of such an audit to organize a site survey, during which to verify in person the characteristics of the products, the possession of certifications (e.g. REACH certification for hazardous materials), the adoption of GMP (Good Manufacturing Practices) or GLP (Good Logistics Practices). During the audit, it is also possible to request the possession of such ISO certifications as ISO 9001 (Quality) ISO 14001 (Environmental Management) ISO 18001 (Safety) or ISO 27001 (Information Security) deemed relevant for the negotiation.

Production by the supplier of references and success stories similar to those of the supply in question may also be subjects of the audit.

The audit is a sort of pass-or-fail filter and serves to screen the range of possible suppliers to become involved in the negotiating of supply contracts through the subsequent Request for Proposal phase.

Request for Proposal

Suppliers who pass such an audit are invited to negotiate. This typically means a structured process known as an RFP—Request for Proposal. The greater the importance of the supply, the more structured the RFP process is: it is absent for supplies of small consumables, while it is essential for the supply of durable and multi-annual supplies of great economic and strategic importance (e.g. plants, software projects, multi-annual services such as transportation and logistics, etc.).

During the RFP, a structured document is prepared, in which the Purchasing Department specifies:

- The object of the supply; the products and services that are the object of the supply, the technical characteristics and any other parameters necessary to avoid making the object of the supply ambiguous are illustrated in detail;
- The SLAs—Service Level Agreements and therefore typically, the maximum permissible defect rates and the methods of verification;
- The service characteristics required (lead time, delivery points, delivery frequency); these can be included in the SLAs;
- Procurement management methods, such as Just in Time procurement through a Kanban system or traditional reorder mechanism, and then the characteristics and waves of orders are specified (min./av./max. number of orders in the period to be fulfilled, order characteristics (min./av./max. of the OL/order; min./av./max. quantities per order line);
- Special requests on expiration dates (e.g. at least 2/3 of the shelf life guaranteed by the supplier to the customer), FIFO management (no deliveries of references with a shorter expiration date than the same reference already received);
- The methods of EDI integration and transmission of information, e.g. of orders, order confirmations and DESADV/ASN;
- Special requests for order preparation and traceability management. For example, there could be a request for the preparation of single-item Handling Units (HUs), to facilitate cross-docking operations, impose restrictions on the dimensions of the HUs, or the adoption of specific standards for identification media, for example by requiring the presence of RFID tags at the HU level, box level, or item level.

The RFP document also contains indications relating to electronic invoicing and payment conditions (bank withholdings or bank transfers; payment dates).

Lastly, in the case of supplying capital goods that require after-sales service, the SLAs relating to after-sales support are specified in the RFP. Participants in the tender are then notified of preventive maintenance requests (number and type of interventions required per year) and repairs of breakdowns. In this case, the methods and times required for the creation and taking charge of a maintenance ticket are specified (e.g. email, portal, dedicated telephone help desk), for triage (triage consists in determining the root cause of the malfunction), for the workaround (temporary solution) and the permanent correction (restoration of a system as-is). The SLAs may vary depending on the level of severity of the incident. Any service credits are also introduced, and therefore the penalties which the supplier accepts in the event of non-compliance with the SLAs, to be assessed over a sufficiently long time horizon to be statistically significant (typically 6 months to 1 year).

Based on the parameters defined in the RFP document, the supplier is invited to make a bid. This can be done using different systems. The simplest way is to collect individual bids on paper (the supplier prepares the tender documents and sends them through a file-sharing mechanism) or via an online portal.

In other cases, the tender can be carried out using online auction mechanisms. In this case, the e-procurement portal makes it possible to proceed top-down starting from an auction base, and at each step, each supplier is informed of how many competitors are left in the game and so they can decide whether or not to confirm their quote. The auction ends when only one supplier remains in play.

Supplier Selection

Based on the RFP and the bids received, the Purchasing Department selects the supplier/s who has/have been awarded the supply contract.

Once again, this process is not automatic but requires the team responsible for the supply to examine the proposals received, appropriately evaluating both the quantitative aspects (usually the price) with the qualitative aspects (e.g. references, service, etc.).

The output of this phase is typically represented by a selection of a small number of vendors, normally two to three, who are admitted to the shortlist and invited to a face-to-face meeting.

During the meeting, the proposal is illustrated, samples or demos can be produced, site visits arranged, any grey areas left unresolved are analysed in depth.

It is also important to check whether the corporate values and culture are shared, since these are essential to establish the necessary empathy for a satisfying and lasting relationship. According to Lewis's Cross Cultural Communication Model (see Lewis 1996), The cultural mindset may also be important, since cultures such as Asian (reactive culture—Introverted, conceal feelings, respectful, harmony-oriented), Latin (multi-active cultures—people-oriented, emotional, talkative, dialogue-oriented) and Anglo-Saxon (Linear active culture—task-oriented, highly organized planners, data-oriented) are at the top of a cultural triangle in which conflicts and misunderstandings can easily arise if not thoroughly understood and managed (see Fig. 9).

Last but not least, a final price is negotiated, if a further margin is available in the negotiation.

The result of this phase is the choice of the buyer/s to whom the contract will be entrusted.

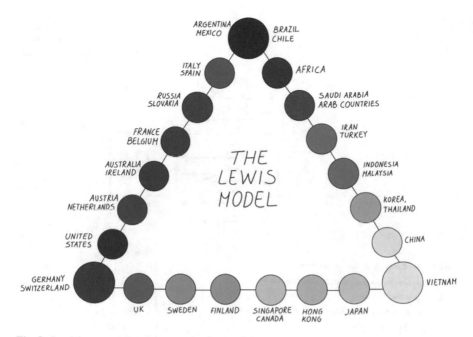

Fig. 9 Lewis's cross cultural communication model

Contract Definition

In this phase, everything covered by the RFP offered by the supplier and agreed in writing and/or orally during the negotiation steps is contracted from a legal point of view. The products and conditions of service, prices, methods of issuing and transmitting orders, guarantees, SLAs, performance mechanisms and service credits, duration, territory, etc. are therefore a part of the contract.

In addition to the various negotiation clauses agreed, it is important that the contract contains aspects of a more generic nature such as (1) safeguard clauses for customers and suppliers, such as the possibility of terminating the contract in the case of a serious breach, impossibility of respecting certain clauses, for example, SLAs, loss of performance beyond certain limits, etc. (ii) regulation of intellectual property—in the event that patents arise from a collaboration, it is good practice to establish the background IP of each party and who will represent any foreground IP that might be generated by the collaboration. (iii) exclusivity—this is a clause which binds a supplier and prevents it from providing the same goods and services to other competitors of the buyer; the exclusivity clauses can also be reciprocal, in the sense of binding the buyer not to obtain the same goods and services from other vendors. These are clauses which may prove necessary when the object of the supply is of a particular technological and strategic value, on which the buyer can build a competitive advantage, so that access to competing third parties needs to be limited. Obviously, the counterpart of such exclusivity is linked to the territory and the quota

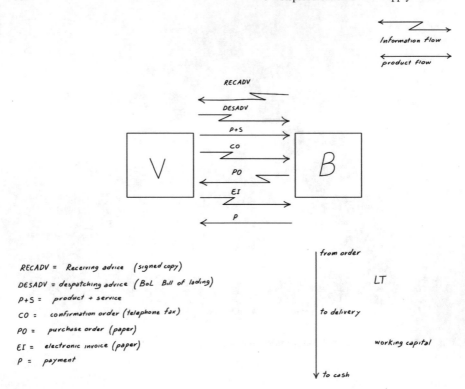

Fig. 10 The procurement process

of business generated, which must also be regulated. (iv) liability clauses, which hold the customer harmless from any third-party claims on the products supplied by the supplier (e.g. violation of intellectual property). (v) indemnification: compensation for direct/or indirect damage caused by the supply, and any liability caps. (vi) dispute resolution methods, which should prefer such streamlined forms as arbitration, rather than a recourse to regular civil justice.

It is essential that the contract be written in a clear and exhaustive form, since this will be the document that will be considered authentic during the entire duration of the relationship between the parties, and to which anyone for the respective parties must/can refer to for any obligation, but above all in the event of a dispute, regardless of whether the people who carried out the original negotiation remain part of the respective teams in the future or not.

Purchase of Materials

Once the contract has been signed, it is possible to proceed with the actual purchasing process, in accordance with the contents of the contract.

The operational part relating to the order cycle is represented in Fig. 10. It begins with the issuing of the Purchase Orders, continues with the order confirmation by the supplier, the preparation and delivery of the order and the issuing of the Transportation Document TD and DESADV by the supplier, the receipt of the compliant goods, the issuing of the RECADV by the buyer, and then the active cycle is completed for the vendor with the issuing of the invoice EI, and the passive cycle for the buyer in paying the invoice.

We talk about *order to delivery* (OTD), to identify the time between the first process and the delivery of compliant goods (3), while the term *order to cash* (OTC) also considers the time for the financial/administrative cycle, which coincides with the payment of the invoice from the vendor. While the former is the basis for evaluating logistical performance, the latter impacts the financial performance, and therefore the ability of the SC to finance itself with equity or not.

This part, as well as the benefits deriving from the use of electronic standards compared to paper systems, will be analysed extensively in the chapter on information flows. In any case it is evident that the use of electronic standards allows a reduction in the times of both OTD and OTC.

It is only necessary to underline that information systems, through a variety of e-collaboration systems including Electronic Data Interchange (EDI) and e-procurement systems, now allow the transmission of information and the exchange of communications in real time, and therefore of automatically and electronically managing various activities related to purchasing. This has allowed Purchasing Departments to shift their focus from daily procurement activities to strategic sourcing tasks, which can help a company achieve market success while at the same time reducing the cost of communications, both in terms of labour and accuracy. Furthermore, the use of ICT combined with collaborative strategies allows the sharing of information accurately and in real time between customer and supplier, and this is essential for optimizing the processes at the overall level of the SC and not merely that of the single player (suffice to think how the sharing of stock and demand levels by the retailer allows the manufacturer to eliminate the Bullwhip Effect, better manage their stocks, and avoid out-of-stock situations).

Supplier Management

The organizations in which the strategic value of the Purchasing Department is included are more in favour of continuous monitoring and evaluation of their supply base. In this way it becomes clear which are the best and worst, critical and necessary suppliers.

First of all, the SLAs and KPIs defined in the negotiation phase are constantly updated and monitored, to keep the situation under control and manage any deviations. In fact, once a collaboration has begun, one of the main risks is that of forgetting the monitoring and measurement of performances, which, on the contrary, must be constantly measured to allow interventions if necessary, as represented in Fig. 11.

Secondly, a periodic review of the supplier base can prove useful, to avoid the classic phenomenon of proliferation. Simply due to the fact that different buyers

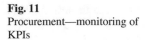

Fig. 11
Procurement—monitoring of
KPIs

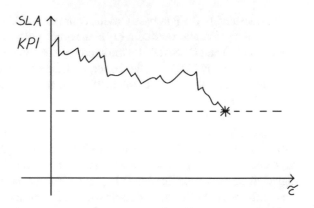

can get on better with different suppliers, or purely with the passage of time, it is more frequent than we might think to find ourselves with an "extended" supply base beyond any objective justification, consequently, it is useful to periodically review the pool of suppliers and see whether there are any margins for rationalizing it (e.g., does it makes sense to keep open the relationship with 10 suppliers for the same component for reasons of resilience or price, or is this situation simply the result of the fact that over 10 years the Purchasing Department has seen 10 people come and go who have opened relations with different suppliers, but there were never any critical issues from the point of view of the procurement of the materials?).

1.4.3 The Kraljic Matrix

The sourcing and procurement process is cyclical and needs constant revisions and iterations for successive loops. Accordingly, at the end of each phase and at the beginning of the next, it may be useful to periodically review suppliers based on the Kraljic Matrix. This matrix was introduced by Peter Kraljic in 1983 in an article in the *Harvard Business Review*, and it represents a milestone in procurement (Kralijc 1983).

It is a strategic purchasing management tool which divides products to be procured into four classes, in a 2×2 matrix, and on the basis of 2 parameters (see Fig. 12):

- *The importance of the purchases* (value and impact on company profit, on the ability to grow the business, on the ability to guarantee high quality standards): The importance of the purchases is determined by the impact that the purchased item (whether in value or volume) has on the company income statement (costs and profitability), or in terms of the percentage of raw material/component costs on the total cost of the final product. But also the impact that a certain component has on a company's business growth capacity. An innovative component could provide a product with unique features which allow the company to gain a competitive advantage (e.g. a high-capacity battery capable of doubling the range for a new electric vehicle). Stationery items are unimportant, since they represent a marginal item of the income statement, while key components to guarantee the functionality

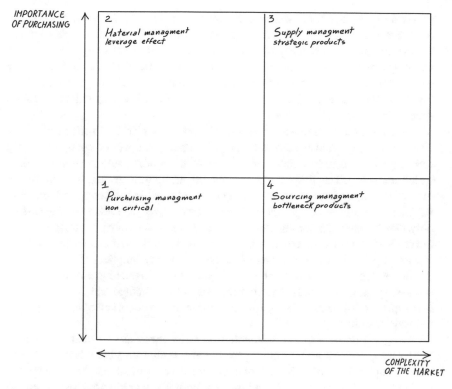

IMPORTANCE
OF PURCHASING

2
Material managment
leverage effect

3
Supply managment
strategic products

1
Purchaising managment
non critical

4
Sourcing managment
bottleneck products

COMPLEXITY
OF THE MARKET

Fig. 12 The Kraljic matrix: from (Kraljic 1983)

of a product (e.g. the new chip for a mobile phone, the high-capacity batteries of an electric car) with a high cost and a high impact on the growth capacity of the business, are certainly more relevant.

- *The complexity of the reference market* (level of availability and supply risks) measured by the conditions of equilibrium between supply and demand, by the number of suppliers (e.g. existence of monopoly conditions), by the availability of substitute/alternative products or services, by the cost of switching suppliers, or a high pace of technological progress, which make supply in that particular market risky. While a supply of detergents is widely available on the market, there could be a single supply capable of producing a critical component, albeit with a low impact on the company's income statement, without which a product cannot be completed.

The 4 classes in which to group products are:

- **Purchasing management—non-critical or not fundamental products**: components which have a low impact on the company's business and are found in abundance and/or in low-risk markets. The real advantage for the company lies not in the price, but in the simplicity of the relationship and in the reduction of the transaction cost. The contract horizon is short, 6–12 months. It is important to standardize products and suppliers, meaning local suppliers, in order to simplify purchasing management, and to optimize the volumes and inventory management. For these types of component, decentralization/delegation of management is generally chosen, precisely to avoid wasting time on negotiating contracts and purchases of non-critical products, either from the point of view of inventory costs, or of supply sources. The goal is to source products that guarantee functional efficiency and do what they were purchased for, without spending too much time on the process.

- **Materials management—products with a leverage effect or with multiplicative effects**: important for any company (with a high impact on costs) but easily found, positioned in low-risk markets and with abundant supply (non-turbulent supply markets). For these types of component, a company tends to make the most of its bargaining power, leveraging high volumes and selecting and reducing suppliers to negotiate on prices in order to lower costs. To seek a leverage effect only, the contractual planning horizon is broader, 12–24 months, but not too broad. Given the high impact on costs, there is a constant search for replacement products at a lower price. In view of the fact that there are many suppliers, if possible they are chosen locally to optimize physical flows, and avoid the high physical costs of stock-keeping and/or market mediation related to shrinkage and OOS. The goal is cost/price optimization by exploiting contractual leverage and optimizing flows to lower inventory costs.

- **Sourcing management—bottleneck products**: these are products subject to bottlenecks, with a low impact on profit in economic terms but where the continuity of supply is at risk. The lack of a product can be critical, but the appeal of the deal in the eyes of the supplier and the negotiating power are both low, given the limited volumes involved. Management of these components is aimed at creating collaborative relationships in the medium to long term between customer and supplier, but also aimed at ensuring supply, while the cost—and therefore the purchase price—is not a problem given the limited impact. Control of the market and suppliers must be continuous due to the uncertain availability, but given the lower volumes/costs compared to strategic items, it is sufficient to ensure the necessary volume, albeit at a higher price. A guarantee of sufficient stocks is therefore one of the main tasks of the procurement manager, who must select suppliers capable of supplying the particular product. Given the potential criticality in the event of a lack, the choice of the supplier is taken at a centralized level, while the management of the relationship is delocalized to a lower level. Such a contract must therefore be negotiated with particular emphasis on guarantees rather than price. Generally speaking, the cost of inventories, given the low impact on the income statement of these products, is not usually a problem. The objective in this quadrant is to secure supply and guarantee volumes over a variable horizon (medium-short term, if short-term availability of the product is expected; medium-long term, if the scenario is that of a scarce resource for a long time).

- **Supply management—strategic products**: important for the company both in terms of economic impact and for the supply conditions from complex and/or risky markets. The company must at the same time develop stable strategic partnerships with the supplier which delivers a scarce resource, in order to ensure volumes, but at the same time optimize costs, in terms of both purchase costs (e.g. by leveraging on volumes and/or monitoring the evolution of the market to evaluate alternative suppliers) and management costs. Attention must be paid to the fact that the physical costs of stocks or market mediation can be significant if the product is subject to obsolescence, and therefore it is best to adopt models of maximum collaboration and information sharing with the supplier (CR, CPFR, VMI) aimed precisely at optimizing flows and lowering costs. In this case, we speak of *supply management* precisely to emphasize this aspect. The planning horizon is long-term, and a decision is centralized precisely because of its strategic nature. The goal is to ensure the availability of high volumes of a scarce resource, but at the same time to optimize the flow of materials.

Clearly, such an approach requires the development of skills within the Procurement Department that go beyond the simple management of the purchase cycle or price negotiation. New skills play a primary role in procurement and can guarantee a competitive advantage: the ability to negotiate long-term contracts, supply chain management (understood as an overall vision, alignment of objectives and optimization of the various components), continuous market scouting in search of new supply opportunities, the design of supply chain resilience, product quality, research and development, up to knowledge of different (and even very different) languages and cultures.

1.5 Operations

The Operations Department has two main business functions: a plant engineering department and an operational department.

The plant engineering part deals with the purchasing and maintenance of production plants.

Often in process companies, where raw materials are physically and chemically transformed into finished products, plants are of a significant size, and this department is given a dignity of function in its own right, with the creation of a plant division. The second function is the operational one, and is therefore delegated to the planning and scheduling, control and management of the production process.

In charge of the department is the Chief Operating Officer (COO).

Precisely in relation to the operational part, (as described in the block diagram in Fig. 13) this business function uses some of the production resources seen in the physical description of the supply chain process—typically process and service plants, manufacturing and assembly lines—but also labour, energy, and consumables. To do this, it uses a series of controls, represented, for example, by essential software for production planning and control, MRP systems, production schedulers, MES systems (Manufacturing Execution Systems), etc..

Fig. 13 Operations: process,
effectiveness and efficiency

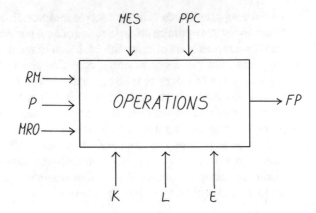

As with all the other business functions in the value chain seen up to now, the typical functional objectives pursued by the Operations business function are those of effectiveness and efficiency. For the Operations, effectiveness means an ability to implement planned and scheduled production programmes and deliver quality products at the planned time, while efficiency aims to minimize the resources used and the waste produced.

Increasing the efficiency of the process typically translates into trying to increase the numerator of the efficiency equation

$$\text{efficiency} = (\text{quality products} - \text{scraps})/\text{resources used},$$

and thus reducing set-up and waste times as much as possible. Scraps affect the numerator, since the quantities produced are in fact the sum of the compliant quantity plus the scraps (scraps decrease the numerator) while set-up times, in addition to generating scraps, are down times when the resources used (machine times) do not lead to any output. Consequently, the plant costs are spread over a lower quantity of finished product, increasing the unit cost of production.

Consequently, with a view to effectiveness and efficiency, for a production manager an optimal scenario is represented by operation of the lines in a continuous cycle, with batches that are as large as possible to minimize set-up times and start-up waste (or at the very least, economical lots if the capital holding costs of the finished products must be taken into account), while format changes to cope with unforeseen productions must be reduced to a minimum, compared to the agreed production schedule.

However, this scenario may well be in contrast with the objectives of the Sales Department (Fig. 14), which, instead, seeks maximum flexibility, in order to sell what is requested at that moment and not necessarily what was planned in the past, based on a sales forecast. Accordingly, the Sales Department could ask for variations in production with respect to the planned one, both in the mix (producing something different) and in terms of volumes (larger, or smaller).

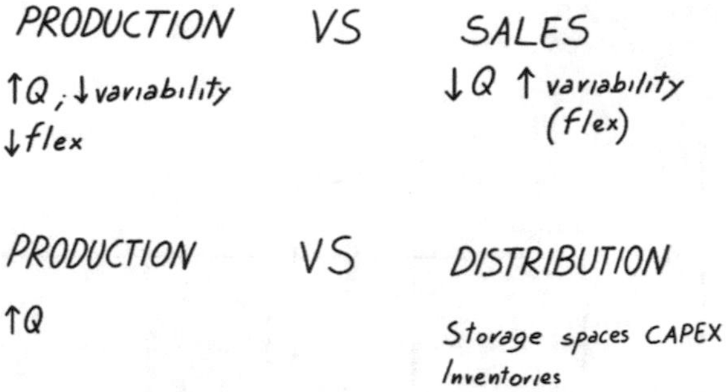

Fig. 14 Operations and
inter-departmental conflicts

PRODUCTION   VS   SALES

$\uparrow Q$, $\downarrow$ variability $\downarrow Q$ $\uparrow$ variability

$\downarrow$ flex   (flex)

PRODUCTION   VS   DISTRIBUTION

$\uparrow Q$ Storage spaces CAPEX

Inventories

For this reason, the performance of the Operations, as well as in terms of effectiveness and efficiency, may be measured by adding an additional flexibility parameter, understood as the ratio between the requests for modification of production plans satisfied with respect to the number of requests received, in terms of both mix and quantity.

The objectives of the Operations Department may also conflict with those of the Logistics Department (see Fig. 14), which may not have the availability to store the production volumes corresponding to the lots of Operations (Logistics business function may have to rent additional warehousing space), or the means of handling and transport necessary to take the finished product from the production lines to the factory warehouses.

Also in this case, therefore, an excellent effectiveness and efficiency must be sought not only from a business function point of view but also from a more global perspective, by evaluating the impact that the choices in the field of operations can have on upstream and downstream costs, which could end up higher than the production savings.

1.5.1 Manufacturing Execution Systems (MES)

MES or Manufacturing Execution Systems are used to measure inputs, outputs and resources used, and therefore Operations effectiveness and efficiency. These monitoring systems manage and monitor the work in progress to increase productivity and reduce cycle times.

An MES has a twofold objective:

- On the one hand, control of the production plants, and therefore the instructions provided to production plants
- On the other hand, the monitoring of parameters and the management of traceability in a broad sense, including everything that happens to a product unit as it moves through the plant during the production process.

These are software systems for monitoring and managing a plant which allow integrated acquisition (via PLC, and Ethernet network protocol, or other network

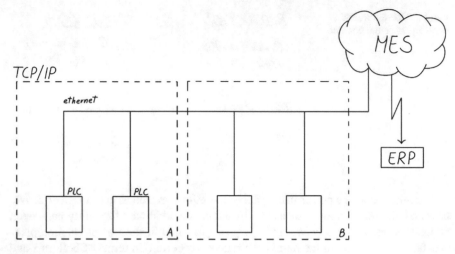

Fig. 15 Systems engineering of an MES

protocols) from the various components of the line of information relating to the resources used, manpower, downtime, dwell time (waiting), raw materials used, finished products made, waste produced (see Fig. 15).

The parts that make up an MES and the information managed are the following:

- Operation planning and scheduling
- Resource management and allocation
- Management of the flow of materials and sorting of production orders on the lines
- Management of process parameters on lines and on individual machines
- Quality management
- Machine maintenance management
- Data collection & acquisition
- Traceability management and document management and control
- Recording and visualization (cockpit) of the current state of the resource machines and
- KPI monitoring and performance analysis.

From an Industry 4.0 perspective, the data gathered from the various machines (typically also in different physical locations) are collected via the network in a centralized system, including the Cloud, where there is a "digital twin" of the single plant and the whole productions system. The digital twin is able to simulate different production scenarios in real time, often applying artificial intelligence algorithms to the massive flow of data coming from the field to search for an optimum condition and send the reconfiguration parameters to the various machines on the basis of the simulations made.

The MES systems are integrated with ERP systems, which deal with the planning and control of production, purchasing and SC management, administrative and management control and CRM (Customer Relationship Management). The

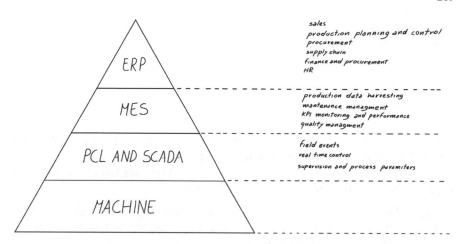

Fig. 16 MES as middleware between ERP machines and systems

image in Fig. 16 shows how the MES fits between the control level (PLC and SCADA) and the planning level (ERP). The MES deals with "production data collection", "Overall Equipment Effectiveness (OEE) and efficiency analysis", "Statistical Process Control", but also with maintenance, materials handling, and machine activation.

All activities which then provide the ERP with useful data for future planning, and therefore for optimization of that continual improvement which is the basis of efficient operations. While the ERP operates schedules with medium-long times (days, weeks, months), an MES acquires data and controls the production process within a very short period—in practice in real time (minutes, hours).

The process to implement an MES is usually quite complex and requires (i) a preliminary analysis phase followed by (ii) the actual implementation, the configuration of the software for the specific case and the development of software integrations with the ERP system. For this reason, to implement the entire process starting from scratch up to full operation can take several months or years, depending on the complexity of the production system.

However, the main advantages connected with the implementation of a 4.0 MES are typically to be found in an increase in productivity, a reduction in cycle times, a reduction in waste and reworking, thanks to an accurate, timely control and optimization in real time of the parameters of the individual steps/processes, thanks to simulation and artificial intelligence.

Ultimately, thanks to the implementation of an MES, significant reductions in Opex operating costs can be achieved, which must balance the Capex associated with the investment, both in terms of implementation and in terms of start-up and training.

The implementation of an MES is in any case the basis for obtaining visibility on the state of the production process necessary for coordinating the supply chain, and therefore the achievement of strategic consistency between the Operations Department and the other corporate departments.

1.6 Logistics and Distribution

This is the business function responsible for bridging the gap in time, space and quantity between production and market demand.

In charge of this department we find the Chief Logistics Officer (CLO), who oversees the block diagram represented in Fig. 17.

- Input: this is the finished product leaving the production lines at that moment τ_0.
- Output: customer orders delivered to the end customer in an instant τ_1; Therefore, the following fall under this department:

  – inventory management,
  – the handling phase, the preparation of orders.
  – the organization of transport.

- As regards the resources used, these are typically:

  – Capex K: materials handling, storage, transport systems, logistics assets, IT hardware and software;
  – labour L: logistics personnel, cooperative, third-party transport.
  – energy E: e.g. for cooling and conditioning in the case of fresh or frozen food, energy for transport, etc.
  – MRO auxiliary materials such as paper, labels, disposable packaging, etc.

- Controls: typically, software systems such as Inventory management systems, WMS for warehouse management, vehicle routing programmes for travel planning, software systems for booking, unloading lots, etc.

Fig. 17 Logistics: process, effectiveness and efficiency

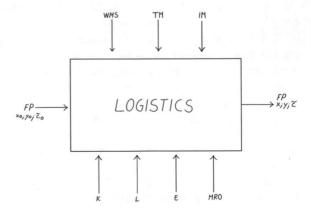

Also in this case, the effectiveness objectives for the logistics business function are measured in its ability to guarantee the process and therefore the transformation of input into output. In logistics, this translates into the concept of the "5 Rs", namely, Right Product, Right Quantity, Right Time, Right Place, Right Time and Right State. This aspect will be abundantly illustrated in Chap. 5 on customer service, to which reference should be made.

As usual, efficiency is measured as the ratio between the output and the resources used. For a logistician, therefore, efficiency means first of all minimizing, among others, the costs of stocks (Opex), logistics infrastructures (Capex) and transport (Opex), etc..

Therefore, precisely in relation to the achievement of these objectives, conflicting needs may arise with respect to the other business functions involved in the SC process and in the value chain. Examples of internal conflicts have already been mentioned in the previous paragraphs. Typically, for example, there may be conflicting needs between Logistics and Production (see Fig. 18). As already mentioned, production may need to produce batches that are as large as possible to minimize set-up times and related waste, to the detriment of the availability in the Logistics Department of storage capacity and holding costs to stock the product.

Typically, conflicts can arise due to conflicting needs between Logistics and Sales. The Sales Department may want high service levels, for example transport based on low capacity vehicles and even dedicated LTT transport, while logistics may be led to delay deliveries to consolidate a sufficient quantity of orders (and therefore volumes) to saturate a means of transport of high capacity and operate according to an FLT logic.

In the example reported in Fig. 19 we can show how transport costs increase in a less than proportional way with the load capacity of the vehicle, so that the unit

Fig. 18 Logistics and operations: inter-departmental conflicts

From A to B		FTL Capacity [pallets]	FTL Transport cost [€/trip]	FTL Unit cost [€/pallet]	FTL Unit cost [€/kg]
1 pallet	Solution A tractor-trailer trucks	33	160	4.84	0,00484
1000 kg	Solution B trucks	15	100	6.66	0,0066
	Solution C van	2	50	25	0,025

Fig. 19 Logistics and Sales: conflicts related to transport unit cost of transport (depending on the load capacity of the vehicle)

	Logistics	Sales
Network design	Centralized	Decentralized
	Economies of scale (↓Capex and Opex)	Close to the customer; ↓ LT
Inventories	Pooled $\dfrac{1}{\sqrt{N}}$	Duplicated N Close to the customer
Transport and LT	FTL; ↑LT	LTT; ↓LT

Fig. 20 Logistics and Sales: conflicts related to the location of stocks (centralized far from the market versus decentralized close to the customer)

cost of transport increases as the size of the vehicle itself decreases.[2] Obviously, with the same saturation of the vehicle, the unit cost is therefore minimal in the case of transport with articulated lorries or tractor-trailer trucks (capacity 30t), intermediate in the case of trucks (capacity 15t) and maximum in the case of trucks of category N2 (up to 12t) or vans of category N1 (up to 3.5t).

Again, as represented in Fig. 20, Sales could be led to keep most of the stocks or to duplicate stocks at local warehouses, close to the customer, to ensure a higher service level, while the Logistics, to optimize the resources used, may be inclined to centralize the stock in order to reduce the management costs of the infrastructures, proportional to the number of warehouses to be managed, and the holding costs of the product, which as we have seen at the end of Chap. 2, vary with a factor of $\sqrt{N}$, both in the cycle component and in the safety stock component, due to the inventory pooling phenomenon.

Also in this case, the choice of an optimal solution for the logistics structure must not be left to a search for the functional objective of efficacy and efficiency of the Logistics business function, which could lead to a partial sacrifice of effectiveness to the benefit of efficiency, but must be taken on the basis of SC's overall competitive strategic choice, according to the concept of strategic consistency.

First of all, it is therefore necessary to move from a concept of logistics (understood as a business function that deals with the distribution of the FP) to the concept of the SC, understood as a set of procurement, operations and distributions, which must

---

[2] The categories the trucks are divided into are:

- Category N. Motor vehicles equipped with at least four wheels and used for the carriage of goods;
- Category N1. Vehicles with a maximum mass not exceeding 3.5 tonnes, intended for the carriage of goods;
- Category N2. Vehicles with a maximum mass not exceeding 12 tonnes, intended for the carriage of goods;
- Category N3. Vehicles with a maximum mass exceeding 12 tonnes, used for the carriage of goods.

According to the Highway Code, the footprint must not exceed 255 cm in width, 400 cm in height and 12 m in length. The tractor-trailer truck is the largest type of industrial vehicle that we commonly see in Italy and Europe. This is a convoy consisting of a traction unit and one or more trailers. The articulated lorry is a vehicle with a cabin without a payload portion, equipped with a fifth wheel on which a semi-trailer rests.

LOGISTICS VS SUPPLY CHAIN MANAGEMENT

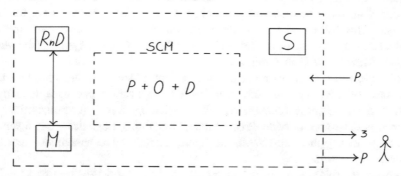

Fig. 21 From logistics to SCM

operate in concert with the other business functions of the Porter's Value Chain to achieve global and non-local objectives of effectiveness and efficiency (Fig. 21).

2 Support Business Functions

The support business functions are those not directly involved in the SC process, but instrumental to it, i.e., without which the process could not take place.

To represent this concept, support business functions are placed above Porter's Value Chain.

The support business functions include:

- IT: Information Technology
- A: Administration
- F: Finance
- HR: Human Resources.

The role of each of the departments described is now highlighted, as well as the possible interrelationships with the core business functions, i.e. how the choices of a particular non-core one can have precise effects on the others, analysing potential inter-departmental conflicts, meaning, as usual, the possible differences that may arise between one department and another due to the pursuit of conflicting objectives.

2.1 Information Technology

The ICT—Information & Communication Technology Department sees the CIO (Chief Information Officer) as its head, and is responsible for providing and managing hardware, software and networking infrastructures.

- Hardware: the ICT Department is responsible for the procurement and management of all the infrastructural part linked to PCs, PDA terminals (e.g. for sales agents who collect orders, for store asssociates who serve customers through tablets, etc.), telephones, printers, BC or RFID readers, AIDC systems (*Automatic Identification and Data Capture*).
- Software: the ICT Department oversees the purchase, management and maintenance of ERP systems but also specific application packages to support the various business functions (R&D, Marketing, Sales, Administration and Finance, Customer Service, Procurement, Operations, Logistics—e.g. inventory management applications, WMS, transport, retail/outlet management—reorders, checkouts, returns, etc.).
- Networking: this area includes both the connectivity (LAN Wi-Fi network) and the server part (physical in-house or hosting on the cloud) the structure of the databases and company security systems—networks and subnets—and related protection from attacks using firewalls and gateways, for example). The networking policies which a company implements can heavily affect the processes. The ability to freely access certain data can enable certain processes or not. The corporate departments also include data security management and disaster recovery policies, both hardware and software.

The ICT business function is the first of the Porter's Value Chain support departments, straddling the core and non-core, since the role of the ICT Department has grown in importance in recent years. In fact, the SC process is increasingly dependent on the resources that fall within the scope of this business function, and in particular on the availability of data which are accurate (linked to the level of error), punctual (available in real time, anywhere), and selective (at the level of a single item). This goes equally for the processes and transactions of a single department (intra-departmental), both between the departments of the same company (cross-functional, intra-company) and at the level of transactions between companies of the same SC (inter-company).

The role of this business functions has gradually become crucial and without the contribution of the ICT, processes at all levels come to a halt. On a down-to-earth level, without the support of the ICT, emails would not work, videoconferencing programmes for meetings would not be possible, just as the repositories on the cloud where company data reside would not be accessible everywhere to everyone, at any time.

Furthermore, in regard to the case of an online retailer such as Amazon, without the support of an ICT Department the web order collection system would not be available, but even if it were, orders could not be transformed into picking missions, the pickers would not have the terminals and software available to process orders, the automatic sorters would not work, the packing processes could not take place, the collections from the courier could not be booked, and the courier would not be able to sort the packages, to program and carry out deliveries. In other words, every SC process depends strictly on the existence of the ICT!

For this reason, in many contexts the ICT can be considered to all intents and purposes a genuine core business function, since it is an indispensable tool of the SC process.

2.1.1 Evolution of ICT Systems

ICT systems have evolved profoundly over the years, moving from non-integrated client systems to integrated client server systems, and subsequently evolving to integrated cloud systems.

Any software application can be schematized in three main macro-elements:

- Frontend: the part of the operator interface software application, e.g. an interface for viewing and processing the order lines of a picking mission.
- Backend: software that supervises the process, but which is invisible to operators. The operational logics of the process are encoded in the backend. E.g. in the case of a picking mission, the integrations for recovery or the logic of creation of the picking missions, the routing logic of the operators, or the verification of the mix/quantity between the picked and the expected.
- Database: this is where the data reside (master data—e.g. picking station register, temporary data storage tables—e.g. the SSCC of a storage bin).

In the 1990s, the frontend, backend and database structures were installed on a single terminal which acted as both client and server and was replicated for one or more stations. In addition, each corporate business function was characterized by its own application, with its own operational logics and its own database with related data and datasets. This is known as a *non-integrated solution* since i) the various applications do not communicate with one another, and also ii) work using different databases and datasets. Sometimes there was a need, for example for integration and/or transcoding tables to allow the data generated by one application to be fed to a second application (e.g. sales within an order-collection software, which had to generate customer orders to be processed in which the references included different terms, management of different units of measurement—one application working in terms of sales units and another of packages, misaligned units for packaging, etc.). The scheme of these applications is represented in Fig. 22.

A non-integrated system is inefficient, not only because of the duplication of data, but also because, as the interactions between the various corporate departments grow, the need arises to create data exchange integrations that become increasingly complex and difficult to manage, with the necessity to update them all once a change has been made. All of which is a source of potential errors, slowdowns in flows, and an explosion of point-to-point integrations that make each change extremely impactful, as shown in Fig. 23.

In the 2000s, thanks to the spread of networking and LAN protocols, first of all there was a shift from an application logic residing on a machine to a client server infrastructure, in which both the backend and database parts were moved to a centralized level. A central structure was therefore created in which there is a

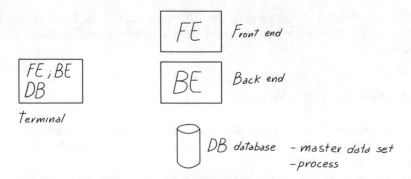

Fig. 22 Software application installed on non-integrated server

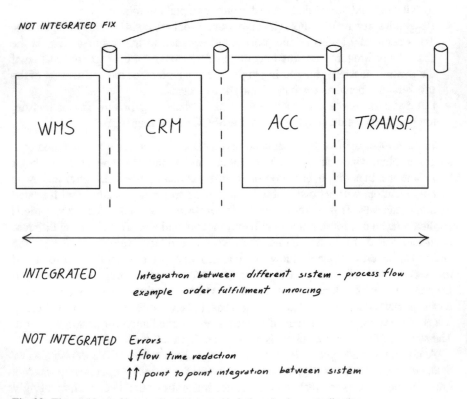

Fig. 23 The problem of integration between stand-alone business applications

common BE and DB on the server side on a single machine, and many instances of client frontends on each terminal connected to the network through a LAN, working on the same low data and datasets levels (see Fig. 24). This is known as a client–server structure because the number of clients can be increased at will since the client resides only on a (eventually web) frontend interface while all the data, the process, the data

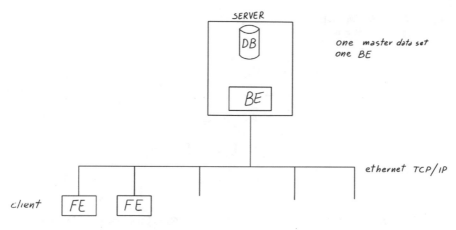

Fig. 24 Client–server structure based on LAN infrastructure

(database and master datasets) and software logics are on a centralized backend. The system can therefore be scaled very simply by enabling any number of frontend stations.

In this way, the data is centralized and aligned, since there is a single management database and there is no need to duplicate the data structure. E.g., a single inventory management system, where availability is automatically aligned for both the retail channel and the online channel, and a sale in either channel decreases availability for both channels.

Furthermore, we have now moved from non-integrated systems, in which each corporate business units had its own software application with its personal data and its database and master datasets, towards an integrated solution, with a single common data structure and with common interconnected processes. In this way no software integrations are necessary, and every update is available for all applications.

Not only that, in an integrated system, the applications of the various corporate departments interact natively with each other, in the sense that the processes of a business unit recall departments and trigger processes on the applications of the other corporate departments. In this case, we speak of ERP systems—*Enterprise Requirement Planning*.

Think for example of the fact that the order collection software for the sales force, or simply a piece of software for online order collection, which, before being able to confirm the order, must communicate with some accounting software, to verify the fact that the customer is in the registry or not, blocked or not; the WMS which has to confirm the availability of the product or the transport planning system that has to confirm the possible dates for the delivery planning, in order to define the lead times. In turn, administrative management software must confirm sales prices, or there is the CRM software which must manage the discounts and promotions for specific customers, for example. Once the order has been confirmed by the order

Fig. 25 Conceptual diagram
of an integrated ERP system

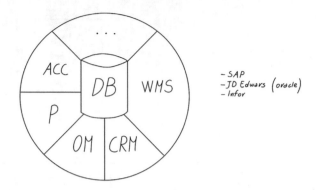

collection software, then the order confirmation triggers processes on the other software applications of the other corporate departments. The order must be processed (WMS), the delivery planned and executed (TM transportation management), while the administration must issue a transport document and then an active invoice, and the customer service (CS) has to update the customer's status (e.g. customer level A, B or C).

An ERP system is therefore a centralized software system capable of managing all the sub-processes making up a complex business processes in an integrated approach, with a single database and single master data, and in which the process workflows are designed with the related interactions between the different processes of the various business functions, based on an installation on a server physically located in a company data centre, where users log in with their credentials via a web window (see Fig. 25).

In the 2010s, with the explosion of the internet, these systems then moved from a local client–server logic to a cloud logic. The backend, database and frontend parts were moved to the Cloud (such as MS Azure, Google Cloud, Amazon Web services which provide a cloud environment), while the frontend became web-based, therefore accessible from any browser screen (either desktop or mobile) without the need for an installation except perhaps a bookmark (see Fig. 26).

When end-users place an Amazon order, they access a web frontend user interface UI of the Amazon cloud system based on AWS (*Amazon Web Services*) which interacts, as any user interface does, with a backend system which manages products, users, and administrative and logistic cycles.

A web repository makes the system much more flexible and independent from a peripheral device, since the system becomes totally independent from the single piece of hardware, on which nothing is installed but a browser. The hardware terminal becomes a transparent object, so that operators, logging in with their credentials from any location, find themselves in their own work environment and can operate with maximum efficiency (e.g., this is what happens with regular mobile phones). With Google Cloud or Apple Cloud, the phone has become transparent; operators log in with their Google or Apple account and find all their address books, contacts, emails, bookmarks, and passwords, etc.

Fig. 26 Cloud-based
infrastructure scheme; the
BE and the DB go to the
cloud and the front end
becomes web based

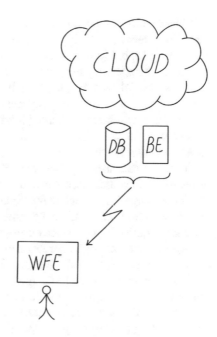

2.2   Administration and Finance

Alongside the ICT business unit, the support departments include the Finance and Administration. These are often combined, but in reality these two business units have two very distinct purposes.

In particular, the Finance Department, headed by the CFO (Chief Financial Officer), has the purpose of finding the resources which allow the other business units to operate the SC process.

Once the budgets of the various corporate business units and therefore the R&D, Marketing & Sales, Purchasing, Production and Logistics budgets and the related investments have been defined, the CFO designs the cash flow trend in income (typically deriving from the sale of products and other types of income) which will correspond to outgoing cash flows. These may have different components, in particular:

- Purchases of raw materials, components, MROs, and services
- Staff (salaries and contributions—with the exception of severance pay which does not involve cash outflow but only a reserve fund)
- Financial liabilities (VAT, taxes)
- Expenses
- Capex investments made during the year.

On the basis of these flows, he/she designs the trend of working capital over time, and identifies possible sources of financing.

These sources may be internal, if the company has sufficient cash to finance itself, or external, if the company, following a negative cash flow, needs to resort to third-party capital (credit lines, mortgages, lines of finance, bond loans, and so on).

Hence, also in this area, tensions can arise deriving from conflicting objectives between a CFO who has negotiated a credit line in advance and who wishes to maintain the status quo, and any business unit manager who, following a change in boundary conditions, requires additional resources to make, for example, new investments, or who needs to change conditions with financial impacts in order to meet particular needs.

Tensions can also arise between Finance and Sales, especially when the sales forecasts of the Sales Department are disregarded. An underestimated sales forecast can have a negative impact on the Finance Department, since it may generate unnecessary costs for borrowing. In the same way, an overestimated sales forecast can have a highly negative impact, since during construction the financial resources necessary to support the company's operations are lacking and it is therefore necessary to resort to buffer solutions, typically a bank loan, and normally high cost.

The Administrative and Management Control Department, on the other hand, has more accounting purposes, and deals with the registration of payable invoices, issuing of active invoices, management of accounting and financial statements according to accounting principles, providing the management with trends and changes with respect to the budget (*management control*).

The Administrative Department also deals with compliance with payment terms, both for active and passive invoices, and can therefore generate delays in payments to suppliers or determine the blocking of customer credit lines, generating tensions with the other corporate departments involved. (Respectively Procurement—a supplier who does not see his invoices paid could block deliveries, generating production delays—and Sales—blocking a customer could prevent goods from being shipped, exacerbating commercial relations with that customer who could contact the competition, therefore losing business).

2.3   HR Human Resources

The Human Resources department has the most important function of all the business units: it is responsible for making the people necessary for the management of supply chain processes available to the other corporate departments.

The HR department deals not only with recruiting, but also with people's professional and economic growth, in addition to corporate welfare policies.

It therefore exercises a support function as we increasingly come to understand the correlation between corporate wellbeing and working conditions, and the motivation and commitment that people put into their daily work, as well as dedication to the cause when the company has to face moments of crisis, such as during the recent Covid-19 pandemic.

Sinek 2019 highlighted the 5 ingredients receipt to create a resilient corporate culture: (1) Advance a just cause; (2) Build trusting teams; (3) Study your worthy rivals; (4) Be ready to pivot on a fundamental level; (5) Demonstrate the courage to lead.

3 Strategic Consistency

3.1 Definition of Strategic Consistency

Each company competes by defining a set of consumer needs/requirements which the company tries to satisfy through its products and services—possibly by differentiating itself from its competitors—to which the consumer is willing to attach a value represented by the purchase price of the product.

E.g., in mass consumption, an organized mass distributor which sells through hypermarkets such as Walmart, focuses on service performance in terms of the breadth of a range (both horizontally in terms of products/departments, and vertically in terms of references by category), working hours and availability 24/7/365, travel distance (a Walmart Supercentre is always a maximum of 10 miles from any other supercentre), at the lowest price.

Conversely, a normal trade shop might focus on the high quality of the product, and its proximity service for which consumers are willing to accept a higher price. In turn, a cut-price supermarket (hard discount) focuses on a reduced range of products (service) of medium–low quality (product) but consistently at a low price (price).

In the case of PCs, Dell focuses its competitive strategy on great variety through customization of the product (service), delivery times that are not immediate but in any case the fastest possible (service), and a price in line with the competition.

Vice versa, Apple has established its own competitive strategy based on the brand name (intangible characteristics of the product), customization reduced to a few models with the possibility of configuration that is in any case limited, sales through retail or wholesale stores, typically corners in consumer electronics retail chains (In Italy Mediaworld, Comet, Euronics, Unieuro, etc.), in which the product is in stock and can therefore be touched and bought, and in which the staff can provide assistance, immediate delivery (service), for which the consumer is willing to accept a high price. Furthermore, the customer values the intangible of feeling part of a world containing people who love the Apple world as much as he/she does.

Finally, in textiles and clothing, a fashion brand focuses on the material characteristics of the product (quality) and on the intangible ones (brand image) on the in-store shopping experience (service) for which the consumer is willing to accept a high price.

Conversely, fast fashion brands such as Oviesse, HM, and Zara do not focus on product performance (medium or medium–low quality) but service performance in the breadth of the range (in terms of high turnover of collections which change every fortnight) and a low price.

The examples above show how each competitive strategy is therefore based on the definition of a mix of priorities given to the product (core product—in terms of tangible and intangible functionality) and service (service surround), which corresponds to the need of a consumer target who are willing to accept a purchase price for that particular product/service mix.

Therefore, there is no right and wrong competitive strategy. Certainly, for every competitive strategy there will be a larger or smaller audience of consumers who attach a value, and are therefore willing to reward the SC with the purchase of its product at that price.

The task of defining the strategy is the responsibility of the Board of Directors, while it is up to the CEO to implement it, orchestrating the different business functions of the Porter's Value Chain analysed in detail in the previous paragraphs.

Each of the core and non-core corporate departments seen in the Porter's Value Chain plays a fundamental role in being able to achieve the strategic objective of product/service/price which the focal company or, more generally, the SC has set itself. In fact, each business function, on the one hand allows completion of a process that adds value to the product, on the other adds a cost to the product according to the scheme represented in Fig. 27.

The unit cost of a product is affected by:

- Indirect costs, typical of R&D, Marketing & Sales, ICT, HR, Administration and Finance
- Direct costs (physical costs—raw materials, components, transformations, transport, handling, stock maintenance, production scraps and waste), deriving from the SC processes: Procurement, Operations and Distribution. The various SC processes add costs, and at the end of the SC process there will be a total cost, the sum of all the costs of the processes, phases and activities that make up the entire SC process

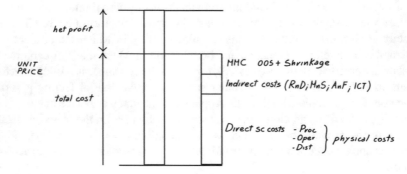

Fig. 27 The impact of corporate departments on unit costs and profit

- Direct costs (market mediation costs) deriving from the opportunity costs of lost sales (OOS deriving from underestimated forecasts) or product shrinkage (obsolete and expired goods deriving from overestimated sales).

The BoD sets the strategy and therefore the characteristics of the product, service and price that it is intended to pursue to target a certain group of consumers willing to attach a certain value (price) to that product/service. It is then the task of the BoD and, through the CEO, to orchestrate the various corporate business functions so that they move in unison, not for the pursuit of their own functional objectives, but of those strategic objectives capable of generating the expected product and service; and secondly the cost structure (understood as total costs, sum of direct costs—physical and market mediation—and indirect costs) so as to generate the total expected cost and therefore the expected margin to be able to sell the product/service for the expected price.

It should be emphasized that the fact that fixing the price means setting an upper limit on the total cost that the SC can afford to keep generating that product and service. The greater the cost at the same price, the lower the margins that the SC can guarantee itself, hence the competitiveness of the SC itself, because the greater will be the "cake" that the SC players will be able to share and reinvest in to generate competitiveness. Competitiveness needs to be assessed both in absolute terms and in comparison with other SC competitors who, for the same price, are able to generate greater margins, or who, with the same margins, are able to offer the same product/service at a lower price and therefore automatically attract more customers.

Limiting ourselves for the moment to looking around within the company boundaries, in general, each of the corporate business functions could try to pursue its own strategy, meaning by strategy the *functional objective*, what the corporate department tries to do particularly well, in terms of effectiveness (what—which product, which service) and efficiency (minimizing costs). Typically, each department could identify its functional objective in the product/service combination that allows it to minimize its own cost component, this is particularly the case if functional MBOs are linked to funcional costs. The problem is that, as we have seen, the choices of a business function are intimately correlated with the effectiveness (*product/service*) and efficiency (*costs*) of the others, and of the SC in a broad sense, and therefore they have an impact both on the product/service and on the different components of the unit cost, as we have seen in the many examples above.

Accordingly, it is a question of understanding, once the competitive strategy has been defined, what the strategy of the various corporate departments needs to be in order to ensure that the strategies are aligned, both with one another and with the competitive strategy in general, in order to obtain the strategic objectives set, both in terms of product/service/price and therefore of turnover, and in terms of total cost and therefore of profit.

Clearly, if the strategy is to maximize profit, on a par with the selling price of the product, it is clear that the total cost must be minimal and therefore the various departments must be orchestrated to achieve this goal. However, it is absolutely not definite that a minimum total cost coincides with minimum costs of the individual

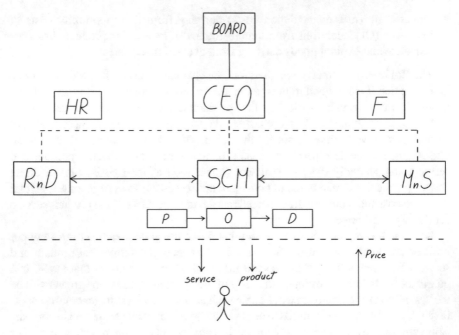

Fig. 28 Strategic consistency: the role of the BoD, the CEO, and the various corporate departments

departments, indeed, this is rarely the case. Usually, the minimum total cost is in sub-optimal conditions for the various business functions. It is therefore the CEO's task to avoid each department pursuing its *local optimum*, unless this coincides with the *global optimum* and is in line with the overall competitive strategy.

Strategic consistency (see Fig. 28) therefore means alignment between the objectives/competitive strategy (product/service/cost) set by the BoD, the SC strategy (procurement, operations and distributions) and the strategy of all the corporate departments of Porter's Value Chain, towards the same objectives of effectiveness (product/service) and efficiency (total cost), which guarantee the turnover (given by the number of customers willing to accept that price for that product/service) and the margins (difference between turnover and total cost) set strategically by the BoD.

3.2   The Causes of Failure to Achieve Strategic Consistency

Basically, failure to achieve strategic consistency can be down to two fundamental causes:

- Lack of alignment between the objectives of the various corporate business functions and the competitive strategic objective. If the success of the company does not depend on any particular department, the failure of just one can compromise the overall success. The goal of a single business function in terms of product/service (effectiveness) or minimization of cost (efficiency) could generate extra costs for other business units which do not minimize the overall total cost.

E.g., it could happen that the Marketing/Sales Department promotes the immediate availability of products (reduced lead times), while the SC (Procurement, Production and Distribution) has been designed and run to pursue cost reduction objectives in favour of consolidation and low price means of transport.

Or again, a retailer that focuses on a high range of products in terms of collection rotation, low stocks, but finds itself with a SC and Procurement Department which, to minimize its functional costs, has selected low-cost suppliers, but not ones that are very reactive and have long delivery lead times.

- Lack of resources and skills: Even if strategies are aligned, business functions can fail in putting the strategy into practice due to inadequate resources (either people or assets) or inadequate processes and skills.

The CEO's task is therefore very simple: on the one hand, as mentioned above, it is to implement the directives of the BoD in terms of competitive strategy and therefore in terms of the product/service/price mix which the company intends to pursue. However, on the other hand, it is the CEO's task not only to ensure that the strategies of the various corporate departments are aligned and integrated with the competitive strategy established but also that the corporate business functions have the necessary resources, not only in economic terms but also in terms of processes and skills—both hard and soft (people)—to pursue these strategies.

3.3   The Different Levels of Strategic Consistency

When the term "*integration*" is used in Supply Chain Management, we mean precisely the alignment between strategic and functional objectives in the terms described above to achieve strategic consistency. Indeed, the term "management" must be understood as meaning *integration*: strategic alignment and consistency between the various corporate departments, inside each single player and outside in the SC.

There are therefore several levels of integration, as represented in Fig. 29:

- No integration. The objectives of effectiveness and efficiency are aligned and orchestrated within a single business functions by the respective CXO, but not between the various corporate departments of a single player. This represents the minimum level of integration;
- Internal cross-functional integration: the objectives of effectiveness and efficiency are aligned between the various business functions, and orchestrated by the CEO, but not with those of the other corporate departments of the other players in the SC process;
- external or inter-company integration: there is a strategic consistency not only between the business functions of a single player, but also between the strategy of the various corporate departments of the various players involved within the

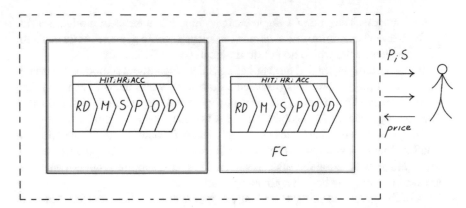

Fig. 29 The different levels of strategic consistency: intradepartmental, internal intra-company, external inter-company

SC, upstream with suppliers and downstream with customers and the third parties involved.

Clearly, internal integration is a necessary but not a sufficient condition to achieve strategic consistency, which can only be fully achieved with external integration, i.e., when all the players involved in the process are aligned towards the same goal and are orchestrated as a single subject. This is indeed a situation where costs can be minimized and SC profits raised to a maximum.

3.4 Strategic Consistency—The Dell Case

Before generalizing how it is possible to achieve strategic consistency and inter-company integration, it is useful to preface some ideas in this regard from a case already analysed, that of Dell.com, to understand how Dell has built its competitive advantage lever on integration, and therefore on the alignment of objectives between the various corporate business functions both inside and outside the company.

In Dell's case, the competitive strategy set by the BoD is as follows:

- Product.

  - PC in line with those of the competition, in terms of design (*intangible performance*) and functionality (*core performance*);

- Service:

  - An extremely wide range of products that can be customized according to a very high number of combinations by the final consumer;
  - direct sales through a web portal; products not available to be touched and seen—and therefore intended for an experienced, confident public;

- Lead time: as low as possible, but the product is not available for immediate purchase. Any PC is inevitably made to order.
- Service: help desk available immediately and after sales support in few business days

- Price:

  - on average in line with the competition.

  Consequently,

- Research and development evolves products with performances in line with those of the competition, but which are highly customizable thanks to a modular design. Each PC can therefore be assembled quickly from standard components. All of this, in addition to expanding the range according to the modules chosen, enables a Pull (make-to-order) and non-Push (make-to-stock) production system as adopted by the competition;
- Marketing promotes the product through call centre channels and especially the web (online);
- Sales do not involve recourse to retail chains or electronic stores, but only sales through direct channels (web or telephone). The prices are in line with those of equivalent products from a technical point of view, but the lack of intermediaries ensures Dell all the margins; physical costs are shrunk to a minimum; market mediation costs are zero thanks to the Pull (make to order) model.
- To be able to keep costs down and maximize margins, the SC strategy is organized so that products are *not made to stock* but *made to order*. In this way the costs of the stocks are zero, as are the costs of obsolescence:

  - Procurement: To be able to make lead times compatible, suppliers are close by; they can see the stock data in order to improve their ability to forecast demand and further reduce their inventory costs and lead times. These suppliers are not necessarily able to provide the lowest price, but they are certainly the most responsive. Key suppliers are managed with a view to partnerships in the development of new products;
  - Operations: The production processes are not necessarily those that guarantee a lower production cost compared to the competition which produces make-to-stock batches, but they are flexible and fast (Dell invests heavily in the flexibility of its lines to lower production costs); a PC can be assembled and customized within 24–48 h;
  - Distribution: the same applies to deliveries, which are made through the use of express couriers/cross-docking to try and achieve economies of scale in transport, reaching the consumer directly at home. If the consumer purchases two components manufactured at different factories, Dell tries to synchronize the flows to make a one-time delivery to the customer.

- The information systems develop the portal for the acquisition of orders and the tools for transforming the customer order into a fulfilled order, as well as for visibility and integration downstream with couriers and upstream with suppliers.

- Human resources promote a cultural climate in which *customer obsession* is at the centre and everything is organized to try to reduce order fulfilment times while maximizing value for the customer. When faced with a problem, everyone knows which direction to head in—the customer!

3.5 Achieving Strategic Consistency

Once the strategic consistency concept has been defined and described like that of a successful player such as Dell which has developed its business currently ranked 27th in the Fortune 500 around this concept, it is now a question of conceptualizing and seeing how it would be possible to achieve strategic consistency within an individual company, and then extend it externally, throughout the entire supply chain.

Fisher, 1997 pointed out that to achieve strategic consistency it is necessary to follow three major steps.

1. First of all, understanding the characteristics of the product/service/price it is intended to pursue. As will be seen in detail in the following paragraphs, there are basically two metrics to analyse the characteristics of the product: on the one hand *demand variability*, on the other, *supply variability*.
2. Secondly, to understand what the possible characteristics of the SC are in terms of the performance that the SC can guarantee. In fact, there are supply chains which are extremely heterogeneous in terms of cost and service performance, and it is therefore necessary to know them in order to choose the one which best matches the specific circumstance.
3. Finally, understanding what the areas of strategic consistency are, in which the performance characteristics of the SC correspond to the characteristics of the product/service, and therefore of supply and demand variability and in which they do not. In fact, strategic consistency means adopting and implementing a supply chain that is suited to the characteristics of the product/service to be provided.

3.5.1 Understanding the Characteristics of the Product

The characteristics of the product/service can be summarized in a metric which considers:

- The demand variability of the product/service which the supply chain intends to provide for
- The supply uncertainty of the product/service.

In the next two paragraphs, each of these two factors are analysed in detail.

Demand Variability

In order to achieve strategic consistency, it is necessary to start from the demand variability which one intends to deal with.

Demand variability must be seen in relation not only to the characteristics of the product itself (how much the demand for the product is variable *tout court*), but also in relation to the characteristics of the service, with which the product is to be sold to a certain tranche of the market (the tranche of consumers who attach a certain value to that product/service and are therefore willing to accept a certain price for it).

The same product can therefore have an extremely stable demand for a certain portion, and certain conditions of service and price, but for other portions of the market, in which the same product is sold with completely different service and price characteristics, the demand may be highly uncertain.

Think, for example, of an automotive component. This can be sold for the mass-produced car market—characterized by standard service, low price and stable demand, given that this component is fitted to all the model variants and versions—or sold to meet the demand for spare parts—and therefore for a market characterized by high service and price but also by difficult-to-predict demand. As thoroughly described later, it would be appropriate to deal with the same product with different demand variability and different supply chains.

We talk about implicit demand variability precisely to imply this concept: considering the demand variability not only for the product itself, but in relation to the service with which the product is offered to the final consumer.

In general, as can be seen from the prospectus in Table 1, for the same product (and therefore the variability of the intrinsic demand for the product) an increase in service generates an increase in the demand variability.

According to the implicit demand variability for the product/service, it is therefore possible to distinguish two macro-categories of products, on a scale which ranges from a minimum to a maximum variability: on the one hand, functional products, on the other innovative products, whose characteristics are described below.

Table 1 Service level and demand variability

Increased service level of the SC	Causing an increase in the implicit demand uncertainty due to
Increased variability in the quantity of accepted orders	Implies greater variance in demand
Increase in the breadth of the range	The demand for each reference becomes more disaggregated and therefore more difficult to predict
Lead time reduction	Less time to react to orders
The number of sales channels increases	Total demand is broken down into several channels
The rate of innovation increases	New products suffer from more uncertain demand

Functional Products

At one end of the variability scale we can place the so-called functional products.

These are products with implicit variability of low demand, therefore characterized by stable demand, which are sold to a portion of the market that does not require particular service conditions. Products that meet basic needs, such as milk, gas, and commodities in general. The vast majority of FMCG (expecially those featured by long shelf life like dry foods) can be considered as functional products.

For functional products, the consumer does not perceive particular variations in the product among the various competitors, other than the availability or lead time (service), and the choice is typically directed towards the lowest price.

The number of variants for functional products is usually low (10–20 variants per category).

Functional products are characterized by a high level of maturity, present on the market for several years (2–5 years), no longer covered by patents, whose initial profitability has attracted many competitors and therefore for which there is fierce competition given that several competitors are able to offer the same product, and for which, given that the competition is based on service and price, with the same service, for example, presence on the supermarket shelf or supply lead time, prices must be kept as low as possible to gain more market share.

Consequently, margins are also low, typically between 5 and 20% at the most, and the struggle lies in a reduction of costs to guarantee margins, but also, as mentioned, in the acquisition of high market shares to guarantee high volumes which can generate profitability even with low margins.

If these are the negative aspects of functional products, there are also some positive aspects in managing this type of product. Precisely because they meet basic needs, the demand for functional products remains stable, consequently the errors that can be made in the demand forecast phase are low, in the order of 10%,[3] with OOS rates of 1–2%.

The low demand variability facilitates the planning of the SC, consequently also the market mediation costs associated with underestimated demand (OOS costs) are low, as is the incidence of the share of market mediation costs deriving from overestimated demand, therefore discounts for expired or end-of-life products which is low, if not near zero.

In light of the above considerations, functional products are therefore characterized by a predominance of physical costs (raw materials, production, distribution and logistics, as well as stock holding costs) over market mediation costs (out of stock and opportunity costs deriving from underestimated demand or expired/obsolete

---

[3] MAPE Demand Forecast Error—Mean Absolute Percentage Error.

$$\text{MAPE} = \frac{1}{N} \sum_{i=1}^{N} \frac{|D_i - F_i|}{D_i}$$

F_i = Forecast at i; D_i = Demand at i.

good deriving from overestimated demand), which is why it is on the former that we need to concentrate in choosing the SC.

Innovative Products

At the opposite end of the demand variability scale are so-called innovative products.

Innovative products, as the word itself implies, are generally produced with a low level of maturity, and are normally in their introductory or development phase. Being innovative, the competition is not yet steep and so the margins are high (50–100% or more), since prices can be kept high in the absence of competition.

The competition does not therefore depend on price, since there are few competitors, but on an ability to get the right product (i.e. for apparel, that model/variant/version) to the customer with the required level of service (e.g. at the right time).

However, being a new product, it is extremely difficult to predict what the market demand will be and the error rates in forecasting demand for innovative products can be 50–100%, or even higher.

Consequently, the MMC part prevails over the PC part. OOS rates can be 10–40% or more, as can end-of-season returns and shrinkage in general.

Innovative products are not only products at the beginning of their life cycle, but also products characterized by multiple variants/versions/models, the demand for which is difficult to predict, since the number of models/variants/versions disaggregates the demand per single item which therefore becomes difficult to predict.

Supply Uncertainty

In addition to demand uncertainty, it is also important to consider the supply uncertainty associated with the raw materials and components from which the product (functional or innovative) is made.

Also in this case, there can be two diametrically opposite situations: low supply uncertainty and high supply uncertainty. Using the terminology of (Lee 2002), the first to introduce this parameter for the achievement of strategic consistency, we can speak of processes characterized by *stable supply* and processes characterized by *evolving supply*.

Stable Supply

In the case of stable supply, the supplier base is large, stable and consolidated, and the processes and technologies for the production of raw materials and components are mature, therefore characterized by a low defect rate and a high level of automation. Consequently, component quality problems are also limited.

Procurement processes are equally stable and consolidated, as are lead times which are affected by low OOS rates.

In a stable supply process, variable seasonality (e.g. availability of raw materials following particular weather conditions) plays a negligible role.

Most functional products such as FMCGs, mass textiles, oil & gas, but also innovative products such as fashion, automobile, or certain food products characterized by difficult-to-predict demand (e.g. seasonal products in which demand is strongly dependent on meteorological variables, such as ice cream).

Evolving Supply

The diametrically opposite case is that of processes characterized by evolving supply.

Usually, we find ourselves in this area with highly innovative products. Since the product does not yet exist on the market, the supplier base too is limited to a very few players or even none. In this case in particular, the procurement processes are all to be built, as are the technological processes for the components and raw materials which do not yet exist. Consequently, lead times can be long and extremely variable, as can the defect rates. Quality problems in the case of an evolving supply system are therefore significant.

It is possible to be in an evolving supply situation not only for particularly innovative products but also in the case of highly functional products. This is the case of continuous food or textile products, in which demand is stable but where raw materials depend on a small number of suppliers, which may be affected by particular weather conditions, for example. Drought, famine, pandemics or temperature can determine an extremely variable supply of raw materials and therefore a rapidly evolving supply situation.

In short, food products dependent on agriculture are functional products with high supply uncertainty, while new electronics products are innovative products with high supply uncertainty.

The Uncertainty Framework

The uncertainty framework can therefore be summarized by the diagram detailed in Fig. 30, taken from (Lee 2002), in which the products are characterized according to the demand uncertainty (functional/innovative) or the supply uncertainty (stable/evolving).

3.5.2   Understanding the Characteristics of the SC

Once the characteristics of the product/service have been defined through a metric (demand variability and supply uncertainty), it is a question of characterizing the SC through a metric which can identify the performance that the SC itself is able to provide.

Demand Uncertainty

		Low (Functional Products)	High (Innovative Products)
Supply Uncertainty	Low (Stable Process)	Grocery, basic apparel, food, oil and gas	Fashion apparel, computers, pop music
	High (Evolving Process)	Hydro-electric power, some food produce	Telecom, high-end computers, semiconductor

Fig. 30 The uncertainity framework according to (Lee 2002)

Fig. 31 The vision of stocks in a lean approach

In an SC performance scale, efficiency (leanness), responsiveness, resilience (risk-hedging) and agility can be positioned at the extremes. The characteristics of four SCs are described in the following paragraphs.

The Lean SC

The term "lean" literally means "without fat". In this context, "fat" is to be understood as everything that means more in the functioning of processes, typically stocks on the one hand and waste on the other, and which must therefore be eliminated/reduced since they represent a cost.

In a lean approach, the SC is seen as a boat sailing over a rough seabed with large rocks, and it therefore needs a lot of water to be able to navigate safely (see Fig. 31). Similarly, in order to guarantee a product and service in the face of inefficiencies, the SC needs a lot of stocks. The lean approach is therefore to progressively reduce water (inventories), bring out rocks (inefficiencies), eliminate them, and make the SC work at minimum water, that is, at maximum efficiency and consequently with the minimum possible level of inventories.

An efficient or lean supply chain, in addition to rejects and inventory, seeks in particular to minimize the overall cost component relating to physical costs. A lean SC schedules in advance and plans all processes to get the required product and service downstream for a minimum overall physical cost.

In the Procurement part, in a lean supply chain, suppliers are chosen according to a price logic, in order to try and reduce the cost of raw materials. The supply can be negotiated on a worldwide scale, and lots are optimized on the basis of procurement costs, quantity discount, and stock maintenance costs. The transport of raw materials is always fully loaded and low cost, precisely to contain the component of physical cost linked to transport costs, at the expense of medium-long lead times.

As for the Operations part, in a lean supply chain, production takes place in advance and by economical lots, in order to generate production economies of scale and optimize stock levels. The product is therefore typically made on forecasts and held in stock pending receipt of demand.

Production is highly automated wherever possible, in large facilities where investments in automation are concentrated, and which produce one or a few models to meet a global demand. In the case of labour-intensive processes such as textile manufacture, production takes place in countries where labour costs are small, albeit at the expense of high lead times and low flexibility in production planning. Also in this case, production is concentrated in a single facility to meet a global demand, in order to enjoy economies of scale or scope.

Also for the distribution part, in a lean supply chain the goal is to reduce the physical costs associated with the logistics. The main strategies therefore concern the network and the means of transport. As regards the network, in a lean SC the attempt is to centralize the stock in a few DCs, which therefore serve large areas, in order to obtain economies of scale in the logistics infrastructure and reduce inventory costs thanks to the pooling effect, despite the increase in lead times. As far as transport is concerned, economies are sought by favouring the logic of consolidation at the expense of speed. Economical choices (sea, rail), albeit slow, are also favoured in choosing the mode of transport.

The Responsive SC

The term "responsive" here means reactive, flexible, and therefore indicates the ability of an SC to adapt quickly to changing surrounding conditions. In this context, in particular, the boundary conditions are represented by the market demand, which can be extremely variable. Consequently, a responsive supply chain can respond quickly to changes in market demand.

A reactive supply chain, unlike a lean SC, tries to minimize the part of the cost related to market mediation costs and therefore opportunity/OOS and shrinkage/unsold costs.

A reactive supply chain therefore tries to supply, produce and distribute exactly the requested product when required, in order not to waste opportunities and not to generate the unsold goods typical of make-to-stock processes, at the expense of physical costs which can also be high.

As for the procurement part, it must be premised that it is often the suppliers who do not allow a supply chain to become sufficiently reactive. The reasons which cause a supplier to be unable to quickly follow a change in demand and make the SC reactive are many. On the one hand, the supplier, being upstream, may not see the variation in demand, and may not keep in stock a component for which a high supply lead time may be required. Then the lead time itself can be very variable precisely because of the innovativeness of the new component. A new component may also be characterized by a high defect rate, and consequently it cannot go directly online, thereby slowing down the downstream processes.

In a reactive supply chain, suppliers, in this case those who supply key components, that is those on which the ability of the SC to supply the new product required by the market depends, are first of all rationalized in order to establish fruitful partnerships. It is in fact unthinkable to be able to establish a structured partnership with all suppliers.

Then in a reactive SC, key suppliers bring value from the start since they are involved in the design of the new product. Production plans are shared, so that they have visibility on the expected volumes, and the supplier can better plan its activities.

In a reactive SC, the key suppliers are located nearby, they keep the product available thanks to the sharing of data via Electronic Data Interchange EDI (both stock and demand data), to be reactive and quick in supplying, even at the expense of price. The lots are small to achieve what is required at that time, to the detriment of supply economies of scale. Stocks are often managed from a Vendor Managed Inventory VMI perspective to make the process more efficient, eliminating the Bullwhip Effect and avoid duplication.

On the operations side, in the SC, production is flexible and typically takes place on demand and not on forecast, or on forecast if postponement logics are applied, which makes it possible to customize a neutral product, even if the product is made through assembly of standard modules (modules that are make-to-stock) or customization with kits. Format changes are the order of the day, and the batches are small thanks to flexibility of production facilities or ultimately, at the expense of production costs. An attempt is made to absorb the related costs by investing in flexible production and assembly lines. The goal is not to optimize production costs through economies of scale and economical lots, but to reduce production lots and production times through the process used. Therefore, precisely to reduce the timelines, the plants are flexible but small and local at the service of a limited market, even if the labour may be more expensive, while the flowtime to travel the production line is the key parameter. The goal is to produce what is required in the shortest time possible, and as quickly as possible. Obviously, at the same time it is necessary to keep production costs under control, which could explode in the search for a unit lot, but which are not the main driver. The main driver is speed, on which the ability to produce make-to-order and therefore to eliminate market mediation costs depends.

Also on the Distribution side, in a reactive SC the key word is again flow and speed. Distribution takes place once again trying to reduce the transit times from the warehouse to the customer. The management of the DC takes place in a flow logic, as in the case of fulfilment centres or transit points, with rapid and targeted processes and transit times of the goods kept to a bare minimum. Where possible, cross-docking techniques are adopted through transit points, precisely to eliminate the time component linked to stock, since any day spent as stock, increases the likely-hood of shrinkage and decreases the product value. As far as transport is concerned, fast modes of transport (air or road) are favoured at the expense of cost, trying to keep costs under control through economies of scale linked to volumes, which can derive either from a synchronization of flows (cross-docking, consolidation and sorting) or the use of a 3PL. Sharing of information (demand and stock) is also essential to ensure successful reactive processes.

Leanness Versus Responsiveness

The representation shown in Table 2 summarizes the characteristics of the two types of SC and the performances that lean and reactive SCs are able to deliver.

Placing the performance in terms of responsiveness on the ordinate axis and those in terms of leanness on the abscissa axis, there is an inverse correlation between the two as shown in Fig. 32.

It is therefore possible to identify a "performance boundary", in the sense that for a given product there are no more reactive SCs with the same efficiency, or more efficient ones with the same reactivity.

The SCs on the left of the curve are SC followers, which must fill a gap to reach the frontier, while the SCs on the frontier are best competitors which, at a certain time interval $\Delta\tau$, represent excellence in performance for the reference market, and which therefore enjoy a competitive advantage.

It sometimes happens that supply chains which are on the frontier or even those that are at a competitive disadvantage to the left of the frontier are able to innovate their supply chain processes. This typically occurs after the introduction of new technologies in the production or distribution processes. Think for example of the use of industry 4.0 technologies (artificial intelligence, autonomous robots, the Internet of Things, for example, by applying an RFID tags to products to reduce OOS and increase efficiency), or thanks to business process reengineering, or to innovations in the way of doing things and in the relationships between corporate departments, internally or externally with individual companies (e.g. VMI, CPFR, cross-docking, etc.).

When this happens, and a supply chain therefore moves to the right of the frontier, the frontier itself automatically shifts in correspondence with the supply chain that has innovated through technologies, processes and/or both. This happened for example when Walmart introduced cross-docking, Dell started its make to order direct sales model, or Zara its fast-fashion model. All the competitors who were in prime position suddenly found themselves at a competitive disadvantage compared to the SC that had innovated. This prompted them to also innovate in order to bridge the competitive lean/responsive performance gap.

That of innovation is an endless game, which over time drives the performance frontier increasingly to the right, thanks to the innovative action of rival SCs competing for the market in a specific product sector.

Table 2 Characteristics of a lean SC and a responsive SC

	Efficient SC	Responsive SC
Primary goal	Meeting demand while minimizing physical cost (procurement, production, logistics, stocks)	Satisfying demand by minimizing market mediation costs (out of stock; shrinkage)
Operational strategy	On forecast (Push) Procurement, production and distribution processes are carried out before demand arises and optimized thanks to economies of scale/scope linked to lots (procurement, production and/or distribution)	On demand (Pull) Procurement, production and distribution processes are carried out following demand and optimized thanks to flexibility (e.g. production and/or logistics flexibility) and economies of scale linked to volumes or deriving from the synchronization of flows (e.g. cross-docking)
Pricing strategy	Reduced margins since the price is the first reference lever for the customer; the competition is high (there are many who can guarantee the same product/service)	High margins, the leverage for the customer is the product (the competition is low) and/or the service (*The right product at the right time*)
Product design strategy	Guaranteeing performance at the lowest unit cost	Creating modularity to allow make-to-order processes, differentiation as far downstream as possible (postponement) and wide differentiation The unit cost is not a problem because the margin is high
Sourcing and procurement strategy	Worldwide Supplier selection on price and quality Economies of scale in procurement	Local Supplier selection on service performance (lead time, punctuality, availability, flexibility) and quality Just In Time (JIT) procurement
Production strategy	Productive economies of scale Low flexibility	High mix and volume flexibility to cope with demand variability; unit lots
Distribution strategies	Reduction of physical costs (Capex, stocks and transport) thanks to centralized warehouses, inventory pooling, load consolidation	Speed, dedicated deliveries; Economies of scale thanks to flow synchronization (e.g. cross-docking) or the use of 3PL/express couriers
Lead time strategy	Reducing lead times but not at the expense of costs	Lead times reduced even at the expense of costs

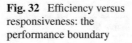

Fig. 32 Efficiency versus responsiveness: the performance boundary

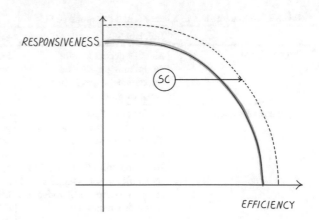

Risk-Hedging Supply Chain

When we talk about a risk-hedging supply chain, we mean one that seeks to minimize supply risk. This risk can be linked both to internal supply chain processes (for example a supplier's inability to provide a certain component due to the breakdown of its internal processes or controls; unavailability of production capacity on the operations side; unavailability of distribution capacity or FP on the distribution of its internal processes or controls), and to causes outside the supply chain (e.g., weather, catastrophes, pandemics, or other socio-political events).

To do this, supply chain risk-hedging typically uses:

- Upstream stocks: stocks of raw materials, components, key semi-finished products make it possible to eliminate the supply uncertainty deriving from the upstream part. In this way, the supply risk can be mitigated, since stocks make it possible to satisfy demand even in the event of a lack of supply of some components subject to supply risk (see the case of Procter and Gamble described in Sect. 4.4).
- Downstream stocks: strategic stocks of FP downstream are added to cope with the distribution risk, linked to a failure of downstream distribution processes and the consequent inability of the SC to deliver the finished product

Consequently, the SC strategy is based on:

- Procurement side: The development of a flexible and surplus supplier base. The attempt is to activate and maintain different suppliers for certain key components subject to potential supply risk, quickly switching from one supplier to another in the event of a lack of the product, adopting multiple sourcing or parallel sourcing strategies (see relevant paragraph) even to the detriment of product costs. The strategy is similar to the case of a lean SC, but in that case the multiple sourcing mainly had the objective of decreasing the purchase price of the component; in a responsive SC, on the other hand, single sourcing is used to try to increase responsiveness through partnerships, at the expense of price and the risk of interruption.

- Operations side: SC risk-hedging copes with supply uncertainty by using extra production capacity, typically oversizing the network, so that in the face of the impossibility of a plant to produce a product, there is always another plant capable of making it. Or else extra capacity from third parties is used, in the event that subcontractors are unable to supply semi-finished products, or certain types of raw materials/components.
- Distribution side: Also in this case it is possible to resort to distribution over capacity according to the principle of redundancy, to mitigate the supply risk on the demand side. The Supply chain manager of an important player in the Italian pharmaceutical industry such as, for example, decided to duplicate the number of DCs in Italy by creating one near Milan and one near Rome. This choice was also dictated by the need for responsiveness, but above all for risk-hedging. This decision was indeed far-sighted and strategic. During the first period of the Covid-19 pandemic, during which the Milan DC area was in full lockdown and therefore unable to work, Bayer continued to supply its customers (Distributors and Hospitals) by diverting all volumes to its Rome platform, maintaining and indeed increasing its present and future share of turnover.

Agile Supply Chain

These are SCs which seek to combine the need for reactivity in the face of demand uncertainty with the need for risk-hedging to cope with supply uncertainty.

A supply chain is "agile" in the sense that it is able to evolve rapidly and simultaneously cope with demand variability through responsive performance and with the risk of supply interruptions using the above risk-hedging strategies.

According to (Christopher 2000), an agile SC is the opposite number to a robust SC. The term "robust" is in fact synonymous with "lean", therefore a robust SC can minimize physical costs and do more with less (less than usual is identified with stocks, hence the "fat" from which the "lean"). All within a scenario of stable demand and supply. Faced with variations in demand or supply, a robust SC is unable to respond in terms of product/service/price, since it does not possess the attribute of agility nor can it reconfigure and do something different quickly and at low cost (responsiveness connected with variation in demand), nor activate different supply channels in the face of interruption of the ordinary ones which were active and optimized but were interrupted (risk-hedging linked to variation in supply).

An agile supply chain, in contrast, is flexible both upstream and downstream, in the sense that can evolve rapidly to find new assets as surrounding conditions change. This is true both for a change in demand (the product/service/price requested by the market changes) and in the face of uncertainty in supply to meet this demand (the procurement scenarios for raw materials and/or semi-finished products change).

The 4 Elements of Agility

According to Martin Christopher, who first introduced the concept of agility in supply chain management in the early 2000s, an agile SC is characterized by

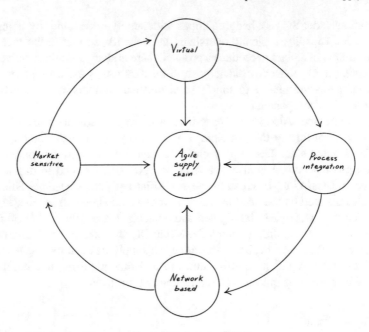

Fig. 33 Ingredients of agility, according to (Christopher 2000)

4 fundamental elements: demand data (*market-driven*), computerization (*virtual*), integration (*process integration*), mindset (*network-based*) as shown in Fig. 33.

- **Market sensitive**: The keyword is POSD—Point of Sale Data. The SC is able to read market demand data in real time and respond quickly to it. This is therefore a highly reactive SC (order-driven-Pull) rather than speculative (forecast-driven-Push), which works without stocks, and is consequently extremely fast in providing the market with whatever is required. It is therefore essential that the market demand data (POS Data in reality) are available immediately and at all levels. In an agile SC, great attention is paid to the concept of flow and the reduction of lead times, precisely to avoid having a flooded pipeline of semi-finished and FP raw materials that may be obsolete; nonetheless there may well be extra stocks of key parts to cope with supply uncertainty.
- **Process integration**: The key word is integration of logistics processes in a broad sense (*procurement, operations, distribution*). To be really order-driven rather than inventory-driven, an agile SC is characterized by a close cooperation between the various links of the supply chain in terms of the processes.

  For example, the development of new products takes place in a partnership with suppliers, to absorb as much as possible the impact on defects and quality instability. The attempt is to leverage the supplier's core competence to create highly innovative products, in which not only is the demand known, but also the few suppliers who are capable of manufacturing the components. On the distribution side, the processes are integrated and based on trust and transparency.

It is recognized that the coordination of the process can achieve higher levels of efficiency (e.g. a VMI process between the focal company and the retailer to coordinate the flow of finished products—the customer transparently shares stock and consumption data; or coordination and distribution by express courier in which everyone focuses on their core competence). There is no need to duplicate stocks and outbound/inbound processes at the boundaries of SC players, that typically witnesses a lack of integration.

- **Virtual**: The key word is ICT integration: extensive use is made of information technologies in an extensive and integrated way among all the players, in which the demand and inventory data are shared, the data is transferred at all levels in an integrated way via computers (EDI). This is an SC that is information-based rather than inventory-based, in the sense that stocks are not used but information is shared to reduce variability (e.g. reduction of the Bullwhip Effect through the sharing of POS Data and not through stocks).
- **Network-based**: In this case there is a mindset aspect. It is recognized that the supply chain is made up of a network of companies connected by flows of products, information, services and money, and that the competition is between networks and not between companies within the same network. Hence the concept of *extended enterprise*. Collaboration with suppliers and customers, upstream towards suppliers and downstream towards retailers and third parties, as opposed to a muscular commercial relationship, is part of the corporate culture and is recognized as a necessary condition to create more efficient processes.

Nowadays, To be able to achieve this performance, agile supply chains make extensive use of 4.0 technologies.

On production and order fulfilment lines, highly flexible autonomous robots work side by side with people to automate and optimize manufacturing and assembly processes, but also logistics processes from picking to delivery (e.g. Kiva robots, or autonomous drones in the case of Amazon).

The data is collected with accuracy and in real time from the field by adding sensors, e.g. RFID tags, which allow independent communications with the machines while collecting data on flows and stocks in real time at any point in the supply chain, in a speedy, selective, and precise manner.

The huge amount of data is gathered on the Cloud based data warehouses, and made available to various stakeholders, according to the concepts of virtual and process integration looked at previously. The same data are also available for AI (*artificial intelligence*) systems which, thanks to machine learning programmes, are able to analyse them, interpret them, and simulate predictive scenarios (through the so-called "digital twins" of machines and processes), recommending to the process manager possible production, logistics and retail scenarios which the decision-maker need only validate or correct. In this way, it is possible to fine tune and optimize the processes.

Lastly, through the integration of systems, software systems and machines in general, data are exchanged in real time between machines via EDI (e.g. *forecasts, demand, inventories, orders, order confirmations, despatching and receiving advices, invoices*), eliminating the time once needed for manual activities, which were inaccurate and also laborious. The data, typically product master data, are unique and shared in order to eliminate all the inefficiencies due to the misalignment of information.

3.5.3   Achieving Strategic Consistency

To achieve strategic consistency means aligning the characteristics of the product in terms of demand variability and supply variability with the performance that the SC is able to provide, by choosing appropriately between responsiveness, efficiency, risk-hedging and agility.

The following paragraphs illustrate in detail the areas of strategic consistency, areas in which the characteristics of the supply chain are optimally adapted to the characteristics of the product. Other combinations are, conversely, inconsistent, in the sense that the characteristics of the supply chain do not optimize the priorities to be managed, given the characteristics of variability and demand and supply uncertainties which characterize the product.

Strategic Consistency in the Case of Stable Supply Processes

Functional Products with Stable Supply Processes

If the demand variability is low and therefore the product is functional, and also the supply uncertainty is low and therefore the supply processes are stable, physical costs prevail over market mediation costs. The SC must therefore be lean, i.e. oriented towards minimization of physical costs (see Fig. 34 for details).

The SC will be programmed on forecast, taking advantage of the low demand and supply variability without incurring high MMCs. The stability of lead times, even

Fig. 34 1st quadrant lean SC for low supply and low demand variability

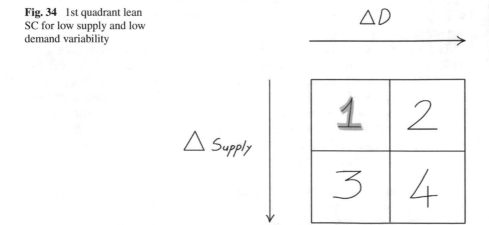

if long, allows application of such rolling pipeline programming techniques as DRP (*Distribution Requirement Planning*).

Economical lots of Procurement, Production and Distribution will allow the generation of economies of scale of both a production and logistic type, necessary to keep the physical cost, and therefore the prices, low.

The supply base is stable in terms of suppliers and the quality of raw materials and components. Availability is extensive and worldwide. Suppliers are selected on price and service, and the planning of procurement and distribution takes place through techniques such as MRP and DRP, exploiting the low demand and supply variability (in terms of lead time, as mentioned) which allows production and distribution plans to be frozen, even for a long time, without causing major problems.

The low demand variability means that forecast errors are few and therefore also the opportunity costs associated with OOS or shrinkage—deriving from underestimated or overestimated demand, respectively—are negligible.

Innovative Products with Low Supply Uncertainty

First of all, it must be emphasized that here we are in a situation where the supply process is stable, and therefore the broad supplier base, quality and lead time are stable, and so the risk of interruption is not a critical issue. It follows that it is unnecessary to adopt risk-hedging strategies in this case.

As for demand, if the demand variability is high and therefore the product is innovative, then MMCs will prevail over physical cost. The SC will need to be responsive (see Fig. 35).

It will be necessary to eliminate OOS and unsold goods (demand is unpredictable and product life cycle is short) by doing exactly what is required at that point in time. If the supply base is stable and the processes are stable, in order not to generate unsold goods, the SC is activated in response to product demand, if possible reading

Fig. 35 2nd quadrant responsive SC for high demand and low supply variability

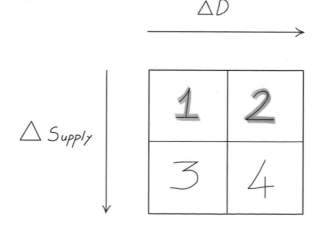

the market data in advance and, in any case, as soon as possible, in order to take timely action.

To be able to deliver on time, the flow must be fast (lead time minimization). The attempt is to minimize the component related to the physical cost through the synchronization of flows (cross-docking and economies of scale linked to volumes).

Lead Time Gap

When dealing with extremely variable demand and there is a need to produce Just in Time, therefore adopting a Pull perspective, in order not to incur MMCs the issue is to understand whether the SC lead time is compatible with the lead time of the final consumer. The consumer lead time (CLT) is shown in Fig. 36 and it is understood to be the time that the consumer, in the face of a purchase price p, is willing to wait between the moment he/she issues the order and the moment he/she receives the compliant goods. CLT is zero in the case of physical stores, where the consumer goes to the location and leaves with the goods purchased, while it can be non-zero in the case of an online transaction.

In other words, the product can be procured, manufactured, and distributed on demand, completely eliminating market mediation costs, only if the time that the supply chain takes to procure, produce and distribute the product is less than the time that the consumer is willing to wait between ordering and receiving the goods. Otherwise, it is necessary to manage one part of the supply chain in a Push on-forecast mode and another part in a Pull on-demand mode, and then keep raw materials and semi-finished or finished products in stock, depending on the mismatch between the values of the two lead times.

According to (Christopher 1992), in particular, indicating with SCLT the lead time of the SC to procure, produce and distribute the product (Fig. 37).

The Lead Time Gap (LTG) can be defined as the difference between the SCLT and CLT (see Fig. 38).

If the LTG is zero or less than zero, the product can be procured, produced and distributed to order in a Pull logic, otherwise it is necessary to procure/produce/distribute in advance on-forecast, in order to have a buffer of raw

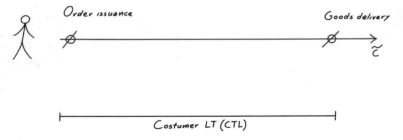

Fig. 36 The customer lead time

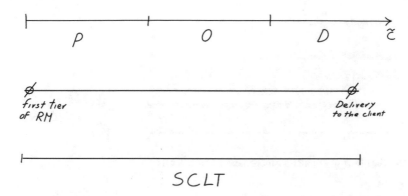

SCLT

Fig. 37 The SC lead time

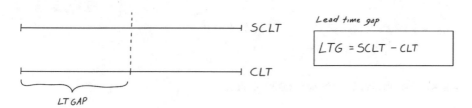

Fig. 38 Definition of lead time gap

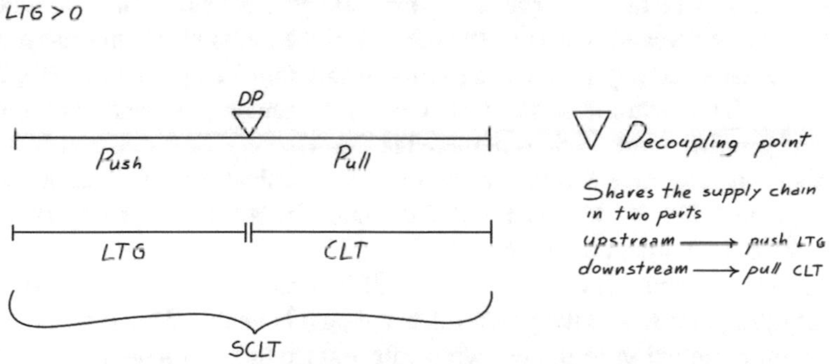

$LTG > 0$

Fig. 39 The SC lead time gap and the decoupling point

materials, semi-finished or finished products available (depending on the extent of the LTG).

Decoupling Point (DP) and Demand Penetration Point (DPP)

We talk of a physical decoupling point to identify the point in the supply chain which divides it into two parts, as shown in Fig. 39.

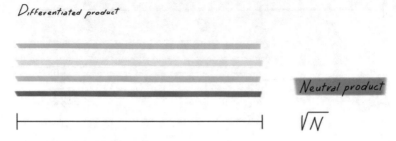

Fig. 40 The effect of product postponement on stock levels

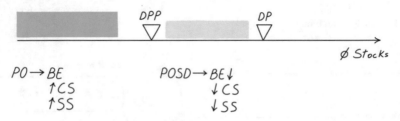

Fig. 41 The effect of the DP and the DPP on stocks

The upstream part is lean, working in a Push-on-forecast perspective and trying to minimize physical costs, creating a "product" with a stable and predictable demand; this product may be a raw material or a component where the raw material is characterized by stable demand thanks to the pooling of the demand for finished products being exploited, or it may be a neutral semi-finished/finished product where, thanks to modularity and customization in view of postponement, demand variability for the neutral product is reduced, as depicted in Fig. 40.

In this Push part, the attempt is to optimize the physical cost of stocks by sharing the demand data. In this way it is possible to reduce the bullwhip effect by estimating the demand not on orders but on real data.

We talk about Demand Penetration Point (DPP) (or *Information Decoupling Point*) to identify the point where the market demand data is visible. Downstream players benefit from a reduction in the bullwhip Effect since they estimate the demand on POS Data; upstream players estimate demand on orders and therefore need to keep more cycle and safety stocks (see Fig. 41).

The more the DPP is pushed upstream, the greater the benefit in terms of a reduction in the stock component on physical costs.

The downstream part of the DP, on the other hand, is reactive, and therefore operates from a Pull point of view. The reactive part of the SC moves on demand and tries to minimize the MMCs, providing the product requested in the time required starting from supply of the upstream part. This part attempts to absorb the physical cost part by means of flexible production lines (if the operations are also Pull), while for the distribution part there is a synchronization of flows—cross-docking—and/or the use of economies of scale on the volumes of 3PL—consolidation/deconsolidation.

Closing the Lead Time Gap

Closing the lead time gap effectively means aligning the SCLT with the lead time of the final consumer, therefore allowing the supply chain to work in a Pull perspective, eliminating the part that must anticipate demand and work in a Push perspective.

To try to reduce the LTG and operate as much as possible in a pull rather than a push perspective, two possible ways exist:

- **Business Process Reengineering**: as depicted in Fig. 42, processes, phases and activities are systematically analysed in terms of time and value, identifying those processes which add time without increasing value, trying to: (i) eliminate a process if it does not add value; (ii) reduce the time required through BPR techniques. In this way, operating phase by phase, activity by activity, it is possible to impact the overall SCLT and therefore to reduce the LTG.

  A Pareto Analysis may prove useful, since the 80/20 rule is often valid, i.e., 20% of the processes consume 80% of the lead time of the SC LTSC, and therefore by acting on these it is possible to substantially reduce it.

  Also because it often happens that some inefficient processes are due to "inertia": a certain activity or process always continuing to unfold in the same way over time; perhaps it had originally been designed efficiently, but changing surrounding conditions have led it to being inefficient (see Annex 4.3 "*Why Did We Design Inefficient Processes?*").

- **Data sharing**: If the part that works in a Push perspective upstream of the DP knows the demand and stock data, then it can see the demand trend in advance of the order and so plan to be ready for an appointment with the order. Which is why the sharing of demand and stock data downstream of the DP is essential. Thanks to the sharing of downstream information, the upstream processes from

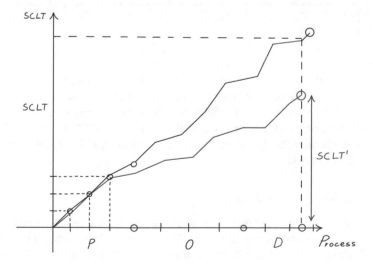

Fig. 42 Closing lead time gap through business process reengineering

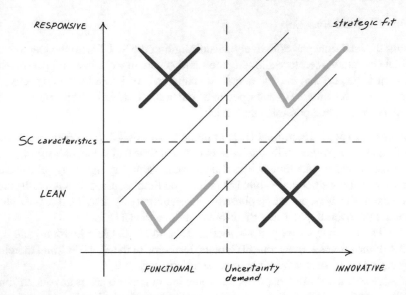

Fig. 43 The fisher matrix: match and mismatch areas

speculative actually become reactive, since the visibility eliminates any forecast uncertainty. Thanks to this visibility, it is possible to know in advance what will be required and when, and it is therefore possible to operate the SC in advance to arrive on time for the appointment, without the risk of incurring MMCs.

Areas of Strategic Consistency

There is therefore a strategic consistency matrix (Fisher 1998), as shown in Fig. 43, which highlights the areas of strategic consistency and those areas which lack strategic consistency.

It is the BoD's task to first of all define the competitive strategy (product/price/service) and therefore understand where the product is positioned on the scale of the demand and supply variability. In particular, in relation to the uncertainty of the question whether the product is functional or innovative. Then, once the strategic consistency characteristics have been identified, steering all the corporate departments which contribute to the Porter's Value Chain towards the same objectives of effectiveness (product/service) and efficiency (cost—MMC or PC?), as represented in Fig. 44.

Evolution of Strategic Consistency

Products evolve throughout their life cycle. At the beginning of its life cycle, a product is typically innovative, and with the passage of time, through introduction and development, it typically ends up in an area of maturity where it tends to become functional.

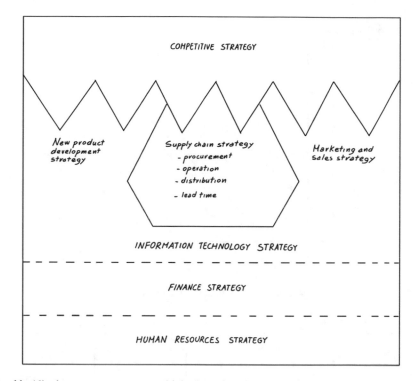

Fig. 44 Aligning corporate strategy with business function strategies

To maintain strategic consistency, at the same time the SC will need to evolve (see Fig. 45) from an initial situation of reactivity during introduction and development (*to cope with an unpredictable demand but also a poor competitiveness that ensures higher margins*) to one of leanness during maturity and the onset of decline, characterized by predictable demand, but low margins due to high competitiveness.

At the same time, a SC can deal simultaneously with functional or innovative products.

The strategy in this case must be mixed, and therefore it is necessary to manage the portion of innovative products with a reactive strategy, and the portion of functional products with a lean strategy, maintaining, wherever possible, certain common traits in order to achieve economies of scale.

In Fig. 46 we represented the Zara case. Zara, simultaneously manages continuous products (blue t-shirts) that are typically functional with stable and predictable demand, and seasonal products (new collections) which are innovative, with difficult-to-predict demand. Zara handles innovative products in a different way from functional products. The innovative products are made in Spain and North Africa with suppliers whose lead times are therefore extremely low, using local suppliers selected on the basis of their reactivity. Functional products, whose demand is easily predictable, are made in the Far East in a leanness logic, since production can be

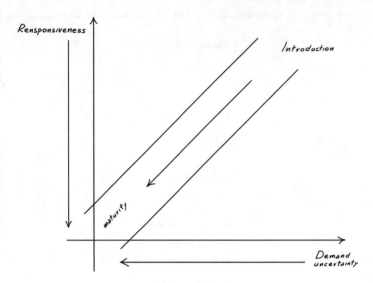

Fig. 45 Evolution of strategic consistency with product lifecycle

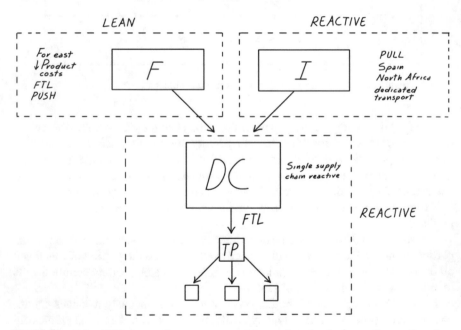

Fig. 46 Zara mixed SC—responsive procurement and manufacturing for innovative products, lean procurement and manufacturing for functional products, same reactive for distribution

planned in advance in an DRP logic, given that the demand is predictable and lead time is predictable. In this case, the suppliers are chosen on a worldwide scale with a view to minimizing the cost of raw materials. The distribution part for the two products is instead shared.

Strategic Consistency in the Case of Evolving Supply Processes

Functional Products with Evolving Supply Processes

When raw materials and components are still characterized by supply uncertainty, in terms of quantity (e.g. production or meteorological shortages that impact quantities) or in terms of lead times, with unstable quality and reliability levels which could potentially introduce high quantities of waste and disruptions in production processes, to achieve strategic consistency the SC must be of the risk-hedging type (see Fig. 47).

In this case, the most efficient strategy for an SC is to create a decoupling element, as the one depicted with a triangle in Fig. 48, capable of absorbing supply uncertainty, so that the downstream part can continue to operate in a lean perspective, since the supply uncertainty is absorbed upstream.

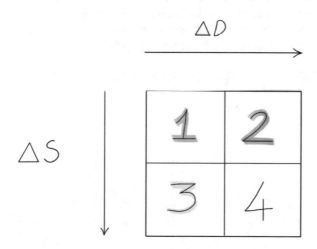

Fig. 47 3rd quadrant risk edging SC for high supply and low demand variability

Fig. 48 Absorbing supply variability allows the downstream part to become lean

The decoupling element can be:

- Stocks—in the event that the uncertainty derives from the lack of a product. The cost of keeping a key component in stock in excess quantities can be high, but in this way the supply risk is absorbed and continuity of flow to the downstream part is guaranteed.
- A multiple supplier base—in the event that the supply uncertainty is represented by a low level of quality, reliability or ability to respect lead times by suppliers (in this way it is possible to switch flexibly from one supplier to another, by reason of a supplier's inability to cope with the required product/service performance). Sourcing, thus activating and managing multiple suppliers can involve higher costs, but it does mitigate the risk of interruption.
- Extra production capacity—a third-party production surplus is used to cope with the inability of the subcontractors (3PM, co-packer, façon) to meet the demand for a specific semi-finished/marketed product. Third parties are always close at hand, i.e. local, to keep lead times shorter and mitigate the risks of non-supply, even at the expense of higher labour costs compared to offshore production.

Innovative Products with Evolving Supply Processes

These are typically highly innovative products, in which neither the market demand nor the ability of the (few) suppliers to supply reliable products within the required time frame is known, given that the product is brand new.

In this case, the SC strategy necessary to cope with the demand variability and supply processes is to adopt a chain that is agile (see Fig. 49).

All the comments brought in for responsiveness to cope with demand uncertainty, and risk edging to manage supply uncertainty apply here.

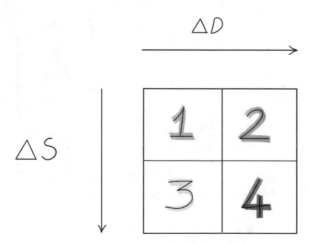

Fig. 49   4th quadrant agile SC for high demand and supply variability

The Areas of Strategic Consistency

In summary, in light of the analysis of demand and supply uncertainty on the one hand, and the characteristics of the SC on the other, the areas of strategic consistency will be as shown in Fig. 50

It is the CEO's task to achieve strategic consistency and then proceed in four steps:

1. Identifying the characteristics of the product/service/price on which it is desired to compete in order to acquire a specific consumer target
2. On the basis of product/service/price characteristics, identifying supply and demand uncertainties
3. Identifying a congruent SC strategy (lean, reactive, risk-hedging or agile) and
4. Organizing the various corporate departments in a manner consistent with the SC strategy.

Expanding Strategic Consistency

However, achieving strategic consistency within the company is not enough; it is necessary to expand it at the SC level upstream (suppliers) and downstream (distributors and retailers), so that all the players of the SC are aligned towards a single objective of minimizing physical cost or market mediation cost.

The goal is to minimize the overall cost and not that of the individual players, so the attempt is made to operate according to a coordinated and shared approach, i.e.,

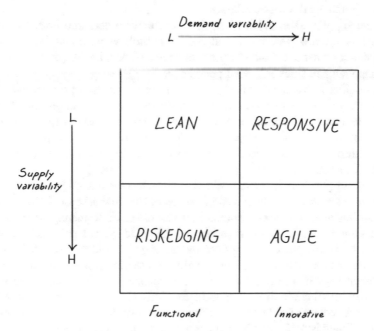

Fig. 50 The areas of strategic consistency/fit (Lee 2002)

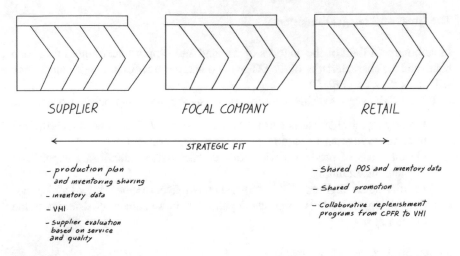

SUPPLIER   FOCAL COMPANY   RETAIL

← ——————————————— STRATEGIC FIT ——————————————— →

- production plan
  and inventoring sharing
- inventory data
- VMI
- Supplier evaluation
  based on service
  and quality

- Shared POS and inventory data
- Shared promotion
- Collaborative replenishment
  programs from CPFR to VMI

Fig. 51 Expanding the strategic consistency throughout the SC

one in which processes are coordinated and information is shared, so that the SC as a whole can achieve an overall goal of excellence, as shown in Fig. 51.

On the retail side, for example, manufacturers and retailers share sales and promotion plans in a coordinated manner to reduce variability, while retailers and manufacturers share demand and stock data and adopt advanced collaboration programmes (e.g. CPFR and VMI, cross-docking).

On the supplier side, production plans, customer orders and inventory data of components and raw materials are shared in the same way, so that the supplier can better predict demand and plan its production in advance, i.e. keeping raw materials and components in stock in a VMI perspective.

The supplier is evaluated not only in terms of price but also of service and quality. In this way, the components do not need to be checked but can go directly online without delaying production or requiring additional reworking (e.g. a low-price supplier could make a mistake in applying identification media—component/garment labels, so that the garments need to be reworked, causing delays and extra costs and potentially lost sales).

Indeed, supply chain management (SCM) involves the integration of all key business functions across the whole chain of stakeholders. In this sense, SCM represents a way for business functions involved in the whole chain to manage their activities to extract maximum value for end users and stakeholders. The process of delivering a product to an end customer involves monitoring and control of the flow of products, inventories and the flow of information and service from production facilities of upstream suppliers to retail stores, in accordance with the competitive strategy set. Production schedule streamlining, inventory management, pinpointing of bottlenecks, improvement of order response time and supply chain lead-time compression, aimed at PC and MMC optimization are among typical SCM goals.

Mentzer et al. 2001, synthesized all the concepts above in SCM definition. They stated that SCM is meant to be *"the systemic, strategic coordination of the traditional business functions and the tactics across these business functions within a particular company and across businesses within the supply chain, for the purposes of improving the long-term performance of the individual companies and the supply chain as a whole."*

This is the SCM definition adopted in this book.

4 Annexes

4.1 Post-it: The Story of an Accidental Multi-million Dollar Invention

Post-it Notes were invented by two ordinary employees, Arthur Fry and Spencer Silver, who were working as researchers at 3M: a company with a truly diversified production, a leader in markets that are often not close to one another in any way.

In 1968, Silver mistakenly made a glue which did not work very well, in fact it did not stick as strongly as the one he had been supposed to invent.

The glue that came out of those experiments was really special, it stuck the surfaces it came into contact with together and then they could be separated without damaging them, but being so weak Silver could not find a use for his formula. He did not throw it away, however, he filed it among the failures awaiting better times. Then he thought that perhaps by speaking to his colleagues a use could be found, and so he began to talk about his discovery in the office.

And so it was that during a company presentation, Silver introduced his glue with a weak grip to Arthur Fry, one of his colleagues present. And it was Arthur Fry, fellow 3M researcher, who figured out how to commercialize Spencer Silver's invention. To do this, as so often happens, he began from a necessity. Because as we know, necessity is what moves people and creativity.

We are in Minnesota, at the North St. Paul Presbyterian Church. Fry sang in the choir there and during the rehearsals at the church, he easily lost the place, because the coloured sheets he used to mark the songs and pages would slip out of his book. Fry came up with the idea that was going to change his life forever. If he had used his colleague's glue to bookmark the pages, the hymn book would not have been damaged and the bookmarks would not have been lost. The idea was brilliant: Bingo!

The development and commercialization of the product were very longwinded. In fact, 3M executives did not believe in Fry's brainwave at all, instead, they were genuinely convinced that the Post-its would be a fiasco because for them, people would prefer traditional bookmarks and would never pay for Post-its. And so for several years absolutely nothing happened.

However, Fry never stopped working on his project. Installing a machine in his apartment he applied Spencer Silver's low-strength glue to some sheets of paper.

Convinced that problems are an integral part of innovative processes, at regular intervals he returned to the office and suggested the company do a test. But without the support of the management, no matter insistent he was, Fry could do nothing. So as soon as he had some Post-its at his disposal, he gave them to the executives who started using them.

In 1977, 3M finally said Yes, and a test launch was made which unfortunately did not yield satisfactory results. At the time, it was thought to use this glue to attach and then detach A4 announcements on company bulletin boards, and so the test was done by promoting the product for this type of use.

However, on analysing the results, which were far from encouraging, the 3M managers realized that the Post-its were being used in the company in a totally different way from the original idea. In fact, colleagues and managers had begun to use Post-its not as bookmarks or for the company bulletin board but as a new way of communicating. Messages were left for one another, written on Post-its and stuck almost everywhere. There were those who used them to mark important points in some documents, putting an arrow on them and those who wrote down personal considerations near the paragraphs where there were things to be borne in mind. At this point, 3M converted the product name to "Post-it Notes", replacing the previous "Press and Peel" and re-scheduled a new test in a Virginia city to see how people would respond to this new cue.

Promoted in a different way, the cute sticky notes were very popular, and the test ended with a very encouraging result. However, it was not until 1980 that Post-it Notes were officially launched by 3M. Over 1,000 different versions have been released since then. A global success which continues today.

From that day on, Fry began to make a career for himself, remaining for over 40 years at 3M until the day he retired. 3M is a company which, in its time, has always invested in research and development, hedging its bets on the ideas of over 7,000 scientists, registering many patents ranging from Post-its to CFC-free dispensers, to adhesive tapes and even adhesives for dentures.

3M was one of the first companies to allow employees to invest a part of their working time in personal projects. A common practice among successful companies which is increasingly being taken into consideration, given the results it is capable of generating. Suffice it to say that 3M has registered 118,000 patents in its history and that it still registers 4,000 patents a year.

So this story can teach us some interesting things.

From an entrepreneurial point of view, if you let people do what they know how to do, the results arrive and can even be unexpectedly interesting. It is necessary to have faith in the talents you have sourced on the market, and give them room to flourish.

The other interesting point, related more to the marketing and promotion of a product, teaches us that success is not necessarily explosive. The spark can take time to arrive. The same product can be promoted in a way that does not produce the desired effect because it does not actually meet any real customer needs. On the other hand, when real problems are solved, when the customer's point of view is taken into consideration, success is not long in coming, without needing to spend millions and millions on promotion and advertising.

4.2 But Does Volkswagen Actually Need a New Phaeton? (Al Volante 2015)

The question is legitimate if we think that the current generation has never sold in sufficient numbers to repay the investment.

AMBITIOUS CONFIRMATION—Although without providing details on the project, the Volkswagen company has confirmed that it is developing a second generation of its flagship Phaeton. According to the Reuters agency, citing sources inside Volkswagen, the new model is designed to compete with the Mercedes S and there will also be a plug-in hybrid version in the range. Its commercial launch has not yet been precisely defined, but is expected to take place between 2017 and 2018.

CONTRADICTIONS—The news triggered a series of considerations on whether or not to invest so much in a Volkswagen-branded flagship. This is because the Phaeton project, in addition to being very expensive, has never proved a success, and indeed it seems uneconomical to propose it again when the company is committed to a cost-containment strategy. In fact, for 2017, Volkswagen had set itself the goal of saving at least 5 billion Euro a year, which meant giving up on models that did not make money. But the Phaeton, it seems, will survive despite being judged one of the three least profitable cars of modern times, with a loss of around 28,000 Euro for each model sold between 2002 and 2012.

The Phaeton dates back to 2002 and was restyled in 2010.

HIGH COSTS, LOW SALES—The costs of designing and developing the model (including the famous "glass factory" in Dresden built specifically to produce the Phaeton) amounted to one billion Euro. And this production reality was accompanied by sales which were constantly far below the estimated 20 thousand units per year. Volkswagen does not provide sales figures for the Phaeton, but production figures in 2013 (latest data available) did not exceed 5,812 units.

4.3 Why Did We Design Inefficient Processes? (Hammer 1990)

"In a way, we didn't. Many of our procedures were not designed at all; they just happened. The company founder one day recognized that he didn't have time to handle a chore, so he delegated it to Smith. Smith improvised. Time passed, the business grew, and Smith hired his entire clan to help him cope with the work volume. They all improvised. Each day brought new challenges and special cases, and the staff adjusted its work accordingly. The hodgepodge of special cases and quick fixes was passed from one generation of workers to the next. We have institutionalized the ad hoc and enshrined the temporary. Why do we send foreign accounts to the corner desk? Because 20 years ago Mary, who spoke French, was in that desk. Today Mary is long gone, and we no longer do business in France, but we still send foreign accounts to the corner desk.

Why does an electronics company spend $10 million a year to manage a field inventory worth $20 million? Once upon a time, the inventory was worth $200 million, and managing it cost $5 million. Since then, warehousing costs have escalated, components have become less expensive, and better forecasting techniques have minimized units in inventory. But the inventory procedures, alas, are the same as always. Of the business processes that were designed, most took their present forms in the 1950s. [*The article in question is from 1990*, ed.]. [...] Nearly all of our processes originated before the advent of modern computer and communications technology. They are replete with mechanisms designed to compensate for 'information poverty'. Although we are now information affluent, we still use those mechanisms, which are now deeply embedded in automated systems."

4.4 Procter and Gamble a Billion Company Made Through Supply Chain Resilience

In 1837, William Procter, a candle maker, and James Gamble, a soap maker, formed the company known as Procter & Gamble. The two men, immigrants from England and Ireland respectively, had settled earlier in Cincinnati and had married sisters. They decided to pool their resources to form their own company.

Today the company has $218 billion in market capitalization and 121,000 employees. The group's worldwide turnover amounted to 83.6 billion dollars last year. An estimated 4.8 billion people worldwide use P&G products. Some examples: Dash, Ariel, Ace for the laundry; Mastro Lindo, Swiffer, Viakal for household cleaning and hygiene; Pantene, Oral B, AZ, Gillette for personal care; Pampers and Tampax for diapers and feminine hygiene; Vicks for over-the-counter drugs; Duracell for batteries; Braun for small appliances. In all, the group has 25 brands which each generate a turnover of over $1 billion a year.

In 1837, starting a business was difficult. Although Cincinnati was a thriving marketplace, the nation was in financial turmoil. Across the country, hundreds of banks had closed their doors and there was widespread concern that the United States was bankrupt. Even so, Procter and Gamble launched their new venture, worrying more about how to compete with other soap and candle makers than about the great financial fear that was shaking the country.

For example, oblivious to rumours of an impending civil war in America, they decided to build a new factory to support their growing business.

In the 1840s Cincinnati was the largest meat packing centre in the United States, The main ingredient for soap and candles was animal fat, which was readily available from the Cincinnati Hog Slaughtering Industry.

At the outbreak of the Civil War, supplies from the south of the country of raw materials for the production of soap and candles were abruptly interrupted, but at Procter & Gamble, thanks to the availability of animal fats deriving from slaughter, the availability was such as not to interrupt the production.

While competitors were interrupting productions and closing their activities, SC resilience allowed P&G not only to continue to operate, but also to won contracts to supply the Union army with soap and candles. In addition to the increased profits experienced during the war, the military contracts introduced soldiers from all over the country to Procter & Gamble's products.

Once the war was over and the men returned home, they continued to purchase the company's products.

Chapter 4
Representation of the Supply Chain Through Information Flows

1 Introduction

A third way of seeing the SC, in addition to representing it through physical flows and through the corporate business functions involved, is through the information flows that characterize it and which the SC players exchange among themselves.

These information flows are fundamental since they represent the 'signals' which allow the SC to carry out its product and service flows. In the same way that the nervous system of an organism carries signals from the senses to the brain and from the brain to the muscles to act accordingly, information flows collect signals from within the supply chain and outside the market, and then carry them to the various players, who act to bring the product/service to the final consumer. And just as in the physical world, where the clearer and faster the signals are, the more prompt and consistent the responses, also in a supply chain the more transparent the information is and not distorted by the transfer mechanisms (e.g. we will also call it Bullwhip Effect), the more the product and service flows are effective and efficient.

The information flows include:

- PO—purchase order
- OC—order confirmation
- BoL Bill of Lading—Transport document TD
- DESADV/RECADV—Despatching Advice/Receiving Advice
- POS—Point of Sales Data
- Inventory—stocks.

The information flows are bidirectional, in the sense that they can move both forwards from upstream to downstream and backwards from downstream to upstream in the SC.

BOLs, DESADVs and OCs are typically forward flows; while POs, RECADVs, and POS Data are typically backward flows; Inventory data is typically bidirectional. Each of these flows will be the subject of a dedicated paragraph later in this chapter.

2   Purchase Order and Confirmation Order

2.1   *Definitions and Functionalities*

The Purchase order is the mechanism by which the flow of a product within the SC is activated. It is a backwards flow which starts downstream at a buyer and goes upstream to a vendor.

The mechanism applies to any level of the SC: retailer focal company; focal company Tier I supplier, Tier I supplier Tier II supplier, and so on.

With reference to the scheme in Fig. 1, in which is represented an example of the retailer and manufacturer, the process takes place as follows:

1. A **buyer** realizes that following the demand for a product downstream, in the case of a retailer, for example, on the part of the final consumer, it is necessary to replenish to restore stock levels;
2. **PO**: The **buyer** prepares and issues an upstream order PO to a vendor (backward information flow) in which it specifies the product and service needs;
3. **CO**: Following this order, the **vendor** issues a Confirmation Order CO with which it accepts the order and its conditions;
4. **P + S; DESADV**: The **vendor** prepares and sets up the order lines for that order and transfers product/packaging plus service downstream, along with related information flows (TD, DESADV).
5. **RECADV**: After receiving goods, the **buyer** signs the TD or sends an electronic receipt confirmation called a Receiving Advice. The purpose is to inform the vendor of the goods received and accepted.

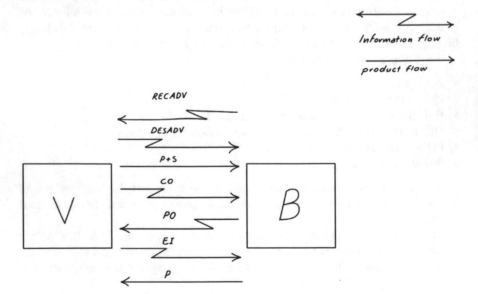

Fig. 1 The flow of information related to a purchase order; in bold the activity's owner

6. **EAI**: Based on this information, the vendor prepares an electronic active invoice EAI.
7. **P**: The **buyer** receives the passive invoice, checks it, and pays it.

2.2 The Structure of the Purchase Order

The order must be formalized through a structured document, on paper rather than electronic, which consists of three essential parts:

*   Heading
*   Body
*   Summary.

This structure contains information relating to what the buyer intends to order from the vendor (which product in what quantity) and the service it intends to receive from the vendor (lead time, place of delivery of the goods, physical state, etc.).

The Heading part contains the general information of the document. Particular parts include:

*   Issuer: the company name of the legal entity issuing the order which will typically be invoiced for the goods, unless specified otherwise in the summary;
*   Recipient: the company name of the legal entity that receives the order and will have to fulfil it.
*   Issue Date: the date when the order is issued and to which reference will be made for exchanges of information;
*   PO number: the unique identification number of the PO. The numbering system is arbitrarily structured by the buyer according to its own internal logic; typically this is a numerical or alphanumeric progression. It is linked 1:1 with Date, Recipient and Issuer. The PO number is important because it is referred to for all subsequent transactions both of a physical type (*e.g. bill of lading for the shipment of goods*) and of an administrative type (e.g. invoicing).

The header part in the body of the order follows. The Body contains information on what the buyer intends to order from the vendor with that particular PO, the quantities, and the relevant prices.

The Body is presented in the form shown in Fig. 2.

It is structured as a recordset of order lines, in which the fields are:

*   Order line number: a unique progression (e.g. 100, 110, 111, 200, 300) at the buyer's discretion, with which the single order line is uniquely identified;
*   Reference code: the unique code with which the single reference is univocally identified.[1] In an open many-to-many system such as FMCG, it is important to adopt a standard identification system, e.g. the GS1 standard, to avoid ambiguity in identifying the reference. In this case, for example, it is possible to use on of

---

[1] Two articles can have the same reference if they differ at most in the expiry date.

ID LINE	REFERENCE CODE	DESCRIPTION	QUANTITY	PRICE
10	$EAN13_{10}$	$Desc_{10}$	Q_{10}	P_{10}
20	$EAN13_{20}$	$Desc_{20}$	Q_{20}	P_{20}

Fig. 2 The structure of the body

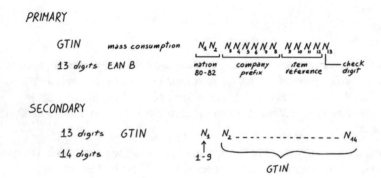

Fig. 3 The GTIN structure for primary and secondary packaging

the GTINs of the sales unit (EAN 13 or EAN 8 if ordering individual items) or of the secondary package (EAN13);

The structure of the GTIN code (GS1 italy 2021a) and its national prefix, company prefix, item reference and check digit structure for primary and secondary packaging is shown in Fig. 3:

- Further insights about the identification of products in FMCG will be given in paragraph.
- In a closed SC, for example in the textile and clothing sector, it is possible to use the supplier's own item code, e.g. the SKU (unique combination of size, variant, model);
- Description: it may be useful to add a description of the item for each order line, to be used in case of ambiguity. In any case, this is a secondary field, not to be used except to resolve any ambiguities, precisely because of the subjectivity linked to the description;
- Unit of Measurement: this field is very useful should ambiguities arise in the quantities. For example, in the FMCG case, the same item (secondary but also primary packaging—suffice to think of multipack products) can be characterized by different units per package. In the event of any ambiguity, it is always a good idea to specify the unit of measurement (pieces, items, packages, kg, l) or the units per package required for each order line;

– Quantity: indicates the quantity required in the unit of measurement specified in the previous field (e.g. 3 packages, two pieces, 1 kg, etc.);
– Unit Price: this is the unit price of the single reference which, multiplied by the quantity will determine the price of the order line. This can be agreed on a one-off basis, or it may have been previously negotiated for the specific transaction to which the PO refers.

The Summary part completes the structure of the PO. The Summary typically contains the buyer's general condition of purchase, as well as information relating to the service that the buyer intends to receive from the vendor. The fields contained in the Summary are therefore typically the following:

• Requested Delivery Date: the exact date (neither before nor after) on which the buyer wishes to receive the goods. E.g., "15 days from order date" or "15 days from order confirmation". The difference between the date of the PO, or rather of the CO, and the delivery date, therefore, represents the lead time expected by the buyer;
• Physical place of delivery, i.e. using GLN standards (GS1 italy 2021c) and Conditions for return of goods: This is an important parameter since it identifies who bears the burden and risk of transporting the goods, and to what extent. Precisely to avoid ambiguity, it is important to use a univocal standard, namely, Incoterms 2021. These are standard methods to deliver goods that uniquely identify who will bear the charges for transport and customs clearance for export and import, as well as who will bear the responsibilities for the transport of the goods in the event of accident/damage/loss. This part is thoroughly addressed in Chap. 5; however it is summarized in Fig. 4. The methods of goods transport can range from EXW

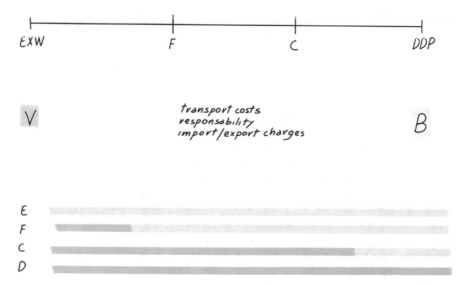

Fig. 4 Incoterms 2021: transport costs, responsibility and import/export charges depending on how the goods are returned

(Ex-Works) in which the goods are made available at the vendor's loading dock, and therefore the import and export transport costs and the risks are totally borne by the buyer, to a DDP (Delivered Duty Paid), in which the goods are delivered to the buyer's unloading dock, and therefore transport and customs charges, as well as the risks are totally borne by the seller. Intermediate situations are type F and type C modes. In the case of type F returns, the goods are cleared for export but not for import; in the case of type C returns, the goods are cleared for export from the country of origin and for import to the country of destination and the cost of transport from the country of origin to the customs of the destination country is covered.

- It is important that in addition to the Incoterms agreed upon, the physical place of return is specified (address, city, country);
- Conditions of the goods: in this section, particular status conditions agreed for the goods are reported, which may include: the type and maximum and minimum dimensions of the packaging used, the identification technology and standards to be used (e.g. GS1 logistics label with an SSCC, BC or RFID tag), traceability information (e.g. special requests on lots—delivery of single-lot pallets) or expiry date (e.g. at least 2/3 of shelf life available to the buyer is the standard requisite from retailers);
- Invoicing Conditions: these contain the conditions under which the vendor will be able to issue an invoice for the order in question. E.g., invoicing upon order or upon receipt of goods, partly upon order and partly upon delivery of the goods;
- Payment Conditions: indicates the conditions (methods and times) with which the invoice is to be paid (e.g. by bank transfer, bank withholding tax, cash on delivery upon receipt of the goods, 30–60 days invoice date at the end of the month, etc.).

2.2.1 Identification of References in the FMCG Sector

The identification of products and packaging is a form of standardizing the information relating to them, so that this information can be understood without difficulty by all the players in the supply chain.

In Italy, and in 100 other countries around the world, the information associated with food products and consumer goods in general is encoded through the GS1 system. This is a two-way coding system, in which each unit (be it a sales unit, secondary packaging, or a load unit) is identified by a single code, and each code corresponds to a single unit, recognizable in all the countries of the GS1 circuit.

With reference to sales units intended for the consumer, the GS1 code used to identify them is known as GTIN (Global Trade Item Number), and can consist of a number of digits equal to 8, 12, 13 or 14. The GTIN-13 code (consisting of 13 digits, is used for the identification of products intended for the consumer, and directly provides information such as the country of origin of the product, the company which owns the brand, product and reference code. In particular, the origin of the product can be recognized by the national prefix, consisting of the first two digits of the GTIN; this prefix is assigned by GS1 to its representatives in the various countries where the code is used. In Italy, the management of GS1 codes is carried out by GS1

Table 1 GTIN structure for variable weight secondary packaging—EAN 13

National EAN prefix	Company prefix	Item reference	Check-digit
N1–N2 (between 80 and 83)	N3–N9	N10–N12	N13

Table 2 GTIN structure for small packs of a fixed weight—EAN 8

National EAN prefix	Item reference	Check-digit
N1–N2 (between 80 and 83)	N3–N7	N8

Table 3 GTIN structure for variable weight secondary packaging—EAN 13

Prefix	Item reference	Price	Check-digit
2	N2–N7	N8–N12	N13

Italy, and Italy is assigned national prefixes between 80 and 83. The subsequent digits of the GTIN include, in order, the code of the company which owns the brand (from the third to the ninth digit of the code—company prefix), the product code (item reference—from the tenth to the twelfth digit) and the check digit (thirteenth digit), according to the scheme illustrated in the chapter on representation through information flows and which is proposed here again (Table 1).

As represented in Chap. 2, the GTIN coded serial is identified by a barcode which makes it possible to automate the checkout, having created the link n:1 between the GTIN (n) of the i-th reference (1). In fact, the same reference may have different GTINs. Suffice to think of a can of coke produced in different countries: the GTIN changes, since the country code is different, but the reference remains the same.

For specific applications, and in particular in cases where the 13-digit GTIN code cannot be used for reasons of space, the GTIN-8 code can be used as an alternative (GS1 Italy 2021a). This code is represented in Table 2, it allows identification of a product and the country of origin, but does not contain additional information.

In many cases, a food product does not have a predetermined selling price, but the price depends on the weight of the product purchased. This is the case, for example, of some fresh products or fruit and vegetables. The resulting type of packaging is designated "variable weight", and is identified through a specific code, the validity of which is limited to the national scope. As detailed in Table 3. The variable weight code is always composed of 13 digits, of which the first, which identifies variable weight packaging in Italy, is 2. The remaining figures indicate the reference (from the second to the seventh digit), the price, expressed in € (from the eighth to the twelfth digit) and a check digit. The corresponding diagram is shown below (GS1 Italy 2021a).

Now consider a case in which it is necessary to identify multiple secondary packaging, containing several units destined for the final consumer. Also in this case, the packaging may be of fixed or variable weight.

In the case of a "homogeneous" secondary packaging (see Table 4), meaning a situation in which the elementary units contained in the packaging all have the

Table 4 GTIN structure for secondary packaging of a fixed weight—EAN 14

Logistic variant	National EAN prefix	Company prefix	Item reference	Check
1–8	N2–N3 (between 80 and 83)	N4–N10	N11–N13	N14

Table 5 GTIN structure for variable weight secondary packaging—EAN 14

Indicator	National EAN prefix	Company prefix	Item reference	Check
9	N2–N3 (between 80 and 83)	N4–N10	N11–N13	N14

same code, and therefore the weight is fixed, a GTIN-14 code can be used. This code is identical in structure to a GTIN-13 for a sales unit, which is preceded by a figure known as an "indicator" or "logistic variant", whose value is between 1 and 8, according to the following scheme:

In the contrary case of a secondary packaging in which the production process does not ensure consistency of weight, size, or length, the resulting packaging is "non-homogeneous" and of variable weight. In order to identify the packaging unit, the quantity of the product contained is relevant information, and, for this reason, GTIN-14 numbering is adopted for these products, recognizable with respect to that of primary packaging since the first digit is 9. The information relating to the quantity of the product contained is expressed using GS1-128 symbology. The code is structured as follows (Table 5).

2.3   Confirmation Order

The Confirmation Order is a structured document which the vendor sends to the buyer following receipt of a PO, and with which it undertakes to fulfil the order contained in the PO itself. It is therefore a forward flow of information, from upstream to downstream.

Like the PO, the Confirmation Order consists of a Header, Body and Summary which contain items similar to those seen for the PO.

Similarly to the PO, the Confirmation Order in the Header section, in addition to the sender and recipient, will have a unique number, linked 1:1 with the buyer's PO, to which the buyer can refer for its communications with the vendor.

In the detail section, the vendor specifies the order lines that it undertakes to fulfil and the related quantities, confirming the prices indicated by the buyer.

In addition, the general conditions of sale are also shown in the summary section, which represent the claims of responsibility according to which the vendor undertakes to fulfil the order and accept any returns from the buyer.

If there are no differences between the CO and the PO, the order can be processed, but if there are differences, these must be negotiated and resolved before proceeding with the physical fulfilment of the order.

2.4 *Manual Orders*

Even today, in many sectors, the order cycle is manual.

This means that all the individual activities which make up the order cycle and therefore: the formation of the order and its structuring in a document (1), the transmission from the buyer to the vendor (2)—by post during the 1980s, via fax in the 1990s and via e-mail since 2000 (3), the insertion into the vendor's information systems, the compilation and sending of the order confirmation (4)—are done manually, as shown in Fig. 5.

The manual process implies certain critical aspects. First and foremost, it is a labour-intensive process, since all the separate are carried out by an operator and therefore generate a labour cost.

Furthermore, even if an e-mail transfer phase, albeit manual, can be quick, the other phases can be long and time-consuming, and therefore the process itself is time-consuming. All of this, on top of the labour cost, mainly impacts the lead time—since the order cycle is lengthened—which in turn impacts cycle stocks and safety stocks. In FMCG, for example, the daily reordering of a department can involve thousands of order lines; it is therefore unthinkable to use a manual reordering process, in which an operator manually formalizes the order on the terminal and then transfers it to a Distribution Centre. A process like this could take days.

Finally, the manual process, by its very nature, suffers from poor accuracy, both in phase (1) of order formalization (therefore on the buyer's side) and in phase (3) of order entry and confirmation (therefore on the vendor's side) errors can be generated. These errors can be both related to the body part and therefore typically errors of mix and quantity, but also errors in the Heading and Summary sections, such as errors in the returns of goods, status errors, etc. Once an order has been processed, these errors will need to be dealt with later, resulting in an increase in administrative

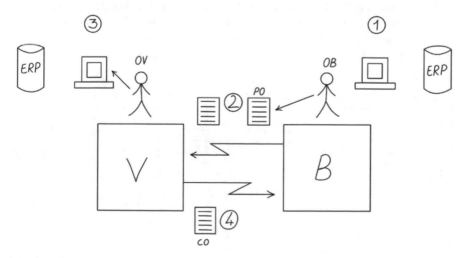

Fig. 5 The order cycle. Manual process

management times and related costs, which impact the total logistical cost of the product. The costs associated with errors in the manual management of orders can be both physical and of a market mediation kind. The additional physical costs are typically those necessary to deal with the error, such as the costs to return the goods or to send an additional shipment. Market mediation costs, on the other hand, can be much higher as in the case where if fewer lines/quantities are ordered, there may be a product OOS and therefore a loss of turnover. Current and future turnover, since automatic reordering systems will interpret OOS as a decrease in product demand, and will reorder less in future. Conversely, if more lines or quantities are ordered by mistake during the insertion phase, or in the order confirmation phase more lines or quantities are inserted, there may be shrinkage.

2.5 Automatic Orders—EDI—Electronic Data Interchange

To overcome the limitations of manual processes (labour-intensive, time-consuming, low accuracy) it is necessary to move over to automated processes.

In an automated process, all the phases of the order cycle—formation of the order, transfer of the order, insertion into the vendor's information systems and formation and transfer of the order confirmation—take place automatically between the buyer's ERP information systems and the vendor, as represented in Fig. 6.

In an automated process, the buyer's ERP system, after putting together a PO based on the internal reordering rules—e.g., every night, based on product sales and available stock, the ERP issues an order proposal for references which the department head will simply need to validate or correct in the morning based on factors which the ERP cannot know. The validated order is transformed into a standard electronic format, which allows information systems to interface with one another.

We talk about EDI—Electronic Data Interchange—to identify the languages and standards that allow transfer of structured documents in an M2M system (many-to-many but also machine to machine). The EDI file can be simply a text file that is transferred via FTP between two folders or created in a shared folder on one of the

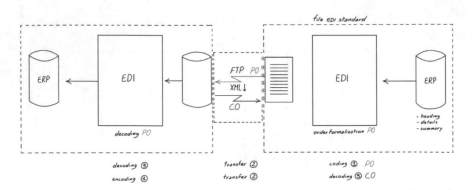

Fig. 6 The order cycle. Automated process

two systems to which both systems have access, or a restful Json, a Web service with which buyers and vendors systems automatically communicate. The vendor's server in turn, through a specular decoding system, is able to decode and correctly interpret the fields of the EDI layout (heading, details and related order and quantity lines, and summary) to insert it into the information system, process it, and proceed in a similar way to the preparation and transfer of the order confirmation via EDI.

In an M2M system—as in FMCG—two requirements are fundamental for the functioning of the EDI.

1. It is essential that the language and the EDI format with which electronic documents are encoded and transferred is standard, so that the information systems can correctly interpret the fields of the document.

    An EDI file is made up of segments (lines) which contain elements within them, separated from each other by special characters, typically asterisks *. The succession of elements and segments therefore constitute the characters, words, and lines of which an EDI document is composed. In order to be interpreted by machines automatically, also the position of the order fields within segments and elements is standardized. The most popular EDI standards in mass consumption are EANCOM (GS1 Italy 2021b) or EURITMO (Gs1 italy 2021d)—which define how to structure segments and elements to transform the various sections of the order (Header, Detail and Summary) into a standard structured file (see Fig. 7).

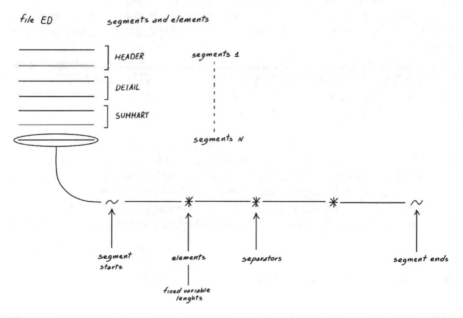

Fig. 7 The structure of segments and elements in an EDI file

Fig. 8 Forward information
flow connected with the BoL

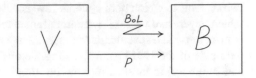

2.   It is essential that the personal data with which the references and the related
     attributes are identified (e.g. units per package, weight, etc.) are standard, so
     that there is no ambiguity in the face of a PO product code. Even today in
     this area—and in particular for the FMCG industry and retailers—difficulties
     are encountered in having a univocal system for aligning personal data. The
     objective of projects such as Allineo of GS1 (https://allineo.it/) is to align product
     data and related attributes in a single system so that a manufacturer can have a
     single data update point which all retailers can access, and, vice versa, a retailer
     can have a single point of reference for all references from different suppliers.

3   Information Flows—Bill of Lading

3.1   Definitions and Features

The Bill of Lading (BoL), the "Transport Document" or simply TD, is a document
which accompanies goods and moves from upstream to downstream. It therefore
represents a forward information flow from the vendor to the buyer, as represented
in Fig. 8.

   The TD is important in the order cycle for two reasons: first of all through it the
vendor informs the buyer of what it is receiving. As will be seen in detail in the
following paragraphs, the earlier the information is received in advance, typically in
electronic format, the more streamlined the reception process can be in terms of both
time and costs.

   Secondly, as schematized in Fig. 9, the TD is the tool that enables the receipt of
goods by the buyer. Once signed by the buyer, the TD represents the document that
certifies receipt of the order lines by the buyer; the vendor will subsequently issue
an invoice on the basis of the signed document. The TD comes in two copies: one
retained by the buyer and one which returns to the vendor, signed for acceptance. Any
non-conformities must therefore be noted on the BoL and dealt with administratively.

3.2   The Structure of the TD

Like the PO and the CO, the TD too is a structured document, which therefore consists
of standard sections: Heading, Detail or Body, and Summary.

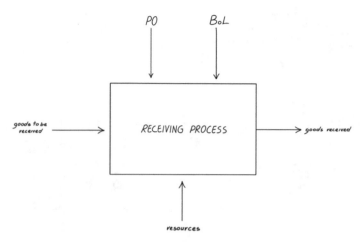

Fig. 9 The TD (also known as BoL) as a control in the receiving process

3.2.1 Heading

In the Heading section there is some standard information that can be found in all transport documents:

- Sender/recipient—departure and destination: first of all the indications of the sender (legal entity) and recipient (legal entity) of the goods as well as the physical place of departure of the goods (company name and address) and physical place of destination (company name and address). The sender is the legal entity of the owner of the goods, while the physical place of departure is the physical place from which the shipment originates. Likewise, the recipient is the legal entity of the recipient of the goods (the one who issued the order) while the physical place of destination is the place where the goods are delivered.

  To avoid ambiguity, it is good practice to use shared standards to identify locations. In the FMCG sector, the GS1 standard for identifying locations is the GLN—"Global Location Number" (see GS1 Italyc, 2021). This is a numerical identifier N1, …, N13 in which N1 N9 represent the company prefix, N10, …, N12 identify the location, normally structured in a hierarchical structure (e.g. country, facility, building/department, floor, office).

  The sender and the physical place of departure do not necessarily coincide, just as the recipient and the physical place of destination do not necessarily coincide. In the example shown in Fig. 10, which refers to an order issued by Italian subsidiary of Auchan Retailer to one of its fresh product suppliers, Müller, the sender is Müller, and the recipient is Auchan, while the physical places of departure and destination of the goods are the sites of the respective Stef TFE logistics operators in Tavazzano (LO) and Italtrans in Calcinate (BG).

- The TD can be physically issued by the carrier transporting the goods, in this case Stef TFE which is Müller's logistics operator, or directly by the sender (Müller).

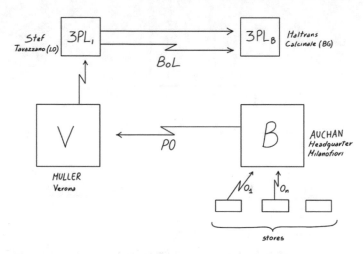

Fig. 10 Example of sender/recipient physical place of departure/destination in the case of a shipment of fresh products in the Italian FMCG sector

- TD No.: represents the unique identifier of the shipment. This is the number to refer to for subsequent communications and administrative requirements. It is linked 1:1 with the transport, while obviously on one means of transport there can be several TDs, depending on the loading plan (*bordereau*). The identification is managed independently by the issuer of the transport document, according to its own proprietary numbering scheme;
- TD date: date on which the TD is issued; it is good practice that this date coincides with the date on which the transport actually takes place (at most a short precedence is permissible, while the TD cannot be post-dated). The TD date, together with the TD No., represent the elements that allow tracking of a shipment;
- PO of reference: the PO to which the transported goods refer to is given in the transport document. In this case, some customers can request only one TD for each PO, and therefore a 1:1 PO-TD-Invoice link to ease the invoicing process, while usually in one TD there can be goods to fulfil multiple POs to allow the vendor to streamline its own processes;
- Reason: identifies why the goods are being shipped. Typically, the reason is "sale", but it may also be work account/return from work account; viewing account/return from viewing account; free gift/sample, etc.

3.2.2   Body

The typical structure of a Body is as shown in Fig. 11.

The Body of the TD includes details of the goods transported. Each line of the Body is made up of some reference fields, typically:

- #line: represents the progressive identifier of the order line;
- Article code: identifier of the supplier's article, e.g. in FMCG, the GTIN of the product can be used (using the EAN 128 standard, reference can be made to the

order line	reference code	quantity	ED	W	LN
10	$GTIN_{10}$	Q_{10}	ED_{10}	W_{10}	LN_{10}
20	$GTIN_{20}$	Q_{20}	ED_{20}	W_{20}	LN_{20}
i	$GTIN_i$	Q_i	ED_i	W_i	LN_i

Fig. 11 Example of the structure of a TD body

application identifier AI 01 to identify the GTIN of the sales unit, or AI 02 to refer to the GTIN of the package);

- Description. Optional field for description of the item;
- Quantity: quantity of items shipped—the unit of measurement must be specified to avoid ambiguity. In the case of fixed-weight products, reference is made to secondary packaging (packages), using the AI 30 in the EAN 128 standard; in the case of variable-weight products (such as dairy products, cured meats, cheese, meat, fruit and vegetables), reference is made per kg, using AI 310;
- Expiry date ED: the commercial expiration date of the goods. Identified by AI 17 in the EAN 128 standard in FMCG for commercial practice; as already said before, in the FMCG sector, it is commercial practice that at least 2/3 of the product's useful life is reserved by the manufacturer for the retailer;
- Lot Number LN: identified via EAN 128 by AI 10, this represents the fundamental information to manage traceability. Using the Lot number, the supplier identifies items that have undergone the same production/logistics process. For example, the items processed by a production line in a certain time interval—timeframe—can represent one production lot. A mixed pallet resulting from several batches can represent a new logistics lot. In the event of a potential hazard for the consumer, once the cause has been identified (e.g. contamination in a production line, mixed pallet affected by a non-compliant batch), the potentially hazardous batches are identified using parent–child links (e.g. batches processed by the line; i.e. "child lots" with which the dangerous "parent lot" may have merged) and the lot numbers to be recalled or collected are communicated downstream. We talk about *recall* when this involves the final consumer, we talk about *withdrawal* when the lot has not yet been put on sale and the recall takes place within the supply chain itself. Typically, in FMCG practice, in order to avoid false negatives in the face of a batch withdrawal communication, the retailer will still withdraw not only the batch but all the product references. All of which is to streamline collection operations (it is not necessary to check the lot for each reference, but the entire reference is withdrawn from the store), and it is precisely the risk of a false negative that is eliminated.

3.2.3 Summary

The summary recapitulates the transport document information, typically:

- No. of loading units/packages in the shipment;
- Details of the carrier carrying out the transport;

Fig. 12 The receiving
process: check including PO,
physical aspects, and bill of
lading

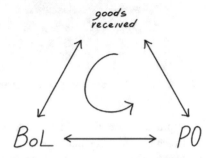

- Note fields for annotations—these are fields in which to note any discrepancies between the mix/quantity marked on the BoL and the mix/quantity physically received. Any such annotations will then need to be dealt with administratively;
- Signature of the recipient: this is the field which, once signed, certifies delivery of the goods and acceptance by the recipient. The signed TD represents an enforceable document in the face of any disputes.

3.3 Manual TD—The Receiving Process

The TD represents the basic document with which to carry out the receiving process. As depicted in Fig. 12, the reception process is essentially a three-way check including:

- Ordered—mix/quantity present in the PO sent by the buyer to the vendor;
- Physical—what is physically delivered by the vendor to the buyer;
- TD—administrative document prepared by the vendor, on the basis of which the administrative processes (invoicing) from the vendor to the buyer will be structured for each line of the TD.

In the FMCG sector, there are different variants on this process, depending on whether the processes are carried out manually or using mobile terminals connected by radio frequency with the management system, typically a WMS. The process flowcharts are described in the following paragraphs.

3.3.1 Manual Process

The vehicle containing the goods is parked at the dock and unloaded; the SUs (shipping units) contained and forming part of the transport document are placed on the ground in a staging area in front of the dock to be checked.

Physical control—TD

The operator collects the BoL in duplicate and for each row of the TD (Fig. 13):

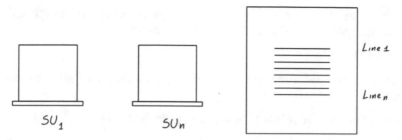

Fig. 13 Manual receipt physical check—TD

- checks the reference code;
- counts the quantities and compares them with what is reported in the BoL;
- checks expiration dates and batches;
- checks the condition of the goods (damage, cold chain, etc.);
- notes on the BoL (physically with a pen next to the line !!) any mix/quantity/status reserves, the so-called "Bill of Lading reservations".

Once the loading units and the TD lines have been used up, the operator signs the BoL with any notes/reservations and delivers a copy to the carrier who can then depart from the dock.

The TD with the annotations is then taken to the office for administrative formalities.

TD—PO check

Here, an administrative operator:

- calls up the relevant PO of the BoL.;

For each line of the TD,

- identifies the corresponding row of the PO and enters the quantities received on the dock door, the expiry date and the lots received (to manage traceability);
- updates the PO with the lines fulfilled completely, the lines to be fulfilled completely or partially;
- at the end of the process, manages any reservations by contacting the supplier and dealing with non-conformities (missing and/or unexpected, damaged, or non-conforming items).

3.3.2 Manual Process—Radiofrequency Evolution-1

The manual radiofrequency process represents an evolution of the process described in the previous paragraph, aimed to (at least partially) reduce the manual activities of checking at the dock and entering data in the administrative offices. In this case, the operator on the dock is equipped with a radio terminal, connected to the ERP information system that can access the administrative information on the order.

Once the vehicle has been logged in and the goods unloaded, the first phase of the reception process consists in the physical-PO check.

Physical-PO check

An operator, equipped with a radio frequency terminal connected to the information system:

- calls up the number of POs to be received from the WMS;

  For each reference to be received,

- opens the secondary packaging and scans the EAN 13 code on the sales unit (this procedure, in addition to identifying the reference, also serves to check the readability of the goods on the checkout barrier);

The WMS checks that the reference is among those to be received for the PO.
The operator then manually enters:

– expiration date—for managing rotations
– the lot number—for traceability management
– the number of pieces physically present in the shipment for sthat reference—with the expiry dates and lot numbers—for the matching of the order;

The WMS checks the correspondence in mix/quantity;
The operator declares (via radio terminal and in writing on the BoL next to the relevant line) any non-compliance of status/preparation/cold chain, etc.

Once the BoL has been signed, the PO is updated with the completely fulfilled lines, the partially fulfilled lines still to be received and the lines yet to be fulfilled, and any non-conformities are then dealt with administratively: missing pieces can be removed or added to the next order, extra pieces can be confirmed or returned, non-compliant pieces can be cancelled.

The TD is signed with reservation, delivered to the carrier who can then leave the Distribution Centre, while a copy is taken to the office for administrative checks.

TD-physical check

In administration, an operator:

- Calls up the relevant PO;
- For each row of the transport document, enters the value reported in the document and compares this with what has been physically delivered by the transporter at the dock and entered in the system;
- Any non-conformities must subsequently be dealt with administratively by contacting the vendor.

Once the administrative checks have also been concluded, the references received are then available in the information system and can be processed for subsequent processes (e.g. storage, retrieving, picking).

3.3.3 Manual Process—Radiofrequency Evolution-2

This is a variant on the radiofrequency process described in the previous paragraph. In this case, the check is made first between the TD and the PO and then a check between the physical situation and the TD. The process lends itself to an evolution towards an automated solution using DESADV.

TD-PO check

Before the vehicle is unloaded, while the driver waits in the parking lot, the TD is inserted into the buyer's information system and linked to the relevant PO.

The TD is taken to the administration department where the lines of the TD are entered into the system, comparing the goods to be received with those ordered and still to be received in mix and quantity. Any serious non-conformities can be dealt with administratively before unloading the vehicle, avoiding wasting time and generating handling costs. Minor non-conformities can be dealt with administratively by contacting the vendor.

As mentioned above, the process lends itself to natural evolution using DESADV. If the TD has been sent in electronic format, this operation can be done before the goods arrive, avoiding having to manage waiting times on the dock for the insertion of the TD, or even before the goods leave, allowing the resolution of any non-compliances.

Once the TD is entered in the system, a check takes place between what is reported in the TD and what has been physically delivered.

Physical-TD check

Once the TD is entered in the system, a check takes place between what is reported in the TD and what has been physically delivered (Physical-TD check).

After unloading the vehicle and signing the BoL with reservation (usually after counting the number of pallets/packages), the vehicle leaves the Distribution Centre, and the actual reception process can begin.

An operator, equipped with a radio frequency terminal connected to the information system:

- calls up the TD number to be received from the WMS

  For each reference to be received:

- opens the secondary packaging and scans the EAN 13 code on the sales unit (this procedure, in addition to identifying the reference, also serves to check the readability of the goods on the checkout barrier).

  The WMS checks that the reference is among those to be received.
  The operator then manually enters,

- expiration date—for managing rotations
- the lot number—for traceability management
- the number of pieces physically present in the shipment for that reference—with the expiry dates and lot numbers—for the matching of the order;

The WMS checks the correspondence in mix/quantity.

- The operator declares (via radio terminal) any non-compliance of status/preparation/cold chain, etc.

The PO is updated with the completely fulfilled lines, the partially fulfilled lines still to be received and the lines yet to be fulfilled, and any non-conformities are then dealt with administratively: missing pieces can be removed or added to the next order, extra pieces can be confirmed or returned, non-compliant pieces can be cancelled.

The references received are therefore available in the information system and can be processed for subsequent processes (e.g. storage, retrieving, picking).

3.3.4 Manual Process—Potential Critical Issues

From the description of the manual processes, both completely manual and with radio frequency terminals, it emerges clearly that the process described (albeit in its variants which tend to reduce some of the criticalities mentioned below), is a highly manual process, slow and subject to inaccuracy, especially in the presence of mixed pallets and different expiration dates/batches. While a load of 33 mono reference, mono lot pallets—e.g. water, milk—can be received quickly with the procedures just described; a load of mixed pallets—as is typically the case for fresh products— prepared in layers in which each layer represents a reference potentially with different expiration dates and batches, can take hours to be received and lends itself to very frequent errors, since expiration dates and lots need to be manually inputed.

Any non-conformities are also detected a posteriori, after the transporter has already left the dock some time ago. Consequently, the vendor has evidence of any reservations in the BoL only after hours if not days, when contacted by the buyer to manage non conformities or when the bill of lading with the related reservations is returned by the carrier and inserted into the vendor's ERP system. For example, if a reference is missing because the vendor has mistakenly inverted a pallet during shipment, or a pallet has not been accepted because its expiration date is too short or prior to a pallet already received, or the cold chain claimed to be poorly managed, the vendor will have evidence of it only hours (or even days) after receiving the goods. The consequence is a product being potentially out of stock at the store, with related impacts on current and future sales, both due to product replacement and due to the fact that in the face of an OOS and consequent drop in sales, the automatic reordering system interprets the drop in sales due to product shortages as a drop in demand and consequently decreases the quantities to be ordered for future periods.

To overcome all these drawbacks and improve efficiency and accuracy, dealing with any non-conformities in real time, it is necessary to switch to automated processes using DESADV/ASN and transport packaging identified by means of the SSCC standard, possibly encoded in an RFID tag (in this way inversion errors can be avoided). The following paragraphs detail how DESADV and identification standards streamline the shipping/receiving process (Youtube 2012).

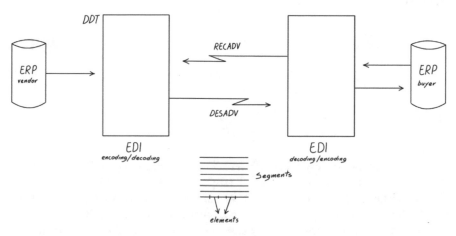

Fig. 14 The flow of goods and information with DESADV RECADV

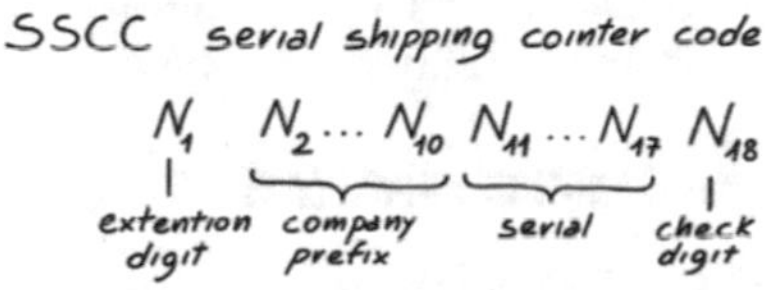

Fig. 15 Structure of the
SSCC

3.3.5 Receipt Via a Despatching Advice (DESADV)

The complete automation of the reception process and the surpassing of all labour
and accuracy limits connected with the manual process and its evolutions in radio
frequency is based on the use of an electronic document in EDI format called
DESADV—DESpatching ADVice or ASN—Advance Shipping Notification, and
its counterpart RECADV—RECeiving ADVice.

The functional scheme and the structure of a DESADV is represented in Fig. 14:

Just like the PO, the DESADV is also structured using standard language in an
XML file or similar, divided into segments and elements. Each segment and element
contains specific information of the TD that can be automatically decoded by the
buyer's information system.

As represented in Fig. 16, DESADV typically contains the list of SSCCs (each
SSCC (GS1 Italy 2021e) uniquely identifies the load unit (shipping unit or single
handling units) in a standard way, according to the format shown in Fig. 15) to be
received, and for each SSCC details on reference, quantity, LN, ED and any additional
information agreed.

The DESADV is sent in advance of the goods, either via and ftp file or WS (web
services) integration, and is therefore available to the buyer (a) in advance of the
delivery of the goods, potentially even before the goods leave the supplier's shipping
dock doors (b) directly within the information systems for the buyer, without the need
for manual entry by an operator, eliminating entry times and accuracy problems.

Fig. 16 Contents of the
DESADV

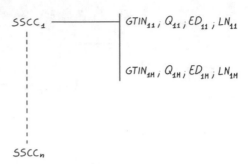

The comparison between DESADV and PO and the related compliance check (Are the references/quantities correct? Are the expiration dates correct?) can therefore be done in advance, even when the goods are still at the vendor's ready to leave; any non-conformities can be resolved by contacting the vendor and deciding what to do even before the goods leave or in any case before they arrive at their destination.

TD-PO check

– As mentioned, this is automatic, and involves sending the DESADV electronically from the vendor to the buyer.

Physical-TD check (Free Pass)

Once the vehicle arrives at the buyer's dock, and the green light to download is given after a successful TD-PO check and the DESADV is present in the buyer's information systems. The process takes place as follows.

The operator responsible for unloading the vehicle:

- unloads the goods and places them in front of the dock to be checked
- calls up the TD at an RF terminal.

The WMS retrieves the list of expected SSCCs for that particular TD.

- The operator scans the SSCCs received (either by reading a BC or by reading the RFID through a gate on the reception gate; in this case the identification process does not need additional time since it is automatically done during unloading activities—experiences of RFID projects in this sense show a 50% reduction in reception times, while completely eliminating errors;
- The WMS checks that the SSCC is among those expected;
- The operator randomly checks the SSCCs from a Free Pass perspective (The Free Pass process is described below);
- Any non-conformities (e.g. damage during the journey, non-compliance with the cold chain, expiry dates to short or shorter than same products received before etc.) can be dealt with in real time by contacting the supplier;
- At the end of the process, the operator closes the BoL on the information system;
- The buyer's ERP automatically sends the RECADV RECeiving Advice to the vendor in real time; this is a structured electronic document, twinned with the

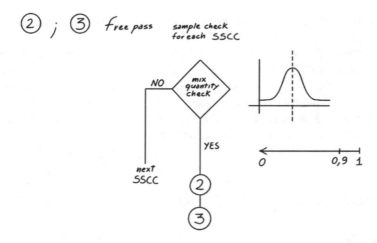

Fig. 17 Operational logic of the Free Pass reception process

DESADV, with which the buyer informs the vendor of the SSCC received and taken in charge, and of any reservations and goods not received due to problems (mix, quantity, expiration date, status, cold chain, etc.);

- The operator signs the document and gives it to the carrier, having noted any non-conformities.

We speak of Free Pass control when the loading units being received are checked only at an SSCC level, while on a sample basis, a single SSCC can be promptly checked at the reference level, lot quantity and expiration date. The process is represented in Fig. 17.

It is the WMS which decides which/how many HUs to subject to punctual checks, through a random process based on a threshold value that changes based on the results of previous checks (as end customers may be randomly checked at a self-checkout kiosk in a retail store). Controls which give a positive result (correspondence of mix/quantity between physical aspects at the level of a single expected SSCC checked) raise the threshold of the sample control, making checks less frequent; In contrast, checks which give negative results increase the probability of precise future checks.

The advantages of using DESADV/RECADV and Free Pass receipt procedures compared to the manual process are summarized below point by point.

- The TD is entered directly into the buyer's information systems, eliminating the Buyer's time for entries and related errors;
- All the quality (Mix/quantity/DS) and traceability (LN) data are available directly in the DESADV, and do not have to be manually checked at the dock by the operator (often by physically typing the values on a PDA keyboard) or in the office from a printed BoL;
- The information of what is being received is available to the buyer in advance, and therefore the matching with the order and the management of non-conformities can be done before the goods arrive or even before the goods leave, solving in

advance the problems that would otherwise lead to rejection of the goods, with all the consequences described above;

- Thanks to the RECADV, the results of the deliveries are known in real time, and not days later, when the BoL is returned by the carrier and inserted in the information systems; in the event of delivery rejections, the vendor can intervene when the vehicle is still parked at the dock, avoiding the product not being delivered;
- The goods can be received in a Free Pass perspective, drastically reducing the time for receiving goods and the related costs.

4 Information Flows—POS Data

4.1 Definitions

POS—Points of Sale Data are a fundamental information flow to manage a supply chain.

POS represent the data of the actual demand by the final consumer, they are measured in pieces/day and are represented by the daily sales of the product at the checkout barrier, as represented in Fig. 18.

4.2 The Transmission of Information on Demand in the SC

In a traditional SC, the POS Data are known only to the retailer, who uses them to elaborate their sales forecasts and consequently the upstream orders with which they procure from the SC. On the other hand, all the other players of the SC represented in Fig. 19 use data from orders downstream to estimate the demand and to process upstream orders in turn. In general, the generic N level then uses O_{n-1} orders to estimate the product demand, and issues O_n upstream orders with which the $n + 1$ level will estimate the product demand.

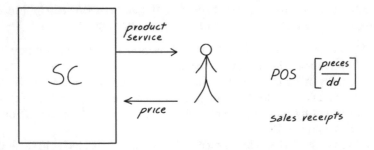

Fig. 18 The information flows: POS data

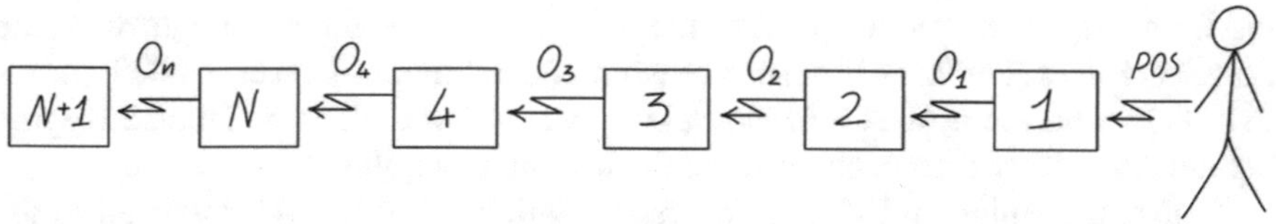

Fig. 19 The reordering mechanism in a traditional SC

This upstream information transmission mechanism of product demand generates two types of problems: Delay in transmission of information, and distortion of information.

4.2.1 Delay in the Transmission of Information

Precisely in relation to this latter aspect, each level transmits information upstream via the order, with a delay which depends on the level of stocks present, the supply lead time, and the frequency of reordering. If the demand of the final consumer varies, it is in any case necessary for the downstream level to decide to issue an order (either because the stock falls below the reorder point—reorder point—or because the reorder time has been reached—order-up-to level) to transmit upstream the information that the demand has changed. And so on and so forth, this is what happens across all levels of the chain, wholesaler, distributor, manufacturer, Tier I supplier, Tier II supplier etc., up to the highest levels. If the reordering takes place with low frequency, as in the case of very high reorder costs (e.g., products that require large batches to be able to saturate a means of transport, or in the presence of large quantity discounts); or in the case of long lead times, where it is necessary to order well in advance, again in the case of class C products for which buyers tend to order less frequently and order quantities to cover adequately—the information of the variation in demand will take a long time to spread to the higher upstream levels, which will therefore be penalized. In fact, even in the face of variations in the POS Data and therefore in demand, they continue to behave in the same way, as if the demand had not changed. If demand falls, upstream levels will continue to produce/supply until a sudden decrease in order quantities causes them to discover that demand felt. Conversely, in the face of increases in downstream demand, the upstream levels, since they have continued to produce supplies as if demand had not increased, will not be able to cope with increased inventories, generating out-of-stock (and then backlogs or cancellations of orders).

4.2.2 Distortion of Information—Definition of the Bullwhip Effect

A second aspect related to the mechanism to transmit information through the order is that of *distortion*. As will be seen analytically in the next paragraphs, the standard deviation in orders is greater than the standard deviation in POS Data, and is greater the further upstream the recipient of the order is. This phenomenon, known as the Forrester Effect (Forrester 1961) or Bullwhip Effect (Smitchi-Levi et al. 2020) means that those who are upstream in the SC not only see variations in demand lagging behind those downstream, but also see them strongly amplified.

In fact, the Bullwhip Effect at the generic i-th level of the SC is defined as the ratio between the standard deviation in demand seen from the i-th level, represented by the orders received from the $i - 1$ level, and the standard deviation in effective market demand, represented by the POS Data.

$$BE_i := \frac{\sigma_{O_{i-1}}}{\sigma_D}$$

Ultimately, it is as if those upstream in the supply chain had to drive a car wearing glasses which lag in transmitting information about what is happening externally on the one hand and transmitting distorted images in which objects are larger or smaller than they actually are.

Distorted information from one end of the supply chain to another can lead to huge inefficiencies: oversized stocks, or low service levels (stockouts and backlogs) and therefore lost sales/turnover, incorrect production capacity planning, inefficient transport, deeply impacting on SC profitability.

4.3   The Effect of Information Flows on Safety Stocks

While cycle stocks CS are used to meet product demand in the supply period (reorder point or order up to level reorder interval), safety stocks are used to cope with stochastic phenomena. Thanks to safety stocks, it is possible to cover oneself from variability in demand and from variability in supply lead times. To determine safety stock, the reorder level case will be reviewed later, the case of the order up to level being conceptually similar (safety stock must cope with variations in demand in the reorder interval T plus the procurement lead time—LT).

4.3.1   Deterministic Demand and Lead Times

If demand and lead times are deterministic, no safety stock is needed. Called D [pieces/days] and LT [days] the average value of the daily demand and of the supply lead time, respectively expressed in units/time and time, the order can be issued when the stock falls below the level D * LT [pieces], so that when the goods arrive (after LT days) the stock is zero (Fig. 20).

The quantity ordered will be equal to the economical lot, able to minimize the costs of stock maintenance and reorder costs. In the simplified case in which the economical lot coincides precisely with the demand in supply lead time, we will simply order every LT day, reordering when the goods arrive exactly the quantity delivered, equal to LT * D. It is in this case that SC cycle stocks are worthwhile

Fig. 20 Representation of
stocks in case of
deterministic demand and
lead time

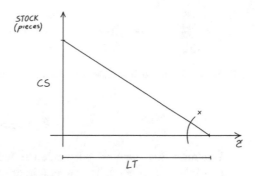

$$CS = \frac{LT * tg\alpha}{2} = \frac{LT * D}{2}$$

If, for example, the lead time is one week, and the demand is one piece/day, a quantity of 7 pieces is ordered weekly, to meet the needs for the entire supply horizon, therefore the average CS is 3.5 pieces.

Instead, in a more generic case:

$$CS = \frac{EOQ}{2}$$

However, in a real case, neither the demand nor the lead time are deterministic, but vary stochastically, so that it becomes necessary to introduce safety stocks, simply to cope with the variability of one and the other. The case of stochastic lead time and deterministic daily demand and the case of stochastic daily demand and deterministic lead time are examined separately, to then analyse the more complex case, a combination of the two—lead time and stochastic demand.

4.3.2 Stochastic Daily Demand and Deterministic Lead Time

Therefore, supposing that the lead time is a constant equal to LT while the daily demand D is a random variable X.

$$X(M_D; \sigma_D)$$

We wish to determine the value of the random variable demand Y in LT. The random variable Y is the sum of LT random variables X all equal to one another (the demand on day 1 is equal to r.v. X, the demand on day 2 is equal to r.v. X, the demand per day LT is equal to r.v. X), as represented in Fig. 21:

Therefore, it follows that the random variable Y "Average demand in supply lead time" is random with a mean and standard deviation equal to M_Y and σ_Y:

$$Y = x_1 + x_2 + \cdots + x_{PLT}$$

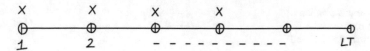

Fig. 21 r.v. demand in lead time as the sum of random variable LTs all equal to one another

$$Y(M_Y; \sigma_Y)$$

Assuming that there is independence and absence of correlation between the values of the individual daily demand, in other words, that the demand for a generic day does not affect that of the other days, which mathematically translates into the fact that i dictates the demand for the i-th day the demand on the generic j-th day, we have cov $(i, j) = 0$, then the mean value and variance are respectively:

$$M_Y = M_{x1} + M_{x2} + \cdots + M_{xLT} = LT * M_D$$

$$\sigma_Y^2 = \sigma_{X1}^2 + \sigma_{X2}^2 + \ldots + \sigma_{LT}^2 + \sum_{i=1}^{n-1} \sum_{j=i+1}^{n} cov(i; j) == \sigma_{X1}^2 + \sigma_{X2}^2 + \ldots$$

$$+\sigma_{LT}^2 = LT * \sigma_D^2$$

For the central theorem of statistics, if the daily demand values are independent of one another and the number of days of the lead time is sufficiently high, the r.v. lead time demand Y tends to be a normally distributed r.v.:

$$Y = N\left(LT * M_D; \sigma_D * \sqrt{LT}\right)$$

With reference to the graph in Fig. 22, it can then be noted that there is a one-to-one correspondence between the safety stock and the probability that the random variable Y—that is, the demand in supply lead time—does not exceed a certain value. For example, if the safety stock is 0, the probability is 50%, if the safety stock is σ_Y, then the probability is equal to $(0.5 + 0.68/2) = 84\%$, if the safety stock is worth 2 * σ_Y, then the probability is equal to $(0.5 + 0.95/2) = 97.5\%$, the safety stock is worth 3 * σ_Y, then the probability is equal to $(0.5 + 0.99/2) = 99.5\%$:

In general, therefore, there is a one-to-one correspondence between safety stock and the probability of having stockouts. Indeed:

$$SS = k * \sigma_Y = k * \sigma_D * \sqrt{LT}$$

Once k is fixed, the safety stock remains fixed on the one hand and, correspondingly, the probability that the demand in lead time does not exceed the value of the safety stock, and therefore of not having OOS situations.

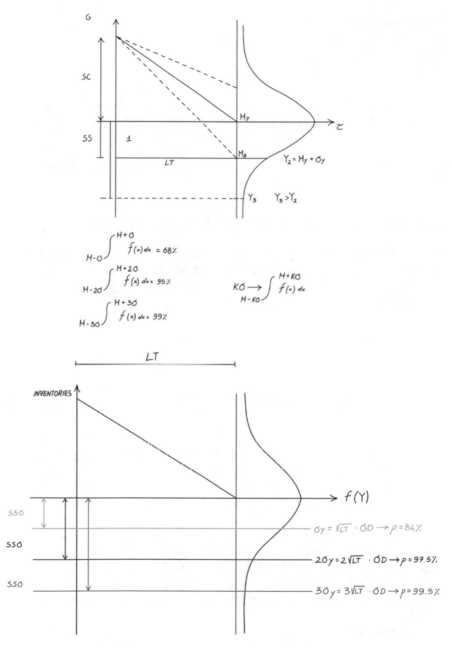

Fig. 22 Representation of the safety stock in the case of deterministic *LT* and stochastic demand and the relationship between safety stocks and probability of OOS

Fig. 23 Representation of
safety stock in the case of
stochastic PLT and
deterministic demand

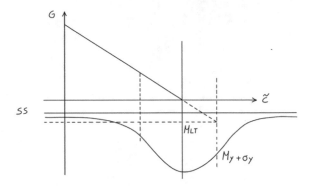

4.3.3   Deterministic Daily Demand and Stochastic Lead Time

This case is the opposite of the previous one. Now the demand is a constant equal to
D, while the supply lead time is a r.v. X, with mean LT and a standard deviation σ_{LT}.

The random variable Y "demand in lead time therefore" is given by a constant D
for the random variable X, and therefore:

$$Y = D * X$$

So:

$$M_Y = D * M_X = D * M_{LT}$$

$$\sigma_Y^2 = D^2 * \sigma_X^2 = D^2 * \sigma_{LT}^2$$

Also in this case, by approximating the r.v. Y to a normal one, it is possible to
make the same considerations on the probability that the demand in lead time will
not exceed a certain value of Y. In this case, the following Fig. 23 applies.

There is a one-to-one correspondence between the safety stock and the probability
that the random variable Y—that is, the demand in supply lead time—does not exceed
a certain value. For example, if the safety stock is 0, the probability is 50%, if the
safety stock is σ_Y, then the probability is equal to $(0.5 + 0.68/2) = 84\%$, if the safety
stock is worth $2 * \sigma_Y$, then the probability is equal to $(0.5 + 0.95/2) = 97.5\%$, the
safety stock is worth $3 * \sigma_Y$, then the probability is equal to $(0.5 + 0.99/2) = 99.5\%$
(see Fig. 24).

In general then,

$$SS = k * \sigma_Y = k * \sigma_D * LT$$

With k fixed, the safety stock is also fixed and therefore the probability of stockout.

$$if\ ss\ =\ \sigma y\ =\ D\ \sigma_{LT}\quad P(Y \le D\ M_{LT} + \sigma_{LT}) = 84\%$$

$$if\ ss\ =\ 2\sigma y = 2D\ \sigma_{LT}\quad P(Y \le D\ M_{LT} + 2D\sigma_{LT}) = 97,5\%$$

$$if\ ss\ =\ 3\sigma y = 3D\ \sigma_{LT}\quad P(Y \le D\ M_{LT} + 3D\sigma_{LT}) = 99,5\%$$

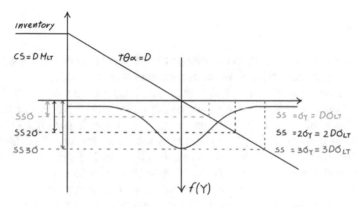

Fig. 24 Safety stock in the case of stochastic LT and deterministic demand relationship between safety stocks and probability of OOS

4.3.4 Stochastic Daily Demand and Stochastic Lead Time

In this scenario, the stochastic effects of lead time and demand are combined.

The random variable Y "demand in lead time" is given by the product of two random variables D * LT:

$$D(M_D; \sigma_D)$$

$$LT(M_{LT}; \sigma_{LT})$$

Assuming that the two random variables are independent, and that therefore the value assumed by the demand has no influence on the possible value assumed by the lead time and vice versa, it follows that the average value of the variable produced is given by the product of the average values:

$$M_Y = M_D * M_{LT}$$

While the variance is given by:

$$\sigma_Y{}^2 = \sigma_{Y1}{}^2 + \sigma_{Y2}{}^2 = \sigma_D{}^2 M_{LT} + \sigma_{LT}{}^2 M_D{}^2$$

$$\sigma_Y = \sqrt{\sigma_D{}^2 M_{LT} + \sigma_{LT}{}^2 M_D{}^2}$$

And therefore also in this case:

$$SS = k * \sqrt{\sigma_D{}^2 M_{LT} + \sigma_{LT}{}^2 M_D{}^2}$$

4.3.5 Conclusions

In all the cases analysed, a direct correlation can be noted between safety stocks and the value of σ_D. In particular, as σ_D increases, the value of SS increases.

It follows that as the Bullwhip Effect increases, there is an increase in the standard deviation of demand which consequently causes an increase in safety stocks, for the same service level k required. Then, as we move upstream in the supply chain, we use an estimator of demand O_{i-1}—given by orders with a level i − 1—which by the definition of the Bullwhip Effect is greater than the standard deviation of effective market demand, is greater the larger it is. It follows that the further a player is upstream in the supply chain, the higher the safety stocks will be at the same service level, or vice versa, with the same stocks, the lower the service level will be, compared to those who are downstream, and has to do with actual market demand.

As will be seen below, the effect of information sharing (POS Data) between all the players of the SC on the one hand and the reduction in LT on the other, mitigate the BE effect, reducing the value of the safety stocks of those located upstream from those downstream.

In other words, a player located upstream in the SC can reduce its safety stocks if it uses as an estimator of demand not the orders of the previous level but the values of the market POS Data, shared by the first player on the one hand, and downstream procurement lead times on the other.

4.4 The Bullwhip Effect

But what is the Bullwhip Effect in practice? How to recognize it? How is it generated? How can it be tackled and managed? These questions are answered in the following paragraphs.

4.4.1 Introduction

In the late 1990s, Procter and Gamble's SC managers were analysing the order trend for one of their high-selling products, *Pampers* diapers. As shown in Fig. 25 the sales in stores showed some variability, but all in all fairly limited. However, by analysing the trend in orders received from upstream retailers, a much greater variability was noted. When in turn P&G observed the progress of their own orders upstream to suppliers, e.g. 3M, realized that the variability was even higher. At first glance, this

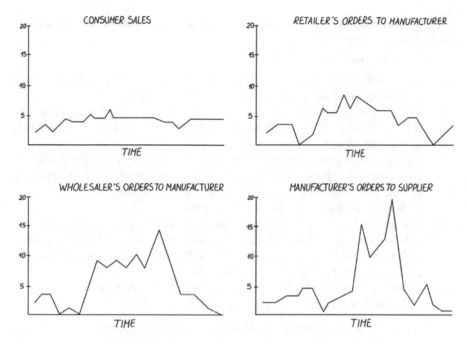

Fig. 25 Increase in the variability of orders when moving upstream in an SC

variability made no sense. As the final consumer was purchasing the product in a stable manner, demand was amplified as it moved upstream in the SC. P&G called this phenomenon the Bullwhip Effect from the shape of the amplification generated upstream in the face of a small variation downstream.

The same phenomenon is experienced in many other sectors from FMCG to the durable goods industry (PCs, cell phones), and from food to healthcare and pharmaceuticals. In the case of the FMCG industry, ECR in the United States has estimated that in certain cases, from when a product leaves the production lines to when it arrives on the shelf, up to 100 days can pass precisely because of the distortion of information which leads every level to create its own buffer, with savings opportunities of 30B USD. In the PC and consumer electronics or pharmaceutical industries, with long multi echelon SCs, where for certain products the supply chain can contain up to one year of demand, the possible savings linked to the optimization of stock levels, the reduction of backlog and OOS could be even greater.

As mentioned in Sect. 4.2.2 the Bullwhip Effect at the generic i-th level of the SC is defined as the ratio between the standard deviation in demand seen from the i-th level, represented by the orders received from the i − 1 level, and the standard deviation in actual market demand, represented by the POS Data.

$$BE_i = \frac{\sigma_{O_{i-1}}}{\sigma_D}$$

The typical symptoms of Bullwhip Effect sufferers are known: high physical costs and high market mediation costs.

In particular, high physical costs are due to:

(i)   high stock levels, since inventories are necessary to cope with extremely variable and difficult to predict demand with high capital costs.
(ii)   the high production costs due to the almost systematic use of extra production capacity in times of high demand, or vice versa, in the case of falls in demand, at low saturation of production lines;
(iii) non-optimized storage and transport costs, also in this case mainly due to the use of extra storage and/or transport capacities to cope with peaks in demand, as well as unused storage and/or transport capacity during falls.
       Instead, market mediation costs are given by:
(iv)   low service level, resulting from stockouts (lost sales) and backlogs (orders processed late entail however a depletion in customer service levels) due to the lack of the product in the face of an order, with deterioration of the downstream relationship between customer and supplier.
(v)   shrinkage costs deriving from the need to keep high safety stocks at all levels which increase the SC flow time and consequently the risks of obsolescence.

Ultimately, in fact, the feeling of any vendor suffering from the Bullwhip Effect and that when there is a product there is a lack of demand, and vice versa when there is no product, the demand is high, inevitably tending to blame the buyer for this with a possible deterioration in supply chain relationships. The same applies to the buyer, that can not fin the product when needed and may be pushed to buy more when he/she does not need it.

4.4.2 The Causes of the Bullwhip Effect

In order to reduce the Bullwhip Effect, it is first of all necessary to understand the causes that generate it, after which we can address them to limit this phenomenon or even eliminate it entirely.

There are four main causes of the Bullwhip Effect in an SC (Lee et al. 1997) in this case:

(1)   The demand forecast
(2)   The reordering mechanism
(3)   Quantity discounts
(4)   Shortage and rationing gaming.

Each of these causes is dealt with in detail in the following paragraphs.

Demand Forecasting Techniques

As already introduced in Sect. 4.2, in common practice, each level of the supply chain uses order data from a downstream level as an estimator of demand. Based on

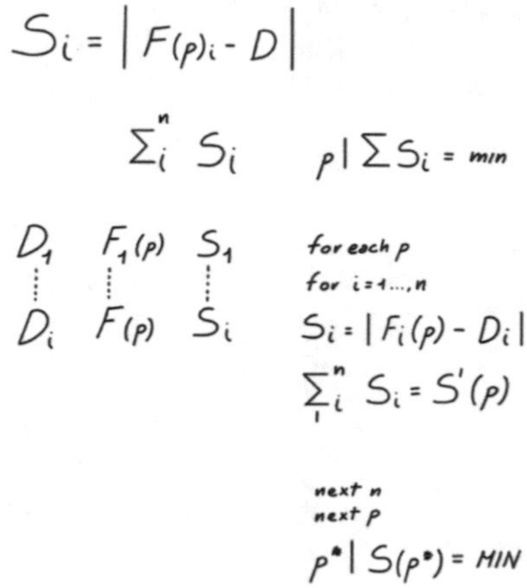

Fig. 26 Optimization of the
moving average interval
finding the p interval that
minimize the S value, being
S the absolute difference
between Forecast F(p) and
Demand D

$$S_i = \left| F_{(p)i} - D \right|$$

$$\sum_i^n S_i \qquad p \mid \sum S_i = min$$

$$
\begin{array}{lll}
D_1 & F_1(p) & S_1 \\
\vdots & \vdots & \vdots \\
D_i & F(p) & S_i
\end{array}
\qquad
\begin{array}{l}
\text{for each } p \\
\text{for } i = 1 \ldots, n \\
S_i = \left| F_i(p) - D_i \right|
\end{array}
$$

$$\sum_i^n S_i = S'(p)$$

next n
next p

$$p^* \mid S(p^*) = MIN$$

the order data, demand forecasting techniques are applied, from the simplest, such
as the moving average, to the most refined, such as exponential smoothing with trend
and seasonality correction. On the basis of the result, the sizing of cycle stocks and
safety stocks is carried out and upstream orders are issued.

The main problem of this mechanism is the fact that demand forecasting techniques must mediate two needs: on the one hand, that of being sufficiently reactive to
perceive a trend reversal in demand; on the other hand, that of not being excessively
variable in order not to chase after insignificant variations. Think, for example, of
the moving average. Reducing the number of moving average periods means having
an excessively variable forecast, which at the very least tends to overlook the trend
and simplistically estimate future demand as the demand of the previous period. On
the other hand, increasing the moving average period too much means perceiving
any change in the trend too late, the more accentuated the more the moving average
period is.

For this, a compromise value is sought between these two needs, determining the
moving average interval value which in the past would have given a minimum MAPE
(Mean Absolute Percentage Error) value, as illustrated in the procedure of Fig. 26.

As a result, in the face of an increase/decrease in downstream demand, those who
are upstream and apply demand forecasting techniques to those data sets, interpret
that variation as a trend and consequently amplify its extent, as shown in Fig. 27.

Period	Demand	Forecast Days of moving average			
		1	2	3	4
1	2				
2	2	2			
3	3	2	1,5		
4	3	3	2,5	2	

(continued)

Fig. 27 Bullwhip effect
induced by demand
forecasting techniques

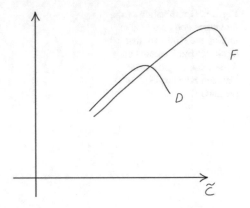

(continued)

Period	Demand	Forecast Days of moving average			
		1	2	3	4
5	3	3	3	2,67	2,25
6	2	3	3	3	2,75
7	1	2	2,5	2.67	2,75
8	1	1	1,5	2	2,25
9	2	1	1	1,33	1,75
10	3	2	1,5	1,33	1,50
11	2	3	2,5	2	1,75
12	1	2	2,5	2,33	2

If those upstream apply a reordering policy (e.g., economical lot), there will consequently be an amplified increase/decrease in both Economic Order Quantity— EOQ (in the economic lot formula, the EOQ value depends on the square root of demand) and stocks (which depend on D^2) and therefore an increase/reduction in the quantities ordered, the greater the higher the supply lead times (see safety stocks formula).

$$EOQ = \sqrt{\frac{2 * D * C_{reorder}}{C_{holdingstocks}}}$$

$$SS = k * \sqrt{\sigma_D^2 * LT * \sigma_{LT}^2 * D^2}$$

Even applying the order up to level policy, the result is the same. Given O as the quantity to be ordered at each Reorder Interval RI interval, this is equal to the target level OUTL (Order Up To Level) minus the stock available at the time of reordering. The Target level in turn must meet the demand for the entire RI reorder interval and the supply LT (whether deterministic or stochastic). So, ultimately the order is a function of D.

$$O = OUTL - G = (IR + LT) * D - G$$

With regard to safety stock, in the case of a reorder interval, the following formula applies, since the safety stock, unlike the case of the economic lot, will have to face possible variations in demand not the LT but in RI + LT.

$$SS = k * \sqrt{\sigma_D{}^2 * (RI + LT) + \sigma_{LT}{}^2 * D^2}$$

It follows that also in this case the safety stock is a function of D and σ_D. Ultimately, those who are upstream and see a variation in demand from downstream, in applying demand forecasting techniques, follow this variation interpreting it as a trend, recalculate reorder lots and safety stocks and transmit the information upstream the more amplified the greater are the LT.

This happens as we move upstream in the SC.

Reorder Mechanism

The reordering mechanism also affects the increase in variability of orders with respect to actual demand. Put simply, the mechanism of the reorder interval or of the economic lot linked to reorder costs means that even in the face of constant demand, orders are concentrated in a period, followed and preceded by periods with zero demand, which therefore increase the variability.

Suffice to think of the simplified case represented in Fig. 28: Neglecting safety stocks (imagining deterministic demand and lead times), a buyer that supplies itself with a certain reference at a reorder interval every week with a one week LT, against a constant demand of 10 pcs/day will order 70 pcs upstream each time, in order to avoid ordering anything in the following 6 periods.

It follows that the supplier, which estimates demand on the basis of the buyer's orders, sees a standard deviation different from 0 (equal in this specific case to 26.14),

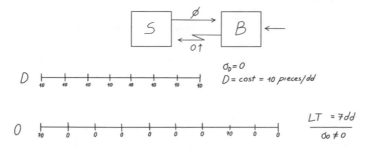

Fig. 28 Bullwhip Effect induced by a batch reordering mechanism

and on the basis of this estimates its own safety stocks and consequently the orders it will issue towards higher levels. Therefore, in the face of a demand Sigma equal to 0, there is an order Sigma other than zero deriving from the batch reordering mechanism, hence a bullwhip effect.

In the case of a reordering interval, the phenomenon is all the more amplified the greater the LT and/or the reordering interval, since the sum between RI and LT is shown in the reordering level formula. In the case of reorder level, on the other hand, the LT affects it by increasing the necessary safety stock, which varies with the root of LT.

The reasons why a buyer procures batch and non-continuous supplies may be many: the buyer is unable to manage frequent reorders due to reorder costs; for example, the administrative management costs, from the issuing of the order to the management of the debit-side invoice to the payment thereof, may be such as to lead to the accumulation of more goods in a single order; the use of production scheduling systems such as MRP, which run on a monthly basis, lead to the accumulation of demand in a single monthly order. Furthermore, the transport costs can lead to the accumulation in a single order of quantities that can generate a transport with full load, thereby optimizing transport costs.

Demand Forecasting and Reordering Mechanism—Quantification of the Bullwhip Effect

One of the fundamental articles in the literature which first quantified the Bullwhip Effect phenomenon triggered by the application of demand forecasting techniques and reordering mechanisms was that of (Chen et al. 1999). Although this work represented a particular case, it has the merit on the one hand of analytically quantifying the phenomenon, and on the other of highlighting the factors on which it depends and therefore how it is possible to mitigate the phenomenon.

Assumptions

The model considers an n-level SC like the one represented in Fig. 29.

In which, in the face of a market demand D (0 level orders), each level, in particular the generic i-th level:

- Uses the orders of the $i - 1$ O_{i-1} level as an estimator of demand (unshared demand);
- Applies a p-period moving average model to estimate demand in the subsequent period;

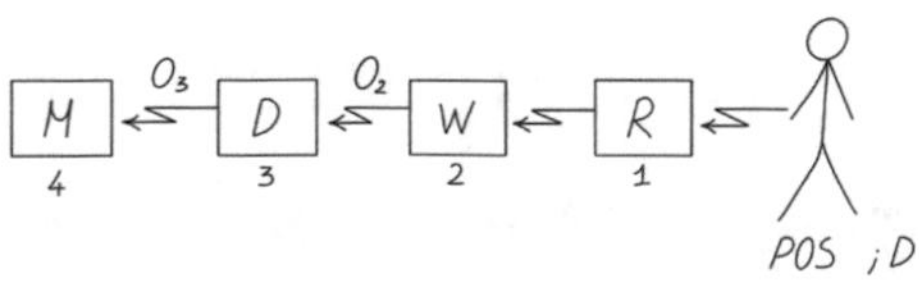

Fig. 29 The SC underlying the model of (Chen et al. 1999) to quantify the Bullwhip Effect

- Procures from upstream following an order up to level policy, by issuing orders O_i at each period towards level $i + 1$.

In their work, the authors provide a quantification model of the Bullwhip Effect following the application by each SC player of demand forecasting and reordering mechanisms. The reordering policy adopted is that of the reordering interval. For each period, each player calculates the Order (order up to level) to be issued, as well as the safety stock:

- The order is given by the Target stock minus the stock in hand G. In turn, the target stock OUTL is equal to the average demand estimated in the lead time $D * LT$:

$$O = OUTL - G = D * LT - G$$

- the safety stock is used to cope with changes in demand in lead time (assuming the lead time is deterministic) and is equal to $SS = k * \sigma_D * \sqrt{LT}$, where k depends on the service level chosen:

$$SS = k * \sigma_D * \sqrt{LT}$$

O and SS are recalculated for each period by applying the moving average to p periods, therefore the estimators of the average demand D_t and of the variance σ_t^2 at the generic instant t are obtained respectively from the following equations:

$$D_t = \frac{\sum_{i=t-p}^{t-1} D_i}{p}$$

$$\sigma_t^2 = \frac{\sum_{i=t-p}^{t-1} (D_i - D_t)^2}{p - 1}$$

Unshared Demand

In accordance with the definition given in Sect. 4.2.2, the authors estimate the Bullwhip Effect as the ratio between the standard deviation of i level orders and the standard deviation of demand D.

$$BE = \frac{\sigma_{Oi}}{\sigma_D}$$

And they show that under these conditions, there is a lower bound to the bullwhip effect equal to:

$$BE = \frac{\sigma_{Oi}}{\sigma_D} \geq \prod_{k=1}^{i-1} \left(1 + \frac{2 * LT_k}{p} + \frac{2 * LT_k^2}{p^2} \right)$$

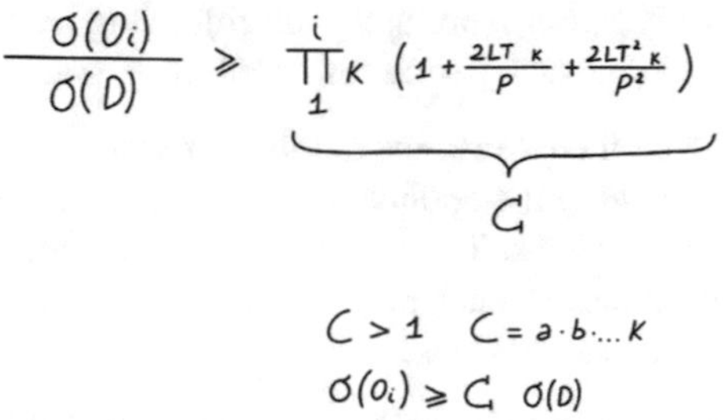

Fig. 30 The bullwhip effect
is greater than a constant C,
which is greater than 1

$$\frac{\sigma(O_i)}{\sigma(D)} \geq \underbrace{\prod_1^i K \left(1 + \frac{2LT\ K}{P} + \frac{2LT^2\ K}{P^2}\right)}_{C}$$

$$C > 1 \quad C = a \cdot b \cdots K$$
$$\sigma(O_i) \geq C\ \sigma(D)$$

The term in Pi (the right-hand member of the equation above) is given by the sum of always positive numbers and therefore is greater than 1; consequently, also the standard deviation of the orders is greater than the standard deviation of the demand by a factor C greater than 1 (see Fig. 30):

The value of C and therefore of the Bullwhip Effect is greater: (i) the higher i is, since as i increases so do the terms in Pi, and therefore the more we move upstream in the SC; (ii) the higher the lead time is, which appears as a numerator in Pi. The value of p, on the other hand, cannot be considered a variable since it is determined (as we have said) to minimize forecast errors.

Shared Demand

The authors then adapt the model to the case in which all the players of the SC, instead of relying on the orders of the lower level to estimate the demand, can use the data of the real demand of the market, the so-called POS Data. In other words, it is assumed that level 1 (usually the retailer), which is the level that sees the POS Data directly, shares them with all the players of the SC, those who use them to recalculate the moving average of average demand at each period and variance and apply the reordering interval technique.

Under the hypothesis of shared demand, Chen et al. demonstrated that the lower bound of the Bullwhip Effect changes as follows:

$$BE = \frac{\sigma_{Oi}}{\sigma_D} \geq 1 + 2 * \frac{\sum_{k=1}^{i-1} LT_k}{p} + 2 * \frac{\left(\sum_{k=1}^{i-1} LT_k\right)^2}{p^2}$$

Also in this case, the terms of the summation are all greater and positive, so that the ratio between the standard deviations is certainly greater than 1. However, the sharing of demand between the players of the SC changes the formula of the Bullwhip Effect from a Pi to a Sigma, drastically reducing the value of the BE in the case of shared demand compared to the case of non-shared demand.

Assuming, for example, a 4-level SC (retailer (1), wholesaler (2), distributor (3) and manufacturer (4) like the one represented above), if $LT = 3$ days and $p = 5$ days, in the case of unshared demand, the manufacturer on the fourth level sees a $BE_{manufacturer}$

$$BE_{manufacturer(4)} = \frac{\sigma_{Oi}}{\sigma_D} \geq \prod_{k=1}^{3}\left(1 + \frac{2*3}{5} + \frac{2*9}{25}\right) = \left(1 + \frac{6}{5} + \frac{18}{25}\right)^3$$

$$= \left(\frac{73}{25}\right)^3 = 24,89$$

while in the case of a shared demand we have:

$$BE_{manufacturer(4)} = \frac{\sigma_{Oi}}{\sigma_D} \geq 1 + 2*\frac{\sum_{k=1}^{3}3}{5} + \frac{\left(\sum_{k=1}^{i-1}3\right)^2}{5^2}$$

$$= 1 + \frac{18}{5} + \frac{162}{25} = \frac{205}{25} = 8,2$$

The sharing of demand involves for the manufacturer a reduction in the bullwhip effect, and therefore of the safety stocks, of around one third.

In conclusion, two aspects can be highlighted from the formulas of Chen et al.

- Firstly, in a two-level SC there are no differences between the unshared model and the shared model, while the benefits in terms of reducing the Bullwhip Effect are greater the deeper the SC is.
- Secondly, the sharing of demand reduces the Bullwhip Effect but does not completely eliminate it. To eliminate the Bullwhip Effect it is necessary to move from models based on the reordering mechanism at various levels to coordinated models, in which the reordering process for all the players is managed in a coordinated manner at a single level, using appropriate techniques such as CPFR (Collaborative Planning Forecasting and Replenishment) and coordinated DRP (Distribution Requirement Planning). These techniques will be thoroughly detailed later in the chapter.

Promotions and Quantity or End-of-Period Discounts

Recourse to the mechanism of unshared promotions or quantity discounts can lead to a significant increase in the Bullwhip Effect.

We speak of an "unshared promotion" when the retailer unilaterally carries out a promotion on a particular product without agreeing it first with the manufacturer, or even notifying it.

During, but especially before a promotion, product demand can increase significantly, even 5–10 times compared to continuous consumption. It follows that the retailer, to cope with the promotion, procures upstream the quantity necessary to cope with the promotion. This is a so-called "installation order", an order needed to supply the product to meet demand in the first week of the promotion. The manufacturer, applying the usual demand forecasting mechanisms, interprets the peak demand of the installation order as a trend and increases production volumes. After one or two weeks, at the end of the promotion, the demand returns to the initial value so that the retailer reduces orders or places any (to dispose of any stock accumulated and not sold during the promo) just as the manufacturer is increasing its volumes following the skewed interpretation of sales data through its demand forecasting models. Volumes which, instead, will not find demand in order to be disposed of.

To limit this phenomenon, it is necessary to move from unshared promotions to shared ones, in which manufacturers and retailers agree on the timing and duration of the promotion, as well as the quantities intended for the promotion. The CPFR (Collaborative Planning Forecasting and Replenishment) programme, developed by ECR in the early 2000s, (VICS 2004) is precisely a collaborative programme between industry and retail created with the aim of sharing promotions, precisely to prevent these from triggering phenomena such as those described above. Through planning, controlling the execution of the promo through appropriate KPIs, identifying problems and their resolution, it is possible to manage promotions in a more efficient and effective way, with lower costs (given that the bullwhip effect has been eliminated) and a higher service level for the final consumer.

Manufacturer discounts based on purchased quantities can also be an incentive to order more than real demand. If the stock maintenance cost is lower than the purchase price differential, faced with a discount, the buyer tends to order more than necessary to meet demand over the period, precisely to take advantage of low prices and cover a longer horizon. So much so that in the literature the formula of the economical lot has been adapted precisely to consider the case of possible discounts linked to the quantities ordered.

The quantities ordered can be one of the aspects which trigger price reductions; these are incentives put in place by sales departments to try and increase sales and reach a sales budget faster. In general, traders, especially in the FMCG sector, have significant budgets with which to incentivize sales through periodic discounts not only linked to quantities, but also at the end of the period, unilateral ones. The effect of discounts on the SC is, however, to increase the Bullwhip Effect, since it further amplifies the phenomenon of variability, represented by high orders in a certain period, followed by long periods with no orders, even in the face of steady demand for a product.

Occasionally, sales managers, in order to reach the end-of-period budget, are prompted to offer discounts in order to reach their target volumes. The effect of these discounts is exactly the same. An increase in orders to take advantage of low prices, followed by a fall in orders in subsequent periods due to product availability. The result also in this case is an increase in variability and therefore a Bullwhip Effect for those upstream.

Rationing and Shortage Gaming

When product demand exceeds supply, the supplier tends to ration orders. For example, faced with an overall demand greater than the quantities available, a manufacturer could proportionally decrease the quantities for each customer in order to be able to satisfy all the customers at least in part. For example, if the available offer is only 50% of the total demand, each customer could receive half of the quantities ordered. However, if the buyer perceives this mechanism, it could be prompted to order double the effective demand, knowing that in this way he will see his real demand satisfied by receiving half. All of this, however, due to the aforementioned demand forecasting mechanisms, generates a Bullwhip Effect for the manufacturer, since in subsequent periods, when the product becomes available again, demand will return to real levels again.

This mechanism is extremely common, especially for innovative products, such as consumer electronics, or electronic components. Whenever a new product is launched, the downstream rings, fearing that the supplier will not be able to produce the required quantities, tend to increase orders beyond the actual demand for the product. Suffice to think of the launch of a new phone model by Apple, for example. A retailer, knowing that Apple will not be able to meet the demand from all retailers for the new iPhone model, will place an order 2–3 times the actual demand, generating rationing and shortage gaming and therefore a Bullwhip Effect. The same applies in the case of a perceived shortage, such as the recent cases of Covid pandemic. Buyers tend to order far above their real needs, hoping to receive the desired quantities, which may actually be lower than those ordered.

4.4.3 Reduction of the Bullwhip Effect

Understanding the causes of the Bullwhip Effect is the first step in putting into practice the correct strategies to mitigate it. For each of the 4 causes of BE identified, the interventions capable of limiting its effects or eliminating it are outlined.

BE reduction initiatives can be categorized in three main strands:

- **Sharing of information**: this category includes all initiatives aimed at sharing stock and sales data throughout the SC;
- **Channel coordination**: this perimeter includes initiatives aimed at coordinating commercial activities (volumes, promotions, prices) and logistical activities (transport, stock management) between buyers and vendors;
- **Operational excellence**: this category includes initiatives aimed at improving logistical performance, such as, typically, a reduction in lead times and operating costs (e.g. reorder costs, transport costs).

For each cause of the Bullwhip Effect, countermeasures are then outlined to limit their effects, trying to frame them along each of the three directions above.

Demand Forecasting Techniques

Normally, each SC player estimates its demand independently, using order data and applying its own demand forecasting techniques. We have seen how the BE is significantly reduced by moving from an unshared model (the one described above) to a shared model, in which, on the other hand, the final consumer demand data, the so-called POS Data, are shared among all the links in the chain and used by all players to estimate demand. A first way to remedy the emergence of BE in an SC is therefore to share effective demand data (POS Data) throughout the SC, so that all the players can update their demand data, not on orders, but on the basis of the final consumer demand data. This information can now be easily transmitted via EDI between company ERPs through a simple software integration. In real time, therefore, the sales data of the cash registers can be made available to all stakeholders, who can use them to apply their own demand forecast models.

Reorder Mechanism

To reduce the BE introduced by this cause it is necessary to operate at the levels of both *channel coordination* and *operational efficiency*. In both cases, the goal is to try to move from a logic of high reordering lots with reduced frequency, to small reordering lots, equal to the instant demand limit, with high frequency.

From an operational point of view, it is therefore necessary to lower procurement lead times and reorder costs.

We have seen that the BE increases as LT increases, given that, as LT increases, both the cycle stocks (in the case of a reorder interval) and the safety stocks (both by interval and by reorder level) also increase. It is therefore necessary to intervene and limit lead times through operational efficiency interventions which can reduce the terms of which the lead time is composed—first of all breaking it down into its various phases, and then working on each of them with a view to BPR.

As for the reorder costs, these may depend on the cost of issuing the order and the transport costs.

If the costs to issue the order are high, the supply batches or reorder intervals will also be high. To reduce reorder costs, it is necessary to switch from manual reordering systems to automatic reordering systems, in which orders are generated automatically and at zero cost by the systems and the operator merely performs a check and manual corrects orders with individual codes. The orders themselves are transmitted via EDI between systems. The fixed part of the fulfillment costs should be reduced as well, to make convenient even small orders prevent aggregation into single batch orders.

To lower transport costs and switch from LTT deliveries to FTL deliveries, it is possible to increase the batches or increase the reorder interval, generating BE towards the upstream rings. To increase the frequency of delivery without increasing transport costs, it is necessary to use 3PLs that are able to ensure fully loaded transport, realizing economies of scale on volume.

From a channel coordination point of view, buyers and suppliers can decide to adopt techniques such as cross-docking. By consolidating orders from multiple stores into a single load, it is possible to make FTL deliveries, increase delivery frequencies, and reduce the batches.

Promotions and Quantity Discounts

Another method to control the BE resulting from the purchase of a product in excess of the necessary quantity in the short term, is to reduce both the frequency of promotions and the level of discount granted.

Among the interventions at the channel coordination level, the manufacturer can decide to incentivize Everyday Low Cost (EDLC) policies. Instead of granting quantity discounts linked to volumes which, as we have seen, generates a BE, the retailer is guaranteed a low constant price which guarantees a supply equal to what is strictly necessary.

In general, however, if the promotions between manufacturers and retailers are shared and coordinated, they do not trigger a Bullwhip Effect. The manufacturer knows exactly when a promo will take place, and to what extent, and can schedule itself in time, knowing that at the end of the promo the demand will return to the continuous consumption value.

Uncoordinated promotions can also generate tactical purchases by the retailer. In other words, the retailer purchases more quantities than the promotion to then sell the product at full price or in other regions where the promotion is not in progress, or at full price once the promotion has ended.

With these policies, the manufacturer tries to eliminate uncoordinated promotions and purchase phenomena of large quantities of product linked to quantity discounts which generate costs linked to the BE higher than the quantity discounts granted. In other words, the benefit obtained in terms of better production planning and inventory control is greater than the loss of margins deriving from the discounts granted. The application of such techniques as Activity Based Costing (ABC) allows the manufacturer to accurately quantify these aspects.

In the same way, the retailer can decide to do the same with the consumer, applying Every Day Low Prices EDLP policies, rather than trying to attract the consumer to the store with the promotion technique. Retailers such as Zara or Walmart were the first to recognize the importance of maintaining constant demand to better plan procurement, inventory and service levels throughout an SC. For this reason, for Walmart, EDLP policies and the scarce recourse to promotions have become a real corporate claim over the years. Conversely, in Italy, promotions are widely used, trying as much as possible (but not always …) to coordinate them between retailers and manufacturers to eliminate the bullwhip effect.

Rationing and Shortage Gaming

To eliminate the phenomenon of gaming in periods of shortage, suppliers can decide to assign the product not on the basis of orders but, for example, based on the importance of the customer. In other words, a manufacturer, in the presence of limited product quantities, can decide to assign the product to customers on the basis of past sales, first meeting the demand of the most important customers and generating backlogs or cancelled orders for less important customers. In this way, customers do not issue orders exceeding the necessary quantities, since they know that in any case if they are important customers their orders will be fulfilled, while if they are not important customers their orders will be fulfilled only in part or not at all. Interventions of this type fall into the category of "channel coordination".

Another method to alleviate the phenomenon of gaming, which instead falls into the category of "information sharing", is to share the levels of production capacity and stocks with customers. This type of visibility, especially in periods when there is a fear of product shortages, relieves the tension on the part of the customer who, seeing the actual availability of the product, is less inclined to resort to excess orders.

In some sectors, the possibility granted to the customer to return unsold products or cancel orders at no cost can induce the customer to issue orders well above the actual demand. In the electronics industry, for example, a customer may order different PC models, then return the unsold ones. Also in this case, to contain this gaming phenomenon, it is therefore necessary to limit the no-cost return policy. Vice versa in online, despite the return at no cost, it encourages gaming, since a customer can be led to order different sizes to then return the ones that do not fit and keep only the right one, there is a tendency to maintain a return policy at no cost to however encourage access to the portal, confidence/trust and the purchasing habit. Therefore, returns must be considered and subtracted from sales to estimate demand.

Conclusions

The Table 6 summarizes the causes and remedies to limit the impact of the Bullwhip Effect, divided into three categories (*information sharing, channel coordination* and *operational excellence*).

4.4.4   Eliminating the Bullwhip Effect—Collaborative Planning Forecasting and Replenishment

One technique which falls among those of "channel coordination" with which it is possible not only to reduce but completely eliminate the phenomenon of BE deriving from all 4 causes, is that of CPFR—Continuous Planning Forecasting and Replenishment (VICS 2004).

CPFR is a business process to coordinate the activities of business partners in planning and meeting customer demand.

CPFR was developed in the late 1990s by VICS—Voluntary Interindustry Commerce Standards. The CPFR guidelines were first published in 1998 by VICS, a US association which brings together FMCG manufacturers and distributors. Revisited by ECR (Efficient Consumer Response) in 2005, they showed that they can bring significant improvements in the reduction of OOS (2–8% less) and at the same time a reduction in stocks of between 10 and 40%.

The goal is to coordinate commercial activities (sales and marketing) with those of the supply chain (procurement, production and distribution) and the players involved

Table 6 Interventions to mitigate or eliminate the causes of the bullwhip effect

Causes of the bullwhip effect	Sharing of information	Channel coordination	Operational excellence
Demand forecasting techniques	Sharing POS Data and using them as a demand estimator		Demand forecasting algorithms Adoption of EDI links as a tool for sharing PoS data
Order batching		Cross docking	Reduction in LT Automatic reordering EDI 3PL for FTL deliveries
Promotions and quantity discounts		Agreed promotions EDLC Everyday low costs EDLP Everyday low prices	ABC to understand the extra cost impact linked to BE and compare it with sales growth due to quantity discounts
Rationing and shortage gaming	Sharing by the vendor of production capacity and inventory	Allocation based on past sales	Including returns in the demand estimate Costs for returns
All	CPFR—collaborative planning forecasting and replenishment		

in the supply chain to increase product availability and minimize out of stock situations, and, at the same time, reduce the costs of production, inventory, transportation, logistics and shrinkage. And all of this not at the level of a single player, but at an overall SC level. In other words, increasing the service level (reduction of OOS; product availability), while cutting physical costs (procurement, production, distribution, stocks) and related market mediation costs (OOS and/or shrinkage).

The aim is to eliminate variability of demand, and therefore reduce stocks and improve the flow of products and services between manufacturers and retailers, through techniques which fall into all three of the BE reduction guidelines seen above: *information sharing*, *channel coordination*, and *operational excellence*. In other words, reducing the variability of demand in order to bring the SC to the case of functional products with low supply uncertainty, therefore manageable with a lean SC, with DRP programmes which minimizes physical costs while guaranteeing the required service level.

In order to function, CPFR needs a potent digital integration. Information must be exchanged, and transactions must take place automatically and not manually, in electronic and paperless formats. The master data must be common to avoid ambiguity, and aligned in real time. Data recording must take place in real time and the information made immediately available to all trading partners. In this way, the SC becomes transparent, and all its players can plan and optimize activities and stocks.

The CPFR model is illustrated in this context with reference to the case of the FMCG sector, in which it was born. In this context, the main players are the manufacturer, the retailer, and the final consumer.

The CPFR is based on 4 fundamental phases:

- Strategy and planning
- Demand & supply management
- Execution
- Analysis.

The phases are shown for display purposes in sequence; in reality they can be present between the coverings and the loops. Furthermore, in some cases, the trading partners may share only certain phases (e.g. strategy and planning) and carry out the others independently. In this case, we speak of "CPFR Lite".

Each phase is in turn broken down into 2 sub-activities, which will be illustrated in the respective sub-paragraphs.

Strategy and Planning

In this phase, the foundations of the collaborative strategy are laid down.

The mix of products and their positioning are defined (stores involved) as well as event plans (continuous sales, promotions) in the reference period (usually one year).

The two activities which make up the strategy and planning are:

Collaboration Arrangement

The activities to be identified in this area are a definition of the business objectives of the collaboration between manufacturers and retailers over the time horizon of CPFR. In this phase are defined the roles, responsibilities, and players involved (e.g. manufacturer side: are production third parties involved? retailer side: are the stores all owned, or there are other players, e.g. franchisees? For both parties: are third party logistics involved?).

Joint Business Plans

In this phase a proper business plan is made, defining the products involved in CPFR (mix and quantity to sell), the Distribution Centres and stores involved and the relative opening and closing days. The introduction and launch of new products is planned. In this phase, the reorder logic and stock levels (cycle and safety stock) are also defined, as are the lead times for preparation/fulfilment of orders, and transport.

One very important aspect is the planning of promotion periods. Promo planning means agreeing for each reference when it will be on promotion, at which stores and for how long, and at what discounts, and consequently what quantity will presumably be sold and must therefore be produced and distributed to meet the demand.

Demand and Supply Management

At this stage, the sales forecasts for the planning horizon are defined, as well as the planning of orders and shipments.

The two activities which make up the Demand & Supply Management are.

Sales Forecasting

Once the continuous and promotional sales periods have been defined, sales forecasts are prepared for each product, based on the POS Data at the various stores during the continuous and promotional sales periods.

Order Planning Forecasting

On the basis of the sales forecasts, orders are planned with which to ensure the restocking of the goods to the stores based on shared reordering policies, lead times, cycle stocks and necessary safety stocks.

Execution

This is the actual operational phase, in which orders are issued and must therefore be processed and shipped. The product received is then stored to be put on the shelf and sold. The cycle ends with the economic transactions and the payment of the sales invoices. We therefore speak of the "Order to Cash" (O2C) cycle to emphasize the fact that the cycle does not end with the delivery of the product but just as important is the economic and financial transactions (issuing of the invoice; payment) which complete the physical cycle, and with which are ensured the correct remuneration and the liquidity necessary for the functioning of all trading partners.

The two activities which make up the execution are.

Order Generation

In this phase, the planned orders are transformed into actual orders which are prepared by the retailer and sent to the manufacturer. Best Practices envisage a flow of orders via EDI between retailers and manufacturers, in order to minimize timeframes and improve accuracy.

Order Fulfilment

This phase includes all the logistical activities of order fulfilment and therefore for the manufacturer, retrieving and/or picking, packing, shipping, transportation and for the retailer, receiving, storage, shelf replenishment, and sale. A particular focus should be placed inorder to avoid duplication of activities that typically take place at the inbound/outbound (i.e. is the double check at shipping and receiving redundant?)

Analysis

The activities of the three previous phases are monitored through a set of KPIs in a logic of continuous improvement (plan/do/check/act). Causes of non-functioning are analysed and relevant adjustments made.

The two activities which make up the Analysis are.

Exception Management

Constant monitoring of exceptions and non-conformities, identification of the causes that led to the non-conformities and definition of solutions to prevent them from happening again.

Performance Assessment

This is the "check and act" phase in the continuous improvement cycle, and therefore the elaboration and calculation of KPIs to evaluate the achievement of the objectives, verify trends, and make alternative decisions in the event that the objectives cannot be achieved.

The Roles of Manufacturer and Retailer in CPFR

In Table 7, for each phase and each activity, the activities carried out by the retailer, and those carried out by the manufacturer in addition to those carried out jointly already described are reported.

From this derives the overall scheme of CPFR reported in Fig. 31.

Table 7 CPFR: activities carried out by the retailer and the manufacturer

Retailer's tasks	Collaboration tasks	Manufacturer's tasks
Strategy and planning		
Vendor management	Collaboration Arrangement	Account planning
Category management	Joint business plan	Market planning
Demand and supply management		
POS forecasting	Sales forecasting	Market data analysis
Replenishment planning	Order planning/forecasting	Demand planning
Execution		
Buying/rebuying	Order generation	Production & supply planning
Logistics/distribution	Order fulfillment	Logistics/distribution
Analysis		
Store execution	Exception management	Execution monitoring
Supplier scorecard	Performance assessment	Customer scorecard

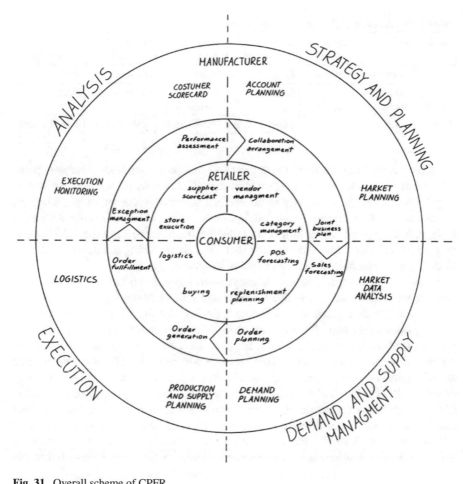

Fig. 31 Overall scheme of CPFR

Traditional Reordering and Quick Response, Continuous Replenishment,
VMI

In the implementation of CPFR, different collaborative scenarios are possible,
ranging from the traditional reordering model to a Vendor Managed Inventory (VMI),
depending on who is involved, manufacturer or retailer.

(1)   Three fundamental processes are involved: sales forecasting, order planning,
        order generation
(2)   Ownership of stocks is maintained: manufacturer or retailer.

We can therefore distinguish between the conventional model of reordering, Co-
Managed Inventory, Continuous Replenishment, and Vendor Managed Inventory (see
Table 8).

Table 8 The different implementation options of CPFR

Alternatives	Sales forecasting	Order planning	Order generation	Stock ownership
Conventional order management	Retailer	Retailer	Retailer	Retailer
Co-managed inventory Continuous replenishment CR	Retailer	Retailer or manufacturer	Manufacturer	Retailer
Vendor managed inventory VMI	Manufacturer	Manufacturer	Manufacturer	Manufacturer

In the conventional model, it is the retailer that processes sales forecasts, plans orders and generates and sends them to the manufacturer, while maintaining ownership of the inventory. A slightly more evolved scenario is that in which, while all the above activities remain in the hands of the retailer, the retailer shares its POS Data with the manufacturer. In this way, the manufacturer can use the POS Data to process its sales forecasts and not the orders, thereby reducing the Bullwhip Effect. In this case, we technically speak of Quick Response.

In the opposite model, Vendor Managed Inventory, in contrast, it is the manufacturer that takes care of the three phases and retains ownership of the stocks at the retailer (we also speak of consignment stock to emphasize the fact that this is actually stock of the manufacturer and managed by the manufacturer on the retailer's shelf or at its warehouse); the retailer becomes the owner only when the product is purchased by the final consumer. This model is win–win, since for the retailer there are no capital costs (the goods become the property of the retailer only at the time of selling them to the final customer), and also the financial leverage is considerable (the customer pays for the purchase while the manufacturer is paid at least 60 days later). The manufacturer directly manages the replenishment cycle, and in this way reduces inventories and is able to better plan production and distribution, since the Bullwhip Effect is eliminated. Furthermore, thanks to the service provided to the retailer, it sees its business share increase compared to competing suppliers. It is evident, in fact, that for the same product and price, the retailer will tend to entrust greater business shares to those suppliers able to provide a better service (e.g. VMI).

Intermediate models of Co-Managed Inventory or Continuous Replenishment foresee, conversely, that the goods are owned by the retailer, as well as the task of sales forecast, while the manufacturer is in charge of order generation. The part of planning the orders can be in the hands of the manufacturer or the retailer as the case may be. In other words, the retailer shares the demand and stock data with the manufacturer, and entrusts the latter with the task of maintaining stock between a minimum and a maximum; however, the stock still remains the property of the retailer.

One of the most know application of VMI started between Procter & Gamble (P&G) and Walmart. Under this system, the vendor based on a pre-agreed level of service with the retailer, decides both on; the appropriate level of inventory to be kept at the retail store, up to the store's sheles and backroom, and the corresponding inventory policies (e.g. delivery intervals, inventory levels) to maintain such targets.

Therefore, there is no need of a sales order to trigger delivery or replenishment process to the retailer. The vendor (manufacturer) automatically receives both inventory level and relative sale data (i.e., demand) direct from retailer and defines the right quantity of goods to deliver. Ultimately, the manufacturer may also assume the ownership of the products within the retail DC and store (consignment) until the retailer briefly becomes the owner when the product is sold. In this system, there is a potential opportunity to eliminate the bullwhip effect. Firstly, there is an elimination of one layer in the replenishment decision-making process. Second, the demand information arrives directly to the manufacturer making it possible react quickly like a just-in-time system, decreasing complexity and reducing demand variability.

In the P&G and Walmart partnership, P&G set up a team to facilitate information-sharing between the two firms and addresses logistics, supply, management information systems, accounting, finance, and other issues. Under this arrangement, Walmart shares point-of-sale information from retail outlets directly with P&G, giving the latter easy access to information on consumer transactions and buying patterns. This improved P&G's on-time deliveries to Wal-Mart and the other retailers while increasing inventory turnovers.

4.4.5 Eliminating the Bullwhip Effect

In an uncoordinated approach, each link of the SC independently manages all the phases of the order cycle, from the forecasting of demand, based on the orders of the previous level (unshared demand) or on POS Data (shared demand), to the planning of orders—OP_i. and to the management of the safety stock level, to the release of orders—OR_i. All of this generates a Bullwhip Effect on the downstream rings.

Through the CPFR technique, retailers and manufacturers coordinate their commercial activities with those of the SC, eliminating the causes of the Bullwhip Effect. For example, through VMI, a manufacturer, in addition to being the owner of the goods, deals with demand forecasting, planning and release of orders, guaranteeing the retailer a stock that can vary between a minimum level (safety stock to cope with variations in demand or supply lead time) and a maximum level, depending on the retailer's availability to stock for the manufacturer at its facilities. The same applies to a Co-managed inventory, but for stock ownership. In this way, the manufacturer eliminates the Bullwhip Effect and plans the replenishment in order to minimize the backlog and at the same time the stocks that are responsible for the related maintenance costs.

In the case of FMCG products, among which CPFR was born, once the variability of demand has been eliminated thanks to CPFR, retailers and manufacturers find themselves working with functional products along with low supply uncertainty (absent possible variables that are difficult to predict, such as meteorological conditions) so that they can apply the lean management techniques of the SC, such as the DRP technique to manage procurement, production and distribution in a coordinated and transparent manner.

The DRP technique is illustrated in the following paragraphs.

DRP—Definitions

Distribution Requirement Planning (DRP) is an order planning technique. It closely resembles MRP (Material Requirement Planning), and in fact represents an extension of it from the production level to the supply level. As seen in the previous chapter, this technique is particularly suitable for managing the supply of products with stable demand and stable lead times, since variability in demand and lead times represent a criticality that is difficult to absorb, especially in the case of long LTs.

The objective of DRP is to satisfy demand D, by minimizing:

- Physical costs, represented by Holding costs in stock I (Inventory);
- Market Mediation Costs, represented by Backlog costs (BL)—the backlog represents the cost of having to manage customer backlogs due to product unavailability. Costs are generally higher as, on the one hand, all economies of scale and scope are lacking, and on the other, a disservice to the customer is generated and therefore a cost may be associated with this.

The data of the problem in DRP are as follows:

- $i = 1,\ldots,$ n periods [w] supposing, for example, weeks;
- GRi—Gross Requirement i-th—coincides with the demand for the product in the i-th period [item/w];
- LT: procurement lead time [w];
- SS_i: safety stock it is wished to keep [item]—to cope with variability in demand or lead time, or both;
- Q_{min}: minimum purchase lot [item]—e.g. to obtain economies of scale in transport or for minimum purchase constraints (often a supplier will not allow purchases below a certain value).

It is necessary to determine:

- OP_i—Order planning for each period;
- OR_i the release of orders for each period.

DRP—Application of the Method

For each period it is necessary to ensure that product availability is able to balance the product demand, as illustrated in Fig. 32.

The product request at generic period i-th is given by:

- SS_i: safety stock it is wished to keep [item];

Fig. 32 DRP: balance between demand and product availability for each period

- GR_i: demand in the i-th period [item];
- BL_{i-1}: represents the backlog demand in the $i-1$ period, to be satisfied in the i-th period. As mentioned, a backlog is assumed to be cumulative, i.e. orders not fulfilled in the period $i-1$ are cumulated with those not fulfilled in the i-th period. In other cases, in contrast, in the event of the lack of a product the customer could simply cancel the order.

Product Availability at generic period i-th is given by.

- SD_i: scheduled deliveries for the i-th period [item]—the scheduled deliveries are children of the orders released with an advance of PLT periods, therefore of $OR_{i-PLT+1}$
- OH_{i-1}: Stock in hand available at period $i-1$.

It therefore follows that the net requirement NR_i for the i-th period is given by:

$$NR_i = (SS_i + GR_i + BL_{i-1}) - (SD_i + OH_{i-1})$$

The orders planned for the i-th period, OP_i, must on the one hand be able to cover the net requirement NR_i of the i-th period, to which the Gross Requirement for the subsequent period GR_{i+1} must be added. Therefore:

$$OP_i = NR_i + GR_{i+1}$$

At this point, it is necessary to evaluate whether the planned orders are greater or less than the minimum lot. In the first case, the planned orders will be equal to the value obtained from the above formula, in the second they will be equal to the minimum lot:

$$if\, OP_i > Q_{min} \rightarrow OP_i = NR_i + GR_{i+1}$$

$$if\, OP_i < Q_{min} \rightarrow OP_i - Q_{min}$$

Orders planned for the i-th period must be released with an advance of $i - LT + 1$ [w] in order to be able to arrive in time to meet the demand in the $i+1$ period, therefore:

$$OR_{i-LT+1} = OP_i$$

$$OR_i = OP_{i+LT-1}$$

If, for example, the lead time is 4 weeks, orders planned in the 5th week, which will generate an SD in the 6th week, must be released at $5-4+1=2$ in the second week in order to arrive in time for the sixth week ($LT = 2, 3, 4, 5$).

DRP—Numerical Example

Suppose we apply the DRP technique to the following case.

Problem Data

- $LT = 4$ weeks; Orders issued at time i, generate availability at time $i + LT = i + 4$
- $SS_i = 5$ units
- $Q_{min} = 5$ units
- $GR_i = (10, 10, 12, 12, 30, 30, 10)$ units
- $OH_0 = 12$ units
- $BL_0 = 0$ units.

For each week, determine the net requirements, planned orders and released orders.

Week 1

Week	0	1	2	3	4	5	6	7
Demand		15						
GRi [pieces]		10	10	12	12	30	30	10
SSi [pieces]	5	5	5	5	5	5	5	5
BLi [pieces]	0	0						
Availability		12						
OHi [pieces]	12	2						
SDi [pieces]		0	0	0	0			
NRi [pieces]		3						
OPi [pieces]		13						
ORi [pieces]		0						

Demand		Availability	
GRi	10	OHi-1	12
SSi	5	SDi	0
BLi-1	0		
	15		12

$$NR_1 = 15 - 12 = 3$$

$$OP_1 = NR_1 + GR_2 = 3 + 10 = 13 \geq Qmin$$

$$OP_1 = 13$$

$$OH_1 = OH_0 + SD_1 - GR_1 - BL_0 = 12 - 0 - 10 - 0 = 2$$

$$BL_1 = 0$$

Week 2

Week	0	1	2	3	4	5	6	7
Demand			15					
GRi [pieces]		10	10	12	12	30	30	10
SSi [pieces]	5	5	5	5	5	5	5	5
BLi [pieces]	0	0	8					
Availability			2					
OHi [pieces]	12	2	0					
SDi [pieces]		0	0	0	0			
NRi [pieces]		3	13					
OPi [pieces]		13	25					
ORi [pieces]								

Demand		Availability	
GRi	10	OHi-1	2
SSi	5	SDi	0
BLi-1	0		
	15		2

$$NR_2 = 15 - 2 = 13$$

$$OP_2 = NR_2 + GR_3 = 13 + 12 = 25 \geq Qmin$$

$$OP_2 = 25$$

$$OH_2 = OH_1 + SD_2 - GR_2 - BL_1 = 2 + 0 - 10 - 0 = -8 = 0$$

$$BL_2 = 8$$

Week 3

Week	0	1	2	3	4	5	6	7
Demand			15	25				
GRi [pieces]		10	10	12	12	30	30	10
SSi [pieces]	5	5	5	5	5	5	5	5
BLi [pieces]	0	0	8	20				
Availability			2	0				
OHi [pieces]	12	2	0	0				
SDi [pieces]		0	0	0	0			
NRi [pieces]		3	13	25				
OPi [pieces]		13	25	37				
ORi [pieces]								

Demand		Availability	
GRi	12	OHi-1	0
SSi	5	SDi	0
BLi-1	8		
	25		0

$$NR_3 = 25 - 0 = 25$$

$$OP_3 = NR_3 + GR_4 = 25 + 12 = 37 \geq Qmin$$

$$OP_3 = 37$$

$$OH_3 = OH_2 + SD_3 - GR_3 - BL_2 = 0 + 0 - 12 - 8 = -20 = 0$$

$$BL_2 = 20$$

Week 4

Week	0	1	2	3	4	5	6	7
Demand			15	25	37			
GRi [pieces]		10	10	12	12	30	30	10
SSi [pieces]	5	5	5	5	5	5	5	5
BLi [pieces]	0	0	8	20	32			
Availability			2	0	0			
OHi [pieces]	12	2	0	0	0			
SDi [pieces]		0	0	0	0			
NRi [pieces]		3	13	25	37			
OPi [pieces]		13	25	37	67			
ORi [pieces]		67						

Demand		Availability	
GRi	12	OHi-1	0
SSi	5	SDi	0
BLi-1	20		
	37		0

$$NR_3 = 37 - 0 = 37$$

$$OP_4 = NR_4 + GR_5 = 37 + 30 = 67 \geq Qmin$$

$$OP_4 = 67$$

$$OH_4 = OH_3 + SD_4 - GR_4 - BL_3 = 0 + 0 - 12 - 20 = -32 = 0$$

$$BL_3 = 32$$

At week 4 it is possible to issue orders since $i - LT + 1 > 0$, in fact $4-4 + 1 = 1$

$$OR_{i-LT+1} = OR_1 = OP_4 = 67$$

Week 5

Week	0	1	2	3	4	5	6	7
Demand			15	25	37	67		
GRi [pieces]		10	10	12	12	30	30	10
SSi [pieces]	5	5	5	5	5	5	5	5
BLi [pieces]	0	0	8	20	32	0		
Availability			2	0	0	67		
OHi [pieces]	12	2	0	0	0	5		
SDi [pieces]		0	0	0	0	67		
NRi [pieces]		3	13	25	37	0		
OPi [pieces]		13	25	37	67	30		
ORi [pieces]		67	30					

Demand		Availability	
GRi	30	OHi-1	0
SSi	5	SDi	67
BLi-1	32		
	67		67

$$NR_5 = 67 - 67 = 0$$

$$OP_5 = NR_5 + GR_6 = 0 + 30 = 30 \geq Qmin$$

$$OP_5 = 30$$

$$OH_5 = OH_4 + SD_5 - GR_5 - BL_4 = 0 + 67 - 30 - 32 = 5$$

$$BL_4 = 0$$

$$OR_{5-LT+1} = OR_2 = OP_5 = 30$$

Week 6

Week	0	1	2	3	4	5	6	7
Demand			15	25	37	67	35	
GRi [pieces]		10	10	12	12	30	30	10
SSi [pieces]	5	5	5	5	5	5	5	5
BLi [pieces]	0	0	8	20	32	0	0	
Availability			2	0	0	67	35	
OHi [pieces]	12	2	0	0	0	5	5	
SDi [pieces]		0	0	0	0	67	30	
NRi [pieces]		3	13	25	37	0	0	
OPi [pieces]		13	25	37	67	30	10	
ORi [pieces]		67	30	10				

Demand		Availability	
GRi	30	OHi-1	5
SSi	5	SDi	30
BLi-1	0		
	35		35

$$NR_6 = 35 - 35 = 0$$

$$OP_6 = NR_6 + GR_7 = 0 + 10 = 10 \geq Q\text{min}$$

$$OP_6 = 10$$

$$OH_6 = OH_5 + SD_6 - GR_6 - BL_5 = 5 + 30 - 30 - 0 = 5$$

$$BL_6 = 0$$

$$OR_{6-LT+1} = OR_3 = OP_6 = 10$$

Week 7

Week	0	1	2	3	4	5	6	7
Demand			15	25	37	67	35	15
GRi [pieces]		10	10	12	12	30	30	10
SSi [pieces]	5	5	5	5	5	5	5	5
BLi [pieces]	0	0	8	20	32	0	0	
Availability			2	0	0	67	35	15
OHi [pieces]	12	2	0	0	0	5	5	
SDi [pieces]		0	0	0	0	67	30	10
NRi [pieces]		3	13	25	37	0	0	
OPi [pieces]		13	25	37	67	30	10	
ORi [pieces]		67	30	10				

Demand		Availability	
GRi	10	OHi-1	5
SSi	5	SDi	10
BLi-1	0		
	15		15

$$NR_7 = 15 - 15 = 0$$

$$OP_7 = NR_7 + GR_8 = 0 + ? = ?$$

$$OP_7 = ?$$

$$OH_7 = OH_6 + SD_7 - GR_7 - BL_6 = 5 + 10 - 10 - 0 = 5$$

$$BL_6 = 0$$

$$OR_{7-LTA+1} = OR_4 = OP_7 = ?$$

DRP Critical Issues

DRP represents a very simple and effective tool for managing replenishment flows in the SC, which, at first sight, may seem to be the solution to all problems related to channel coordination.

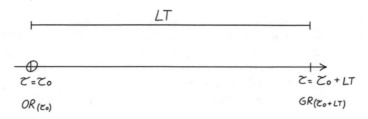

Fig. 33 DRP—LT as the freeze horizon

Only one player, either upstream or downstream, who, through DRP, takes care of the flow management. On the one hand, the Bullwhip Effect is eliminated, and the costs of backlog (market mediation costs) and inventory (physical costs) are minimized on the other.

All of which is true, however, only under certain conditions. As seen in the example, in order to apply DRP and release orders at the generic i-th instant, it is necessary to know the demand data (GR) at instant $i + LT$ as illustrated in Fig. 33. In other words, it is necessary to predict future demand with a horizon equal to LT. But expected demand GR a $t_0 + LT$ could be equal or different to the actual demand D which will actually manifest itself at the instant $t_0 + LT$.

Orders issued at the i-th instant are delivered after LT intervals, which represent a DRP freeze horizon, within which it is not possible to carry out any actions. In other words, any action to modify orders, whether increasing or decreasing, produces its effects no earlier than LT. Nothing can be done within the LT.

The higher the lead time, the less DRP is suitable for managing replenishment, since on the one hand, as the LT increases, the DRP freeze horizon increases (the horizon in which no can do nothing to modify the replenishment plans) and on the other, as LT increases, it is necessary to make forward predictions which, all other conditions being equal on the product, are less accurate the greater the planning horizon is (with other conditions the same—e.g. product characteristics, demand forecasting techniques—predicting demand at one month is easier than in one year).

Think, for example, in relation to the previous example, what would have happened if the actual demand at week 6, instead of being equal to 30 pieces as estimated, had been equal to 60 pieces or 5 pieces. The situation is summarized in Fig. 34

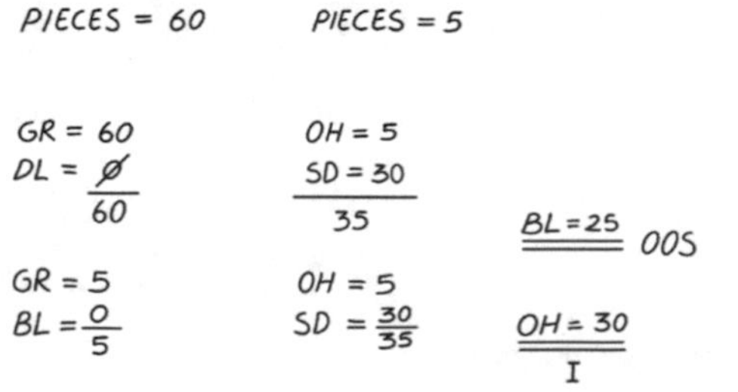

$PIECES = 60$	$PIECES = 5$	
$GR = 60$	$OH = 5$	
$DL = \emptyset$	$SD = 30$	
$\overline{60}$	$\overline{35}$	$\underline{BL = 25}$ OOS
$GR = 5$	$OH = 5$	
$BL = \dfrac{0}{5}$	$SD = \dfrac{30}{35}$	$\underline{OH = 30}$
		I

Fig. 34 DRP numerical example—Effect of demand variability during supply lead time

As represented in Fig. 34, in the first case we would have ended up with a backlog of 25, and therefore a market mediation cost linked to the opportunity cost (margin of 25 product units not sold), in the second case an inventory cost of 30, to be inventoried for later sales (physical cost), unless the product was perishable (e.g. fresh or recurring, so that at the end of the recurrence the product has no market) in which there would instead be an MMC equal to the industrial plus logistics costs of 30 units.

In addition to the problem of the variability in demand in the DRP freeze horizon, it is also necessary to take into account the possible variability in lead time, which produces similar effects, as illustrated in Fig. 35:

If an order released on time τ_0 and which should be delivered after LT periods suffers a delay, even if the delay is noticed with the periods ahead of the SD schedule delivery, this does not allow any room for an intervention.

Again in relation to the previous example, assuming a delivery delay of one week of the 30 units planned at week 6, there would have been a stockout cost of 25 units, and therefore the following week (7), when the delivery would have taken place of the 30 delayed units, depending on the characteristics of the product, there would have been corresponding inventory costs (if the product can be kept in stock) or shrinkage costs (if the product were perishable or if it was a recurring product which, at the end of the recurrence period, is no longer in demand and must therefore be disposed of). Figure 36 summarizes these arguments.

Here then, with reference to the different supply chain models seen in Chap. 3, in a "lean" scenario in which on the one hand the demand for the product (low demand variability—functional products) forecast errors are small, in the order of a few percentage points, and lead times are easily predictable (low variability in supply) the application of the DRP technique works very well. Suffice it to plan the replenishment well in advance and any changes in demand or lead time are absorbed by the safety stock.

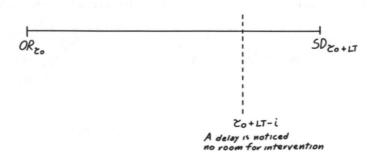

Fig. 35 Effect of LT variability on DRP

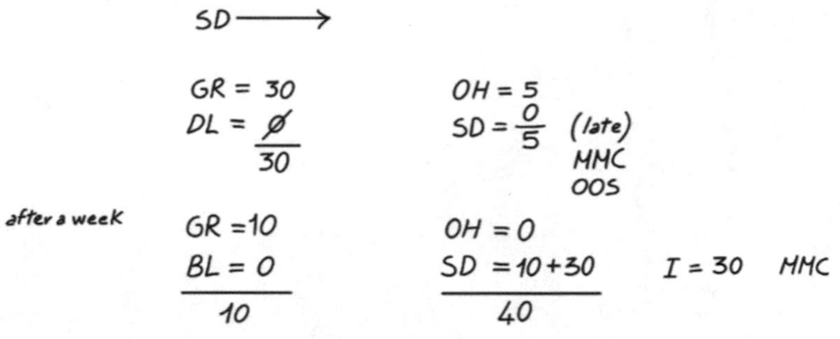

$$SD \longrightarrow$$

$$GR = 30 \qquad\qquad OH = 5$$
$$DL = \cancel{0} \qquad\qquad SD = \dfrac{0}{5} \quad (late)$$
$$\overline{30} \qquad\qquad\qquad MMC$$
$$OOS$$

after a week
$$GR = 10 \qquad\qquad OH = 0$$
$$BL = 0 \qquad\qquad SD = 10 + 30 \qquad I = 30 \quad MMC$$
$$\overline{10} \qquad\qquad \overline{40}$$

Fig. 36 DRP numerical example—Effect of lead time variability

Conversely, in an "agile" scenario, in which product demand (innovative products) and lead times (supply uncertainty) are unpredictable and highly variable, any changes with respect to the planning cannot be absorbed due to the DRP freeze horizon. Indeed, in a scenario with variable demand and lead times.

In conclusion, the application of the DRP technique works much better the more it is applied in "lean" scenarios where demand is predictable and lead time is deterministic. In this scenario, the LT can also be very high, it is sufficient to plan ahead; vice versa, it works really badly in an "agile" scenario, in which demand is difficult to predict and/or the lead time is difficult to predict, and it works even worse the greater the value of the lead time, since the freeze horizon will increase.

Chapter 5
Representation of the Supply Chain Through Service Flows

1 Introduction

In the previous chapters, we saw that it is possible to represent the supply chain in diverse ways.

Initially, emphasis was placed on the physical structure of the SC, analysing the way in which the macro-processes can be broken down and organized: procurement, production, and distribution, players involved and the infrastructures through which the macro-processes take place. The corporate functions entailed in the value creation process underlying the supply chain were then defined, using Porter's value chain, and how the relationships between the various functions influence the efficiency and effectiveness of the SC process. Subsequently, we saw how we can characterize the supply chain through the information flows which the stakeholders exchange, flows which are the basis for managing the process that brings products and services to the final consumer.

A further way to analyse the supply chain process, namely the creation of value through the provision of a product and a service capable of satisfying a set of expressed or implicit needs of the consumer, is precisely to concentrate on the term *service*, recognizing that in the supply chain process defined above in addition to the product flow, with its technological, functional and image characteristics, there is also a service flow (see Fig. 1), which accompanies the product and which goes to satisfying a further set of intangible needs expressed by the final consumer.

On the other hand, any process can be represented through the input, output, resource and control scheme. Similarly it is possible to represent the supply chain process as in Fig. 2.

The process begins with an input represented by a consumer's need in terms of product (material performance) and service (material and/or immaterial performance) and produces as an output a satisfied consumer, in terms of both product and service. Satisfied consumer in this context means two simple things i) a consumer willing to buy again from the supply chain and keep on doing business with it and ii) willing to recommend products and services to colleagues/friends. To carry

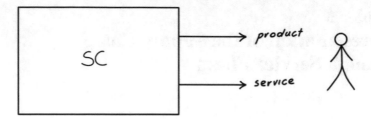

Fig. 1 Value creation through product and service

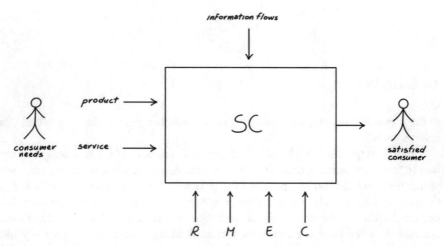

Fig. 2 The supply chain process which brings the product and service to the consumer

out this process, resources are needed, which can be (i) technical resources R (factories, physical equipment, etc. as analysed in Chap. 2), (ii) labour or manpower M (organized into corporate functions, as seen in Chap. 3), (iii) the energy E required for the operation of the processes, (iv) raw materials, semi-finished products, components and consumables. Controls—typically represented by information flows—(studied in Chap. 4) round off the process.

This particular chapter analyses the service flows, since (i) these flows largely depend on how the supply chain is organized and managed, and (ii) for most product sectors, they represent a real competitive factor, on which the competitiveness of the SC is based. In fact, more and more, it is not the product that is the competitive strategic lever, but the service, and the consumer chooses on the basis of the service which SC to buy a product from.

2 Customer Service as a Competitive Advantage

The definition of *customer service* starts from the following consideration.

In many industrial sectors, the overall "value" of a product, that is how much the consumer is willing to pay for that product and therefore how much the consumer is willing to remunerate all the players in the supply chain to satisfy their needs, is not only attributable to the technical characteristics of the product itself, but to a set of technical performances combined with a series of substantially intangible performances requested by the customer, which accompany a product, and which allow an increase in the value perceived by the consumer.

This concept is well represented by Fig. 3, introduced for the first time by (Christopher 1998) in which the overall value of a product as perceived by the customer is divided into two elements.

At the centre is the core product, i.e. the basic product as it leaves the factory. The size of this sector of the area depends substantially on the technical characteristics of the product at the end of the line (i.e. when it leaves the plant which produced it). As a non-exhaustive example:

- Functional characteristics—related to the ability to perform a certain task.
- Technical characteristics—linked to the level of performance.
- Reliability—probability of the product to be in a state of uptime and therefore able to deliver its expected functional and technical performances.
- Quality—maintenance of the technical characteristics as the product changes.
- Useful life—duration over time.
- Intangible performance related to the brand—particularly significant for brand-intensive products such as luxury goods,
- Etc.

According to Christopher, around the core product is a ring, *customer service*, which increases the value perceived by the final consumer (see Fig. 4). This ring

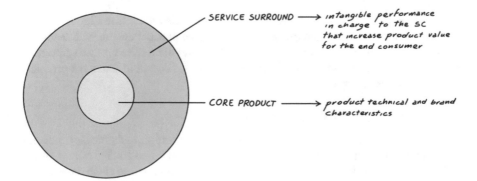

Fig. 3 The concept of overall product value according to Christopher—core product + service surround

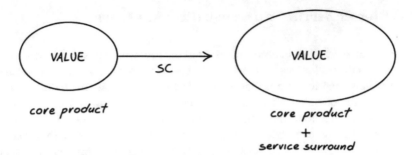

Fig. 4 Customer service increases the final consumer's perceived value of a product

is represented by the ability of the supply chain to increase the value perceived by the customer of the product itself through a series of substantially intangible performances, which concern, for example, the availability in space and time of the product itself downstream of the end of the line, in compliance with certain conditions, as well as the provision of value-added services for the customer. Some of these factors are anticipated here, to be analysed in greater detail below:

- Breadth of the range
- Pre-sales support
- Lead time
- Delivery places
- Punctuality of delivery
- After sales services
- Return management
- Withdrawal of discontinued products
- Etc.

Customer service is therefore defined as "the ability of a supply chain to increase the 'value' of a product as perceived by the customer, by providing intangible services, designed to meet and anticipate the expressed or implicit needs and expectations of the customer, including after-sales assistance," (see Fig. 4).

Only in some sectors is the importance of the core product greater with respect to the service surround. From this point of view we can divide the products into three main categories: (A) technology-intensive products, (B) brand-intensive products, (C) service-intensive products.

(A) Technology-intensive products

For so-called "technology-intensive" products, with a high technological content, such as pharmaceutical products, medical devices, functional food products, chemical industry products or other capital goods such as miniaturized electronic components, the role of service is certainly secondary to technical functionality. In this case, final consumers purchase mainly for the ability of the products themselves to generate a material performance of a functional and technological nature, and up to

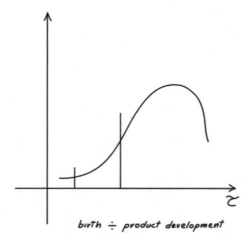

birth ÷ product development

Fig. 5 Technology-intensive products are typically found in the birth/developmental phase of a product's life cycle

a certain point, they are willing to forgo the service (until they find a competitor able to provide the same product with better service). Typically, these products are still in the initial phase of the product life cycle, in the birth and/or developmental phases (see Fig. 5), and therefore their competitiveness is not yet developed; often these products are even covered by patents, and so they are unique, and the customer must adapt to the service which the SC decides to provide. SCs operating in these sectors invest a large part of their resources in research and development (pharma companies invest up to 15/20% of their turnover in R&D) precisely because they recognize that competitiveness is based on the ability to innovate by creating products with a high technological content and high innovativeness.

(B) Brand-intensive products

Also "brand-intensive" products have a clear preponderance of the core product over the service surround. Those who buy products of the main luxury brands, for example from clothes to cars, from jewellery to food products, or certain electronic devices such as certain mobile phones, certainly do so for a mix of technical, functional and/or organoleptic characteristics of excellence, as observed for technology-intensive products, but above all for the intangible performance that such a brand is able to transmit. Manufacturers invest in both research and development to increase the functional performance of a product, but especially in style and marketing, to increase the perceived value of the core product as much as possible, not only in terms of technical characteristics, but in terms of image and status belonging to their brand—which the product transmits. At the same time, the SC invests important resources to protect its brands, trying to contrast as much as possible the black market—of the fake—and the grey market—of the parallel product. If the counterfeiting mechanism with which the black market erodes market shares and brand

Fig. 6 How a parallel
market erodes a share of an
FC's business

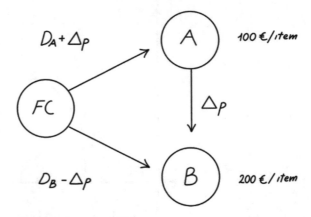

image is clear, the grey market should not be underestimated, with which an importer operating in area A purchases more products from an FC than are strictly necessary to satisfy the area demand D_A, to then resell the surplus quantity Δp in an area B where the FC sells the product at a higher price simply because the market allows it. In this way, the FC sees itself deprived of its sales Δp. The following Fig. 6 schematizes the mechanism described.

(III) Service-intensive products

In all other sectors, and in some ways also in those brand-intensive and technology-intensive where there is competition, the importance of the service level has grown over the last thirty years, and in most cases it has become the real competitive element. In mature markets, where a commercial brand does not represent a discriminant, the competition between SCs is no longer based on the core product, which tends to be similar from a technical point of view for the various competitors and, if anything, to become a necessary prerequisite for competitiveness. Nor is the brand a competitive factor. Rather, the element which increases the value of as product, as perceived by the customer is represented by the combination of service level/price. In particular, service can therefore be exploited in a strategic way to acquire new customers and retain those already acquired. We talk about *customer satisfaction* or even *customer delight*, namely, the ability to generate satisfied or more than satisfied consumers thanks to the product/service generated by an SC. In all these sectors, therefore, the leading companies are those that invest heavily in SC management, which is the main factor responsible for generating service. Thanks to investments in (i) SCM—procurement, operations, logistics and distribution, and (ii) IT infrastructure, companies are able to generate services that can grow more than the competition.

Think, for example, of Amazon, which has based a part of its competitive strategy on customer service. In a nutshell, and summarizing what was explained in detail in Chap. 2, the competitive factors of the Amazon service are precisely linked to service and are: (i) breadth of the range, (ii) delivery lead time in 24/48 h thanks to prime service, (iii) the most convenient delivery point for the customer (home, 24 h locker, collection point), (iv) returns guarantee, and (v) simplicity of the way of shopping). Thanks to this strategy—whereby buying on Amazon is simply more convenient and as cheap or sometimes cheaper than a traditional store—within 20 years it has become the second largest retailer in the world, second only to Walmart. Not only that, by doing so, it has shifted consumer expectations and service needs, putting competitors at a competitive disadvantage, including Walmart itself, which has had to invest tens of billions of dollars to adapt and develop its online infrastructure in order to provide a service similar to that of Amazon.

3 Customer Satisfaction Measurement

Customer service can be measured through two approaches:

(1) Indirect measurement: customer satisfaction is measured indirectly through customer loyalty. The assumption is that loyal customers receive a service adequate to expectations and therefore the measurement of the loyalty rate represents an indirect measure of customer satisfaction.
(2) Direct measurement, the measurement of customer service is conducted directly, through such methodologies as:

    i.   NPS (Net Promoter Score), a synthetic approach, in which the customer is asked if he/she would be available to promote the brand.
    ii.   analytical methodologies based on a set of indicators with which to measure the service provided and the service perceived.

In the following paragraphs we will analyse these two approaches.

4 Indirect Measurement of Customer Service—Customer Service and Loyalty

Back in the early 1980s, one of the most significant treatises on management, i.e. Peters et al. (1982), underlined that the reality of many companies is represented by the fact that *"customers generate sales and that successful companies are those which are able to acquire new customers and retain them."* In particular, this treatise shows that the natural tendency towards favouring the acquisition of new customers over the maintenance of those already acquired is partially incorrect, or at the very least unjustified. Indeed, if new customers are welcome in any industry, loyal customers generate greater profit and offer greater potential for growth in terms of value and frequency of purchase. This slice of the market is what is called "recurring business", which has the advantage of not generating the commercial and marketing costs necessary for the acquisition of new customers.

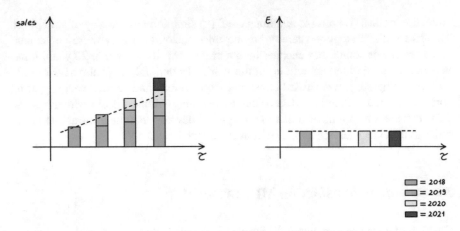

Fig. 7 Analysis of turnover by customer cohort—on the left the case in which a company manages to retain customers already acquired, on the right the opposite

In other words, companies experience growth in turnover if they manage to acquire new customers, but above all if they manage to retain those acquired in subsequent years, especially in sectors characterized by recurring business. This situation is represented in the graph on the left of Fig. 7, which analyses the turnover generated by customers according to their cohort. Conversely, if loyalty is lacking, the turnover curve does not see growth, as shown in the figure on the right.

Given that customer loyalty depends on the service level they are offered, a measure of the ability to build customer loyalty is an indirect measure not only of the product but also of the service provided and perceived.

A first measure of service level is therefore an analysis of the cohorts, or persistence of customers, represented in Fig. 7 if the number of customers replaces the € on the ordinate. To conduct a cohort or persistence analysis, the years are reported on the abscissa and the number of active customers starting from that year on the ordinate. If the curve has a hyperbolic trend like the one on the left, it means that customers who start doing business with the SC are satisfied with the product and service they receive and therefore continue to generate business over time; a business which actually tends to grow. The customers acquired each year are therefore added to those of the previous years and the curve has an upward trend. Conversely, if the curve remains flat, it means that only new customers are active each year, while existing customers are leaving the SC, mainly due to the fact that they are not satisfied with the service they receive.

Some indicators which measure the degree of persistence are shown in the following paragraphs.

4.1 Customer Retention Rate CRR

A first indicator to measure the level of persistence is represented by the CRR: Customer Retention Rate. This is expressed by the following relationship, and represents the customer loyalty rate:

$$CRR(t + nt) = \frac{NRC(t_0 + nt)}{NC(t_0 + nt)}$$

where:

- $NRC(t_0 + nt)$ is the number of loyal customers for the period $t_0 + nt$
- $NC(t_0 + nt)$ is the number of customers overall at a given time $t_0 + nt$.

The closer the value is to one, the higher the persistence level is. On the other hand, values smaller than one indicate a low loyalty capacity of the SC and therefore a low level of perceived service.

4.2 Average Customer Seniority ACS

A further parameter to evaluate the loyalty capacity is represented by the average age of customer loyalty (ACS: Average Customer Seniority). This value can be calculated as follows:

$$ACS(t + nt) = \frac{\sum_{i=1}^{n} NC(t_0 + it) * i}{NC(t_0 + nt)}$$

where:

- $NC(t_0 + it) * i$: number of customers at a given time $t + nt$ with seniority of purchase equal to i periods
- $NC(t_0 + nt)$: number of total customers for the period $t_0 + nt$.

The greater this parameter, the greater the loyalty capacity of the system.

4.3 Customer Lifetime Value CLV

Finally, as noted previously, the importance of customer loyalty in terms of business generation is represented by the concept of a customer's lifetime value—Customer Lifetime Value or CLV expressed in [€]:

$$CLV = ATV \cdot YFP \cdot CLE \ [\text{€}]$$

where:

- ATV: Average Transaction Value [€/ transaction]
- YFP: Yearly Frequency of Purchase [transactions/year]
- CLE: Customer "Life Expectancy" [years].

The importance of loyalty and CLV has been stressed by numerous studies, in many industrial sectors, especially for durable goods (e.g. the automotive industry). Empirical research has shown that a customer satisfied with a service remains with the same brand for decades, and in that period, on average, purchases many other products of the same brand. Consequently, a satisfied and loyal customer, in addition to being characterized by a high CLE value, generates higher values in terms of YFP and ATV, and is therefore characterized by a greater value of CLV than a poorly loyal one. Moreover, as will be seen in relation to the Net Promoter Score, a satisfied and loyal customer is also a source of positive publicity (good word-of-mouth), which leads the SC to acquire new customers. Last but not least, it should not be overlooked that a loyal customer costs less in terms of the cost of the sale and service, since he/she does not have to be constantly informed of the characteristics of the product and the supplier.

On the other hand, a customer who is dissatisfied with a service is very likely to switch to a competing brand, if it is true that for each customer who complains about service, at least 20 others simply stop purchasing from that company. Furthermore, dissatisfied customers spread the word of their negative experience (so-called *detractors*). Everything is even more amplified today, when customers have many social tools to hand to communicate their shopping experience in terms of product and service, and when consumers increasingly use social networks to inform themselves before making purchases, by consulting the reviews from other consumers about their shopping experiences.

The impact of a satisfied and a dissatisfied customer is summarized in Fig. 8.

This aspect is particularly significant in the case of online sales. Think of the same situation of non-service caused by Out of Stock—OOS on the two distinct channels: physical and online:

- Physical channel: A negative shopping experience is perceived as a failure to sell (the customer leaves the store without making purchases because he/she has not found what he/she is looking for), or is again perceived negatively but to a lesser extent because the customer buys a replacement product. In any case, it is unlikely that the customer will stop buying from the particular store or will be a source of negative publicity, except in the case of prolonged out-of-stock episodes.

Unsatisfied customer	Satisfied customer
1 out of 20 complains; 19 leave	Increased YFP; ATV and seniority
Negative word of mouth (web, social)	Positive word of mouth –
DETRACTORS	*PROMOTERS*

Fig. 8 The impact of satisfied and dissatisfied customers

- Online channel: the desired product is purchased because it is available online but cannot be delivered because in reality it is not physically available, typically due to a misalignment between the system stock and the actual stock, or the inability of the SC upstream to deliver that product on schedule (e.g., upstream OOS). In this case, the customer's shopping experience is highly negative. The positive shopping experience that is generated at the time the order is placed, when the SC has made the commitment to deliver a product by a certain date to the consumer, turns into a strongly negative one coupled with disappointment or anger when the order is cancelled. In the face of a single episode of this type, it is a proven fact that the customer will not only no longer make purchases on that online channel, but will most likely be a source of negative publicity through negative reviews on the platform.

A customer service which is to function as a strategic lever therefore has as its first objective not only an increase in the number of customers, but, above all, the retention of those already acquired. Providing an exciting shopping experience (in this sense we speak of *customer delight*, to take the concept beyond mere satisfaction). The customer perceives the SC as indispensable, able not only to provide the requested service, but to invent value-added services which the customer had not even thought of. Think of Amazon Prime, for example, to which a delivery service to home in 48 h has been joined by music, movies and other services. Thanks to these, the customer remains increasingly loyal and willing to entrust a growing share of business.

5 Direct Measurement of Customer Service

5.1 Synthetic Measurement—Net Promoter Score—NPS

A metric recognized nowadays as a measure of service and the ability of an SC to generate customer satisfaction is the Net Promoter Score—NPS.

This is based on the consideration that a satisfied customer is always a source of good publicity for potential new customers. If the Customer Retention Rate is a measure of profitability, then the NPS is a measure not only of profitability but principally of growth.

The Net Promoter Score was first introduced by Frederick Reicheld in his 2003 HBR article "The one number you need to grow" (Reicheld 2003).In this article, Reicheld argued that the ability of the supply chain to generate a service which meets customer expectations and therefore to grow by retaining its customers and acquiring new ones is measurable by asking a statistically significant sample of customers the following simple one question: *"How likely is it that, on a scale of 1–10, you would recommend our company/product to a colleague or friend?"*.

On the basis of the answers, it is possible to calculate the relationship between "promoters" and "detractors". The former, the promoters P, a source of good publicity, are those who answer the question with a score of 9 or 10, while the latter, the

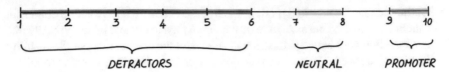

Fig. 9 The net promoter score scale: detractors (1–6), neutral (7–8), promoters (9–10)

detractors D, are those who create negative publicity, those who assign a score to the question of 6 or lower. In between are the "neutrals", who are neither a source of positive nor negative publicity and who assign a value of 7 or 8 to the question (Fig. 9).

To calculate the NPS, it is necessary to subtract the quantity of detractors from the quantity of promoters. The calculation is a percentage of the number of all respondents.

$$NPS = \frac{P - D}{N}$$

An NPS may vary in the interval $-100/+100$; a value of -100% means that all respondents are detractors, conversely a value of $+100\%$ indicates that all respondents are promoters. Values of excellence of the NPS which attest to the SC's ability to grow, are from 80% upwards.

Ultimately, according to (Reicheld 2003); if the SC can generate a product and service that meets customer expectations, it will generate satisfied customers and therefore "promoters". According to (Reichheld 2003) promoters are not just customers who keep buying a product. Also a customer who does not find alternatives or simply buys by inertia is a customer who continues to buy a product, but is not necessarily a loyal customer. When this customer finds an alternative product or simply gets tired and manages to overcome his/her inertia, he/she will switch to a competing product. A necessary and sufficient condition to consider a customer as "loyal", according to (Reicheld 2003); is to "stand up and be counted" by a recommendation to third parties of the SC's products and services. Loyal customers are therefore not just customers who continue to buy a product but those who show their loyalty through recommendations. They are a source of positive publicity, therefore attract new customers and thus generate growth.

The NPS metric is a very simple but extremely effective KPI to photograph the "health status" of the SC given that it measures the level of customers' "endorsement", which is closely linked to the SC's ability to meet their expectations.

From some statistics between NPS and the growth rate of the SC it can be seen that in both the secondary and tertiary sectors, there is a direct correlation between these two factors. The SCs with the highest NPS value are those with the highest growth rate, as shown by (Reicheld 2003) in the bubble diagrams of Fig. 10. In various sectors, the three-year growth rate is correlated with the value of the NPS; the size of the circle is proportional to the revenues.

Fig. 10 Correlation between growth rate and NPS in different sectors according to (Reicheld 2003)

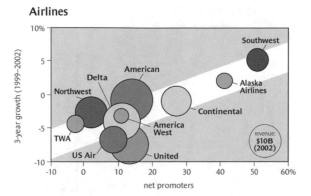

Airlines

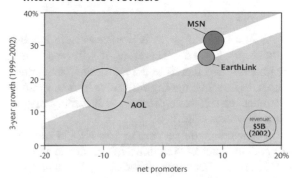

Internet Service Providers

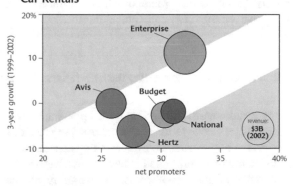

Car Rentals

Conversely, if the SC generates products and services that do not meet expectations, it generates "detractors". Detractors are particularly dangerous not only because they stop generating transactions, but because they are also a source of negative publicity for potential new customers. In other words, in the presence of detractors, the efforts of marketing and sales to acquire new customers will need to be considerably greater.

Fig. 11 Promoters and
detractors as analogy of a
tank

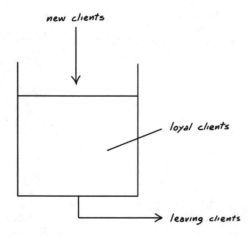

As in a tank, to grow, it is necessary to increase the level of customers or at least keep it constant. Therefore, it will be necessary to increase the flow of new customers—and therefore the marketing/commercial costs associated with it—the higher the flow of outgoing customers is, taking into account that the flow of outgoing customers also hinders the flow of incoming customers through negative publicity from detractors (Fig. 11).

In conclusion, the modern trends of SCM and marketing must go in the direction of creating a sort of "relationship", understood as *customer loyalty*. The basic idea is that a level of customer satisfaction should be created such that the customer does not even consider the possibility of leaving and using competing SCs, especially in highly "promiscuous" markets where customers tend to buy a brand once and the next time another one, quite nonchalantly.

5.2   Analytical Measurement of Service Using a KPI System

If the parameters seen in the previous paragraph can give an idea of some results of customer service in terms of customer loyalty capacity, they are however insufficient to develop a performance evaluation system in terms of customer service that can assist in the monitoring and control of the system.

For this purpose, it is necessary to use an organic methodology, based on analytical indicators of service provision structured in levels, in order to take into account the different dimensions of customer service, where metrics such as NPS, although they photograph the service perceived directly, can provide a global measure of it. This methodology must lead to a quantitative assessment (expressed through numbers by calculating KPIs) and an objective assessment (not subject to personal interpretation) of the service level.

5.3 Measurement of a Service Managed by a Buyer or a Vendor

In order to build and run a customer service performance measurement system, it is first necessary to distinguish case (i) in which the SC performance control system is implemented and managed directly by the customer, from case (ii) in which it is instead the SC that structures and manages this control system autonomously.

In the first case, the customer implements Service Vendor Rating (SVR) techniques for analytical monitoring and quantitative measurement of the services provided by its suppliers. As seen in relation to sourcing and procurement, this control system can be established (for example), to evaluate a possible elimination of those which, on the basis of objective measurements, provide the poorest performance. Once the number of suppliers has been reduced, the SVR system can still be maintained, in order to assign business shares to each of the suppliers that are proportional to the rating. It is important to underline that these techniques are part of the quality procedures envisaged by ISO 9001 and by GXP (good × practices) and absorb a huge amount of resources, both in terms of time, necessary data, and qualified personnel. These techniques are therefore usually implemented by structured companies.

In any case, it is important to discriminate in this case, since if it is the customer who is managing the service measurement process, the upstream supplier SC is more or less passively assessed, in the sense that the rating system is developed and managed independently by the customer. In general, there is a relationship of proportionality between this rating autonomy and the customer's contractual strength, in the sense that the greater the customer's contractual strength is with respect to the supplier, the lower the degree of participation of the supplier in the rating process implemented by the customer. Conversely, suppliers with greater contractual strength can negotiate the rating conditions with the customer.

In contrast, if the customer does not implement analytical rating techniques of the upstream SC, it is the SC itself that can set up a system for measuring its performance in terms of customer service, a system which is therefore organized and managed by the same SC. In this case, the system has a proactive function, in the sense that it allows the SC to control the service level, and to intervene proactively and in real time, according to a logic of continuous improvement (a PDCA cycle—Plan-Do-Check-Act), if the service provided does not meet the expected standards.

The service approaches managed by the buyer or vendor described above are schematized in Fig. 12.

5.3.1 Measurement of Service Managed by a Vendor: Service Provided, and Service Perceived

At this point, it is necessary to make a further clarification before addressing in detail the elements that make up a customer service evaluation system. In order for the

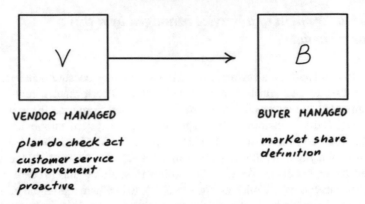

Fig. 12 Measurement of service managed by a buyer or vendor

customer service performance control system to be effective in helping to efficiently design and manage the supply chain by aligning performance with strategic objectives, this evaluation system cannot be limited to a simple measure of the performance "produced" by the supply chain, but must instead be oriented towards a measure of the performance "perceived" by the end customer.

This distinction, which at first sight may appear marginal, is actually extremely important. When the supply chain, or on its behalf the supplier, implements a measurement of the performance "produced", the performance of service to the customer is evaluated from the point of view of the supply chain itself. The performance is evaluated through analytical service KPIs (e.g. punctuality of deliveries, lead time, accuracy, and so on) which are calculated by the supplier from the data in its possession, and are evaluated as results of the SC process. The supplier then evaluates these indices as satisfactory or otherwise on the basis of its own service standards, and can decide which interventions to implement on those parameters which, based on the evaluations, are deemed unsatisfactory.

In some ways, this procedure can be dangerous and counter-productive, when it comes to achieving the strategic objectives of customer satisfaction, if adopted *tout court* for the design and management of an SC. A customer service performance measurement system based on a measurement of the performance "produced" is in fact managed for objectives within the supply chain rather than external ones. In other words, a fundamental point is being ignored: the measured performance parameters are received by a customer, who can judge the performance differently from the supplier and can also consider certain factors more important for his/her needs which the supplier deems secondary (and vice versa). An example is shown below in Fig. 13, taken from (Bryne and Markham 1991), in which the performance provided by the vendor is compared with the performance perceived by the buyer in seven service factors (billing accuracy, speed of response to questions, level of complaints, accuracy of deliveries in terms of damage, completeness of orders, lead time, and level of order fulfilment, the latter being $\frac{\#orders\ fulfilled}{\#orders\ received}$).

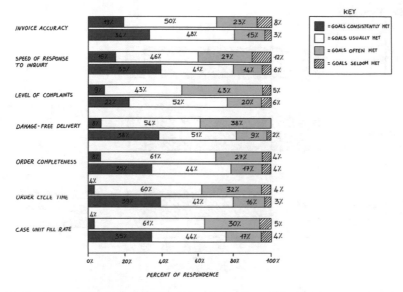

Fig. 13 Gap between the service provided (supplier's point of view) and perceived (customer's point of view)—according to (Bryne and Markham 1991)

As can be seen from the graph, the performance delivered (lower bar) is very different from the performance perceived (upper bar). On a four-value scale ranging from: "Service objectives achieved in a consistent manner" to "service objectives achieved only rarely", it can be noted above all that in the case of cycle time and the level of stock coverage, while 35% of the suppliers believe they are achieving the objectives of consistent service, only 4% of the customers share this point of view!

We tend to note how in each of the service factors analysed, suppliers tend to overestimate the service provided compared to that actually perceived by the customer.

According to (Bryne and Markham 1991) there are also substantial differences in the perception of the relative importance of service factors, i.e., what suppliers consider important for the purposes of the service and what customers actually consider important. This comparison is analysed in Fig. 14, which the discrepancy between the importance of the factors attributed by the vendor and the buyer is evident.

In fact, in this case, a sample of suppliers who had conducted surveys of their customers on the importance of service factors were asked whether their initial perception of the importance of service factors corresponded to the actual one of their customers. For the three fundamental elements of order fulfilment (order processing, delivery cycle, receipt and follow up) only in about 30% of cases was there the same perception of the importance of service factors, while about 60% of the cases showed differences between the importance of factors attributed by the supply chain and by the customer. Finally, 10% of the suppliers observed substantial differences

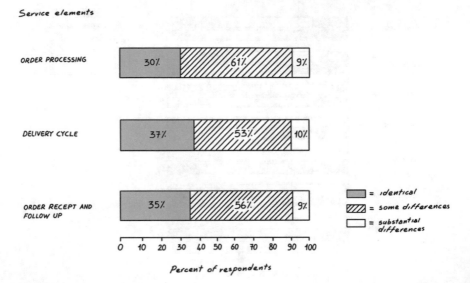

Fig. 14 Gap between the importance of the service factors attributed by supplier and customer—according to (Bryne and Markham 1991)

between what was considered important for the purposes of the service and the actual requirements of their customers.

In general, a vendor naturally tends to attribute greater importance to the factors on which it performs best and less importance to those in which performance is poorer. Sometimes, however, it can happen that these are precisely the factors which the customer considers important to remain loyal and attribute an increasing business share.

In conclusion, a sentence by Drucker (1999) which beautifully summarizes the difference between measurement of the service provided and perceived is the following: *"What people in the business think they know about their customers and the market is more likely to be wrong than right. There's only one person who really knows: the customer."* A supply chain whose strategic objective is customer service, cannot therefore disregard the customer's point of view in evaluating its service performance.

For effective and efficient management of the service provided by an SC, it is therefore necessary to put in the customers' shoes and measure the performance of the system as perceived by them, checking and adapting it accordingly. The indicators used are the same as in the case of measuring the performance produced, but in this case they are not evaluated on the basis of the SC's concept of lead time, punctuality, accuracy, but on the basis of the customer's concept. Without this control system, interventions are only possible in retrospect, when a deterioration of the service level could have already wrought enormous damage, e.g., in terms of loss of customers and image. For an evaluation of the customer service performance of a supply chain, reference will be made below to the latter case, which is certainly more interesting, precisely because it sees the SC actively intervene in the development and management of the rating system.

It must also be recognized that each of these indicators can be given greater or lesser importance by different clients. Consequently, different customers may request different services. To achieve service excellence, it will therefore be necessary to segment the market for customers with homogeneous service needs and to guarantee each of them the required service standards. To this end, there are specific statistical techniques—such as cluster analysis—which allow the creation of homogeneous groups on the basis of customer databases. In this way, each customer, based on certain data, such as income, age, sex, education, can be framed within a cluster whose service needs are known, in terms of the amount of weight given to each analytical service indicator.

5.4   Structuring of a Rating System

According to Ruggeri (1998) the performance of a supply chain in terms of customer service depends, as already mentioned, on multiple factors, each of which can be traced back to a hierarchical tree structure, as represented in Fig. 15.

In other words, the performance perceived by the customer of the service (CS— customer satisfaction) depends on a series of analytical factors of an immediately lower level, whose performance can be measured through summary analytical KPIs of the SC performance, such as lead time, order fulfilment, delivery accuracy, delivery frequency, punctuality, flexibility, returns management, or breadth of range to name a few. But also intangible factors relating to the SC, such as the provision of additional services in the pre- or post-sales phase, such as e.g. support for the design/customization of the product, rather than the presence of a help desk or an after sales service or the option of replacement products, etc.

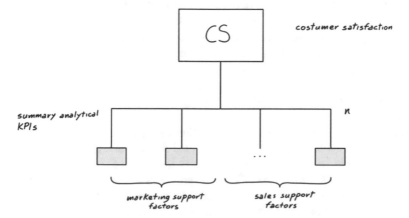

Fig. 15   Structuring of a customer service rating system

5.4.1 Marketing and Sales Support Factors

Level I analytical factors which directly impact customer satisfaction can be grouped into two major categories:

- marketing support service factors
- sales support service factors.

Among the marketing support service factors considered are all those factors which directly influence the service level, but are not expressly linked to an existing economic transaction between customer and supplier. Instead, they are service components which the supplier offers whether the customer buys or not.

Conversely, among the sales support service factors, are all those factors which a supplier offers to meet specific customer needs consequent to the economic transaction in progress, ranging from a phase prior to the actual order up to after-sales assistance. The sales support service factors are therefore linked to an economic transaction between supplier and customer, and are directly influenced by the SC performance and processes.

5.4.2 Marketing Support Factors

This paragraph highlights the main factors and related performance indicators which fall into the category of marketing support service factors. As mentioned previously, we talk about marketing support service factors to indicate those customer service factors that the supplier offers to increase the service itself regardless of the presence of an economic transaction with a customer. These are therefore services which fall more within the area of marketing rather than the sphere of the supply chain, understood as a process which, through a product, satisfies consumers' needs.

From this point on, we want to underline that the indicators presented are certainly not exhaustive for a category since, in some particular product sectors, there may be other determining factors to determine customer service. However, the factors presented have the advantage of being of a fairly general value as the industrial sector changes, and also of allowing an assessment of the service performance of more or less easy factors, using quantitative indicators.

Breadth of the Range

The first service factor attributable to marketing support is the breadth of the product range BR. This factor will be the one analysed in most detail, also due to the fact that it is possible to evaluate it in quantitative terms and for the impact which the breadth of the range has on the SC and its processes, both at the operational level (e.g., simply put, producing a standard product is much easier than organizing the production of n variants and versions) and at the procurement and distribution level

Fig. 16 Breadth of the range as a service factor straddling the marketing and sales support factors

SALES
↑n° products
with different
functional carachteristics

MARKETING
same product but
different service
needs

(managing the stock of n variants versions is much more complex than in the case of a single component or standard product).

This factor could actually be placed midway between a factor capable of increasing the performance of the supply chain in terms of product or service. However, the breadth of the range is considered to all intents and purposes as a service factor when a product is not differentiated by technical or functional characteristics in a range, but by characteristics which vary its value in terms of responsiveness to the needs of different customers. When it comes to FMCG, an example of greater service thanks to the expansion of the range is represented by the presence of products in different formats (e.g., family-sized biscuits—to be consumed at home—or packaged in snacks of 2–3 pieces, for a snack outside), or the creation of multipacks to respond to a promotion. The product is still the same, but the expansion of the range through larger or smaller sizes increases the service level offered, in terms of a response to implicit consumer needs.

Therefore, when the breadth of a range represents a factor which enables a product to meet the needs of a greater number of customers for the service it is able to provide, we speak of the breadth of the range as an external analytical factor supporting the marketing. When, on the other hand, the breadth of the range is characterized in terms of references that are different from a functional point of view, the external analytical factor is closer to those of sales support. Consider the case of Amazon, which has built the key service factor for its success around a range of tens or maybe hundreds of millions of references. This concept is summarized in Fig. 16.

Definition of "Product"

Regarding the concept of "range", it is advisable to adopt a unique terminology. The same product is supplied in different models, different variants, different versions, different sizes/formats, and ultimately, as different references (Stock Keeping Unit SKU). How the differentiation between products is made depends on the product sector.

For highly complex products, such as the automotive industry or electronics, we normally speak of *segments*, *basic models*, *versions* and *items*. Thus, for example, a *segment* might be represented by sedans, SUVs, and station wagons, or laptops, desktops and servers. Within each segment there are then the basic *models* (e.g., in the case of Alfa Romeo the Giulia is a basic model in the sedan segment, in the case of Apple the MacBook is a basic model), the various *versions* can then be identified by a diesel or gas engine, the set-up, by a rear or front wheel drive transmission to a 2- or 4-wheel drive, etc., or by a processor, memory, graphics card, and so on. The *item*, on the other hand, coincides with the single car or with the single personalized

PC, while the serial number (SN) and part number corresponds with the unique item purchased by the customer.

On the other hand, in industries such as apparel and fashion, we speak first of all of *classes* to identify a macro-family of garments (e.g. jackets, trousers, dresses, knitwear, bags, accessories, etc.). Within a class there are then the *models*, to identify a garment defined by an aesthetic shape, a fabric or a cut (a certain model of coat rather than a jacket or other outerwear), while the colour/variant and size subsequently identify each single garment. We can speak of a *line* to identify the combination of model/variant (regardless of size) and of the SKU (stock keeping unit), or code to identify a particular model/variant/size combination. In some cases, serial identification can also be achieved in this sector, obtained by adding an SN to the SKU (SGTIN serialized GTIN, given by the GTIN + Serial Number.

Finally, in the FMCG sector, we first talk about a segment/department such as for instance Food and Beverages; Home care products; Personal healthcare products; Home utensils; Spare-time items; Fashion accessories; Domestic animal foods and accessories and Consumer Electronics. Then, within a department, product categories identify a group of products to meet the shopper's needs. In food for instance we may have meat, dairy products, baked goods, etc. talk about a *product/category* to actually identify a particular family, for example biscuits. Such a product is made in different variants (with or without sugar, shortbread, dessert, dry), up to the single reference item, identified by a particular format and possibly by an expiration date. An item reference definition might be the following: Two items represent the same reference if they are distinguishable at most by the expiration date. Conversely, two references of non-perishable products, in order to be treated as a single reference, must not be distinguishable in any way.

As mentioned in Chap. 4, In the FMCG sector, the GTIN (EAN 13 or EAN 8) is used to identify a reference item (GS1 Italy 2021a). Even in FMCG, this may be pushed at the serial level, distinguishing by SN items belonging to the same reference (same product, lot and expiry date but different serial). In this case an SGTIN is used. In the following paragraph, this topic will be dealt with in detail.

In Table 1 is a classification by product category to summarize what has been proposed above.

Breadth of the Range—KPIs

The breadth of a range is a marketing support service factor which can be quantitatively measured. It is advisable to keep separate, for the measurement of this factor, cases in which one places oneself in the shoes of the manufacturer or distributor, since the measure acquires different connotations.

From a manufacturer's point of view, the breadth of a range can be measured vertically or horizontally. Horizontally, the breadth of a range can be measured by the number of basic models (BM) or basic products (BP) into which the range is divided, and by the number of collateral models and products (CM and CP, respectively). In contrast, as regards the vertical measurement of the breadth of a range, still from

Table 1 Organization of products, articles, models, variants, versions, sizes and formats, up to individual reference items—SKUs

Sector	Durable goods (e.g. Cars)	Clothing fabrics	Food/FMCG
Level I	Segment (e.g. sedan, SUV, SW)	Class (e.g. trousers, knitwear)	Product category (e.g. preserves)
Level II	Model	Model	Products (e.g. Tins, sauces)
Level III	Version	Variant (e.g. colour)	Variants (peeled, puréed)
Level IV	Item	Size	References
Level V		SKU	GTIN (EAN-13/EAN-8)
Last level coding	SN/serial number	SN/serial number	SGTIN

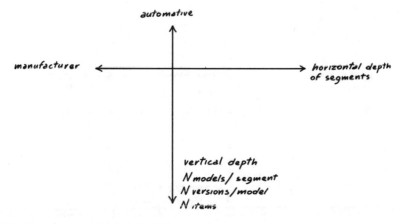

Fig. 17 Measurement of the breadth of the range for a producer of durable consumer goods

a manufacturer's point of view, we mean the number of versions per basic model (VPBM) or the number of items per version (IPV), in the case of manufacturers of high complicacy, or the number of variants per basic product (VPBP) or the number of references per variant (RPV), in the case of FMCGs.

Thus, for example, for an automobile manufacturer the breadth of the range can be measured by the number of segments (compact, sedan, off-road vehicle, station wagon, etc.). The horizontal width of the range can also be measured by evaluating the total number of collateral products or models. For example, a car manufacturer might also produce basic models in the field of industrial vehicles, motorcycles, accessories, and in general in other non-similar sectors. The vertical depth is represented by the number of models per segment, the number of versions per model, and the number of items (see Fig. 17).

In the fashion sector, the breadth of a range is measured horizontally in terms of the number of classes produced per season (e.g., trousers, outerwear, jackets,

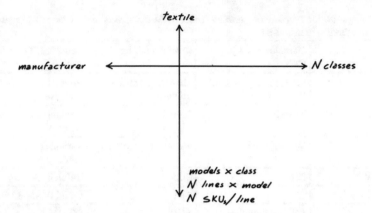

Fig. 18 Measurement of the breadth of the range for a producer of textiles and clothing

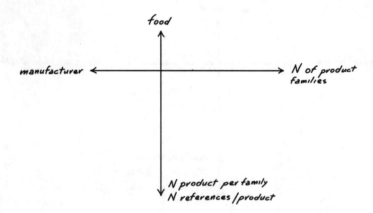

Fig. 19 Measurement of the breadth of the range for an FMCG manufacturer

shoes, accessories, etc.) while vertically, within the class, it is given by the number of "lines", meaning the model/variant combination, or SKUs where the SKU is the model/variant/size combination (see Fig. 18).

In the case of FMCGs, on the other hand, the horizontal breadth is given by the number of product families (in the case of Barilla, for instance bakery, pasta, sauces, etc.), and the vertical depth by the number of products per family and the number of references per product (see Fig. 19).

For all the factors highlighted, the measurements can be absolute, or related to the competition. In the latter case, the values can be related either to the top competitor, or to an average of the competitors.

From the point of view of the distributor (retailer), instead, the breadth of the range is measured through different parameters. However, also in this case we speak of breadth both horizontally and vertically. Referring to the FMCG case, the breadth of the horizontal range for a distributor is measured by the number of product families

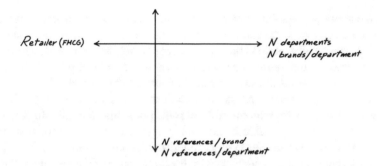

Fig. 20 Measuring the breadth of the range for an FMCG retailer

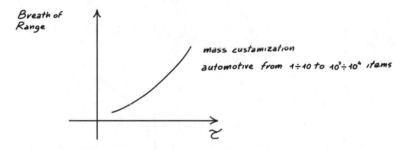

Fig. 21 Explosion of the breadth of a range over time until mass customization

in the assortment (categories), such as fruit and vegetables, fresh and dried food, meat and fish, cured meats and cheeses, health & beauty, homeware, bazaar, multimedia, etc., and for each category by the number of brands per family (for each product family how many commercial brands there are in the assortment) and by the number of products for each brand. As for the breadth of the vertical range, this is measured by parameters such as the number of products per brand, the number of variants per product, the number of references per variant (see Fig. 20).

Breadth of the Range—Strategic Evolution

The analytical factor of customer service represented by the breadth of the range has assumed increasing importance in recent decades. The general trend underway in virtually all product sectors in the last thirty years has been that of expanding ranges, up to the so-called *mass customization*, i.e. the creation of a customized mass product through a maximum extension of the range of products, to be able to create a mass standard product, but customizable to the needs of the individual consumer (see Fig. 21).

The phenomenon that has led to the expansion of the range in recent years is the recognized influence of the breadth of the range on the number of customers and on the average annual turnover per customer. An absolute example of a player for which the breadth of the range is a real competitive element is Amazon.com, as already extensively discussed in the other SC representations proposed, whose breath of range is confidential, but in the range of several hundreds of millions references.

The qualitative trends in the relationship between the number of customers and the turnover they guarantee, and the breadth of the range are shown Fig. 22.

Clearly, the trend is asymptotic, since at some point, also if continuing to increase the range, it will not be possible to indefinitely increase the number of customers or induce the individual customer to buy more than a certain amount.

A further representation which clarifies the importance of the breadth of the range as a service factor but also as a strategic element of success for companies in a given product sector is the one shown in Fig. 23.

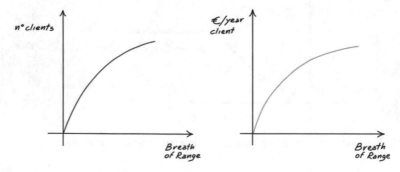

Fig. 22 Relationship between the breadth of a range, number of customers and turnover per customer

Fig. 23 Relationship between the breadth of a range and market share

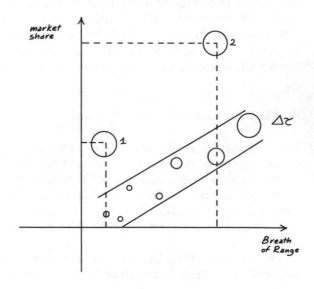

On a system of Cartesian axes Breadth of Range BR—Market Share, the companies of a specific product sector, belonging to a given geographic market, are represented at a certain period. Companies are represented by a circle of an area proportional to the market share held by the company. As can be seen, there is a sort of linear correlation between the breadth of the company's range and the market share it holds, highlighted by the trend lines: the greater the breadth of the range, the greater the market share. Special cases such as that of company (1) and company (2) are an exception.

Company (1) is one which, despite a limited range, can compete with a higher market share than expected based on the general trend of the sector. This happens when such a company owns a global brand, that is, it possesses a product that is so innovative from a technological point of view, or owned exclusively and in any case appreciated by a large population of customers, as to allow the company itself to enjoy important market shares, despite the narrow range. A typical case of such a global brand is that of Tesla in 2015–2020, when electric cars were limited and few competitors shared the whole market. Conversely, companies such as (2) are those with wide ranges of products and market shares predominant in the sector, however more than proportional to the breadth of the range for the sector they belong to. In this case, the company still leverages a global brand, but has a particularly high range of products anyway (e.g., Ferrero in the Italian market has a global brand with Nutella, but still has a particularly wide range).

Companies normally expand their range in the sector they belong to, thus increasing the number of base models (Alfa Romeo a few years ago, through its Stelvio, expanded the range into a new segment, that of compact SUVs, with a new base model), or of basic products in the case of companies producing FMCGs. However, such expansion can also take place in more or less similar sectors, in different ways.

For example, *brand stretching* phenomena can occur. In this case, a manufacturer who owns an important brand in a certain product sector decides to add new basic models or basic products in other sectors. In other words, it seeks to increase the business by leveraging an important brand. This phenomenon often occurs for clothing manufacturing companies, which expand their range through a brand stretching in the field of accessories, leather goods, cosmetics, etc.. In the case of brand stretching, a manufacturer sees its logistical efforts increase as the size of its supply chain increases. We could think of the *Pan di Stelle* biscuit case, with the addition of a spreadable chocolate cream to the product catalogue, which involved the creation and distribution of a brand-new product.

The phenomenon of *licensing* is something different. In this case, the owner of an important brand assigns the use of a brand to companies operating in other sectors, against the payment of royalties. In the case of licensing, the owner of the brand still tries to increase its business by leveraging the brand, but the expansion of the range does not involve an increase in the size of the supply chain. This is typically the case of a fashion company which, for example, sells its brand name to an eyewear company for the creation of eyeware lines, or to a cosmetic company for the creation of a fragrance. The logistics of the fashion company are not affected in any way by this phenomenon, but it receives a premium for the exploitation of the brand.

5.4.3   Other Marketing Support Factors

This paragraph briefly reviews the other factors which, together with the breadth of the range, constitute marketing support service factors, in other words, which contribute to increasing the service level perceived by the customer regardless of whether or not an economic transaction is involved. Unlike the amplitude factor of the range, analysed in detail and for which possible quantitative measurements have been introduced, for the remaining factors we will only provide a qualitative description of how they can contribute to raising the perceived service level.

A first factor of any marketing support service is the presence of written documentation on the supplier's policies in terms of customer service. This information must be written, understandable, and easily visible to potential customers, typically on the web portal or in corporate communications. This document has a dual purpose: on the one hand it informs the customer about the services they can expect, on the other it guarantees the supplier against any unjustified claims by the customer.

Another factor supporting marketing is the accessibility of the organization, understood as the ease of potential customers to receive an answer to their problems. These answers can come in order of involvement of staff, from web pages with FAQs, chatbots, call centres, and also direct contact with internal staff. It is a common experience that being continually bounced on the web or on the phone between chatbots and auto responders without being able to talk to a real person, or being able to talk to several people, to whom, however, you have to continually re-explain your problem, is a highly frustrating experience. A consumer is unlikely to be willing to establish commercial relationships.

Another factor of marketing support service, especially for durable goods, are the forms of free consultancy which the supplier puts in place for its customers, whether current or potential. These consultancies can be of a technical, fiscal or other nature, and are implemented through various forms (toll-free numbers, internet, commercial consultants).

Again, a marketing support service factor is the ability of the supplier to offer potential customers free trial products for a certain period of time, after which the customer decides whether or not to buy a product, or to return it.

In other cases, such as online purchases, products can instead be replaceable/refundable, meaning the possibility, within a certain period of time, to return a product for an exchange or simply to return it. This type of factor increases a consumers' level of confidence in a web portal, knowing that in the event of a product that does not fully comply with their expectations, they will still be able to return it and receive compensation. For example, it is not uncommon for a consumer to order several sizes of the same garment, try them on at home, keep the one that fits and return the others. We have thoroughly takled the importance of this service factor for on line retailers in the chapter related the phyisical representation of the supply chain.

Another factor which falls within the category in question and contributes to increasing the service level perceived by the customer regardless of an economic transaction especially suited for durable equipment, is the possibility of taking advantage of financial benefits, with instruments such as subsidized rate loans, leasing, purchase replacement with a usage fee (especially in the case of packages if we are talking about SaaS—Software as a Service) and other forms.

Finally, the "availability" of a supplier helps to increase the service perceived by the customer. This term refers to the perception that the customer has of a supplier in terms of its predisposition to adapt a service and customize it based on a customer's needs. This factor can play a key role in many industrial sectors.

5.4.4 Sales Support Factors

As already mentioned, those external summary factors of customer service that are implemented by a supplier in the face of an economic transaction with customers, and on which the service level perceived by customers depends, are framed among the sales support service factors. These factors typically fall within the sphere of SC management, in the sense that the way an SC is organized and run influences the performance of the sales support factors which the SC can deliver.

Since these factors are consequent to an economic transaction, La Londe and Zinser (1976), in a study on customer service, suggest framing them in three macro-groups, depending on the moment of the transaction in which they intervene.

We then speak of

(i)   pre-transactional sales support service factors—which intervene during the phase prior to the economic transaction of issuing an order,
(ii)   transactional service factors that intervene during the transaction phase, which ranges from the issuing of an order to delivery of the goods,
(iii)   post-transactional service factors, which intervene downstream of a transaction, during the after-sales service period.

As the reference market changes, these factors may assume a greater or lesser importance. For example, in the online sales sector, pre-transactional and transactional factors are of greater importance (ease of placing an order and lead time for delivery of goods are the keys); in the durable goods sectors, on the other hand, after-sales can certainly play a much more significant role.

We will initially analyse the factors grouped in the pre- and post-transactional phase, and then move on to the characteristic factors of the transactional phase, which typically include performance indicators that depend on the supply chain. For the first two categories, a qualitative description of the factors will be given, highlighting the main characteristics in terms of perceived service, without elaborating quantitative metrics (also because they are often on/off metrics), to move instead to a more in-depth analysis and quantitative of the factors which fall within the product delivery phase, since the latter more typically falls within the sphere of supply chain management.

Pre-transactional Factors

The pre-transactional phase includes all those factors which influence the relationship between supplier and customer in the initial phase of a transaction, which must prompt the customer to issue an order. That is to say, these factors facilitate the issuing of an order by a customer, and are therefore particularly important in the case of suppliers who adopt organizational systems on order (ETO—Engineering To Order or MTO—Make To Order), where some of the main elements that influence the service level perceived by the customer and which can therefore induce the customer to issue an order are typically:

- the frequency of visits and contacts with the customer
- the existence of an information system through which to monitor the progress of orders (i.e., order tracking)
- the possibility of providing the customer with preliminary assistance or appropriately configuring the requested product
- the ability to provide technological insights, intended as an updated point of view on technological development and the new opportunities that technologies can offer the customer
- the ability to analyse the customer's problems/opportunities in advance, and provide adequate solutions.

Clearly, this picture is not valid in absolute terms, in the sense that, in general, in some sectors certain elements can play a negligible role while others can intervene to "complete the picture".

For example, if we consider the first factor mentioned, we see how the frequency of visits and/or contacts by a supplier assumes a marginal role for customers who are connected to the supplier via web portals in which they can see stocks and availability in real time, while it might play a far more important role for other customers.

As for the possibility of providing preliminary technical assistance before the formalization of a transaction and in the compilation of the documents to formalize the transaction itself, it can represent an extremely important aspect. Consider, for example, the advice that IKEA provides with its online tools, where the customer can configure the chosen furniture and place it within a layout of their environment, or the possibility of a contracting company to draw up all the contractual documentation. Clearly, this factor has a marginal role for fast-moving consumer goods.

As regards the presence of a system to monitor the progress of the preliminary phase of an order, this increases the service perceived by the customer who, thanks to this service, has the possibility to act directly based on the progress of the order. Suffice to think of the value that this service parameter has assumed in the field of express couriers, which by now have all adopted an order tracking system with which the stakeholders (senders and recipients) can see an update of the latest position of each individual delivery.

In relation to the last two factors, it is worth underlining that as a consequence of exponential growth in technology, the role of the commercial function is increasingly evolving from a product seller to an expert partner in a specific sector or technology which, in addition to its own products, offers the customer innovative solutions based on the actual products, able to solve a problem or fulfil an opportunity. In other words, the salesperson collaborates with the customer to try and anticipate the customer's needs, solve problems, or suggest technologies and solutions to seize opportunities which the customer might not even have thought of. The customer thus remains focused on their core business, while the partner is the technology expert, able to bring the opportunities offered by technology into the customer's business.

Post-transactional Factors

The factors that fall within the after-sales assistance phase assume a certain importance in sectors where after-sales technical assistance is required, such as IT systems, durable goods, i.e. household appliances and cars, but also in the construction of such one-off works as industrial plants or machinery, or in those sectors in which products must be disposed of at the end of their useful life.

In this phase, the support management service deserves a particular debate, in which it is possible to provide a series of quantitative indicators to measure the service produced and perceived, and which are often subject to contractual obligations between customer and supplier. Instead, for the other kinds, a qualitative discussion will be offered.

Management of Maintenance Tickets

For the management of maintenance tickets, buyers and vendors contract SLAs (Service Level Agreements), which regulate the after-sales service conditions. In particular, the SLAs define the times in which a request for assistance must be taken up and resolved by the vendor. For example, the SLAs may envisage that the vendor must deal with a maintenance request immediately, by providing a telephone help desk, or within a certain number of working hours, for example, following a report via email from the vendor to a specific address, or in a more structured case, when opening a ticket on a maintenance management portal.

As for the workaround and permanent correction, we usually speak of "best effort", since it is not possible to commit a priori to the resolution of an unknown problem. In other cases, buyers and vendors can agree that in any case a certain percentage of tickets must be resolved within a certain period of time—e.g. 95% of requests dealt with and resolved within one working day.

Again on the basis of the SLAs, an in-depth analysis distinguishes between maintenance interventions on two, or even three levels. This request could be Level I, which includes the easiest problems to solve, and which can be solved directly by the customer independently or appropriately guided via tutorials, websites, FAQs, and/or chatbots. Level II and level III requests, on the other hand, refer to more complex cases, in which the customer is unable to conduct the intervention independently. In this case, assistance is provided by the vendor through a specialized operator or escalated to the engineers/developers of the product/system.

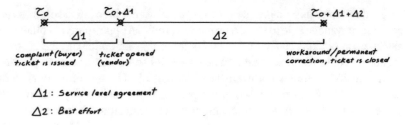

$\Delta 1$: Service level agreement

$\Delta 2$: Best effort

Fig. 24   Management of a maintenance ticket according to an SLA

The scheme to manage a maintenance ticket is represented in Fig. 24 in which the various phases and the relative times are highlighted.

- τ_0 Instant reporting of the problem by the customer; takes place via access to ticket management software, emails, phone calls, must still be tracked, and represents the initial instant of the ticket
- $\tau_0 + \Delta 1$ Represents the moment of taking charge, that is, the moment when the ticket begins to be managed by the vendor; the time $\Delta 1$ is a first service KPI, and precisely defines the time taken by the vendor to take charge of the problem
- $\tau_0 + \Delta 1 + \Delta 2$ Then there is a triage time $\Delta 2$, in which the vendor examines the problem and establishes the cause, which may be attributable to the buyer (an intervention not under warranty, an intervention not falling within the assistance clauses—e.g. a contract which only covers Level 2 interventions, and the request is for a Level 1 intervention), and so on, or which is the responsibility of the vendor. In the first case, the ticket is closed, in the second case it requires a resolution by the vendor
- $\tau_0 + \Delta 1 + \Delta 2 + \Delta 3$ when the intervention falls within those provided for in assistance, the time $\Delta 3$ defines the amount of time required to resolve the problem, restore the system, and close the ticket. In this case, a further intermediate event can be identified, represented by the creation of a "workaround", i.e. a temporary solution while waiting for the system to be definitively restored as it was before (permanent correction) and the ticket to be closed.

In addition to the take-over time $\Delta 1$ seen previously, an additional *a-posteriori* KPI for evaluation of the after-sales service and management of maintenance tickets could be the average duration of the maintenance ticket, expressed through the following relationship

$$\overline{D} = \frac{1}{N} \sum_{i=1}^{N} (\Delta 1 + \Delta 2 + \Delta 3)$$

It is evaluated on the N tickets managed.

As mentioned, buyers and vendors can agree a priori on a performance to be achieved, e.g.,

- 95% of maintenance requests taken up within 4 working hours
- 95% of tickets resolved within a working day
- 95% of tickets with a workaround within a working day; definitive solution within two days.

Or agree on different SLAs depending on the severity of the problem, and therefore typically more stringent SLAs if the severity is high (e.g. the customer's operations are totally compromised) or less stringent SLAs if the severity of the problem is minor or minimal (the problem has a limited impact and customer can operate but with reduced performance).

Other Post-transactional Factors

Other factors which fall into a post-transactional phase include:

- The management of scheduled maintenance phases.
- The temporary replacement of a product (e.g. courtesy car or mobile phone) with a spare device, allows the customer to avoid managing spare parts.
- A service to collect unsold or expired products from its customers. A typical example of the latter type of service is in the case of an attempted sale, where the supplier collects products close to the expiry date from a refrigerated counter without invoicing them to the buyer, newspapers and magazines industries, or in the confectionery industry of highly seasonal products, where the producer of *panettone* or Easter eggs undertakes to collect the unsold items for a variable fee agreed with the customer.
- Similarly, this category includes a service to collect products at the end of their life, or which must be disposed of because they are no longer functioning or suitable for the performance requested by the customer. This typical of household appliances (e.g. televisions, white goods such as refrigerators, washing machines, ovens, etc.).
- Free returns: as mentioned, this is a particularly important service for online sales, which helps to increase the confidence of consumers in the purchase of products that they cannot touch/try before purchasing.

In this case, the factors listed represent one of the possible sets of components of the service level perceived by the customer during the after-sales phase. As the product sector varies, some may assume greater or lesser relative importance, or others not mentioned above might take precedence.

Transactional Factors

The analysis of sales support service factors now focuses on transactional factors, ranging from the management of customer orders to the delivery of finished products to an end customer. They typically depend on SC processes, and directly influence the service perceived by the customer itself (see Fig. 25), be it a player downstream

Fig. 25   Service level as an ability to respect an SLA or shopping experience

of the SC (in this case we speak of the SC's ability to comply with the SLAs), or a final consumer (in this case we speak of the purchasing experience of the final consumer).

The analysis focuses on five transactional service factors, which will be treated analytically, and of which Ruggeri (1998) has proposed quantitative indicators to measure the performance provided:

1. Delivery Lead Time
2. Delivery frequency
3. Punctuality of delivery
4. Accuracy of delivery
5. Delivery flexibility.

Lead Time—Cycle Time

The Strategic Role of Cycle Time

A first definition of cycle time is Vendor Lead Time (VLT), that is, the time which elapses between the issuing of an order by a customer and the receipt of the goods by the same customer, compliant to the order issued (Fig. 26).

Cycle time is arguably the most important of the transactional factors. In fact, it directly influences the costs of the SC, on the one hand in terms of physical costs related to stocks (both cycle stocks and safety stocks) on the other in terms of MMC Market Mediation Costs, whether backlog and/or shrinkage.

For physical inventory costs, cycle time affects both cycle stocks and safety stocks. In the case of a reorder level, the cycle stock will need to meet the demand in supply lead time, and so it is evident that the reorder point, and therefore $CS = \frac{OP}{2}$ will need to be kept higher the higher the delivery lead time is. In the case of an order up to level, the target stock, which represents ½ of the cycle stock, will cover the demand in the reordering interval plus the supply lead time. Also in this case, as the LT increases, the cycle stock increases.

On the other hand, as the lead time increases, safety stocks also increase (see Fig. 27), which, as we have seen, depend on the root of the lead time.

Fig. 26   Vendor lead time

Fig. 27 Safety stocks are proportional to the root of the supply lead time

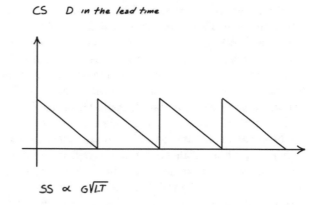

$$SS \propto G\sqrt{LT}$$

As for the impact of lead time on market mediation costs, high lead times require progressively producing forward forecasts, and therefore the accuracy will be less. It is clear that, with the same algorithms, it is easier to make short-term forecasts than long-term ones. Suffice to think of the application of the DRP technique, for example. We have seen how the LT defines the freezing horizon of the DRP. If demand increases during LT, no action is possible, and the result is, at best, a backlog or a missed sale. As can be seen from the drawing in Fig. 28, compared to a lead time of one week, an error in the sales forecast for the following week can be minimal; in the face of a lead time of one month; however, the forecast error could be much more pronounced, thus generating high market mediation costs.

Which is why cycle time has taken on an increasingly strategic role in winning over new customers. In all industrial sectors, the life cycle of each product is shorter and shorter, the demand is increasingly variable, and the physical and market mediation costs associated with inventories are higher and higher. Accordingly, no one wants to hold stocks today, and customers in all industries are increasingly sensitive to lead times. Customers tend to source from suppliers with the shortest lead times possible, who are able to supply the requested product at an acceptable price. Frequently, many customers are even willing to partially ignore the price, as long as there is a reduction in lead time.

Fig. 28 Prediction error as a function of the LT

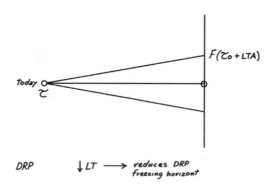

In the case of online sales, lead time has become the competitive factor upon which companies like Amazon or Dell.com have built their success. Dell guarantees deliveries in 40–72 h while Amazon, with initiatives such as Amazon Prime, guarantees the consumer a cycle time of 24–48 h. Initially, the web was struggling to take off because cycle times were in the order of days or even weeks. In some cases, the portals required the closure of a sales campaign (BuyVIP, Privalia, etc.) which had lasted a few weeks before shipping the goods. Only when suppliers were able to deliver products to the consumer quickly, even within 24 h, did e-commerce definitively take off. This has been called "instant gratification", and it means a compulsive purchase resulting from a consumer's need to find instant gratification of a sudden desire. It induces dopamine, an addictive neurotransmitter, and is the element with which the giants of the web keep its customers hooked. It works only if the supply chain is able to guarantee a lead time (order-receipt of goods) of a few hours.

In the FMCG sector, only when the lead time is zero are the products purchased, which means that only the products on the shelf are sold, while in the face of OOS, the consumer increasingly turns to alternative products.

The following Fig. 29 summarizes everything said about the LT as an indicator to measure SC performance.

Vendor and Buyer Lead Times

There are two values of cycle time, a cycle time of the supplier—VLT (Vendor Lead Time), and a cycle time "seen" by the customer—BLT (Buyer Lead Time).

While the Vendor Lead Time VLT depends solely on factors that can be controlled by the supplier itself, the cycle time "seen" by the buyer (BLT) also includes, but not only, the cycle time of the supplier (VLT), including the time to issue the order (Order Issuing Time—OIT) and the time to receive a delivery (RECT) (see Fig. 30).

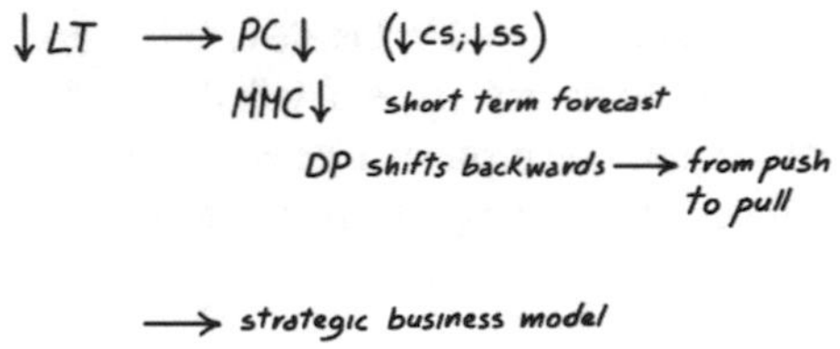

$\downarrow LT \longrightarrow PC\downarrow \quad (\downarrow cs; \downarrow ss)$
$MHC\downarrow$ short term forecast
DP shifts backwards $\longrightarrow$ from push
                                            to pull

$\longrightarrow$ strategic business model

Fig. 29 The impact of lead time on physical costs and market mediation costs

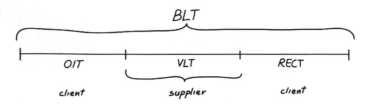

Fig. 30 Definition of BLT and VLT

Therefore, while a Vendor Lead Time can be controlled and managed in full by the supplier itself, the cycle time "seen" by the customer can only be partially managed by the latter. The term "seen" in quotation marks therefore wants to emphasize the fact that on some elements of this time (in this case the Vendor Lead Time), the customer has no active control but must limit itself to "seeing" it passively, nevertheless it is on this term that the customer bases its perception of service.

The cycle time "seen" by a customer Buyer Lead Time—BLT is defined as the total time (time interval expressed in the appropriate units of measurement according to the specific situation in question—hours, days, or weeks) which elapses between the perception of the state of scarcity and the moment goods conforming to the order become available for the end customer.

The following equation applies:

$$BLT = TDOQ + OPT + OFT + VLT + TRT + ULT + RECT + DET$$

These terms are analysed analytically below.

- $TDOQ$: time to define the order quantity. Once the reorder system has noticed that there is a shortage of one or more items for the items managed at the reorder level (typically the stock reaches an order point), or is at the end of the reorder interval for the items managed according to the reordering period (order up to level policy), the reorder quantity Q_i for each reference i must be determined. The time needed to determine Q_i is calculated within the definition time of the quantity to be ordered. This can be instantaneous in the case of automatic reordering systems, or more or less significant if the quantity to be reordered needs to be defined manually. Even in the case of automatic reordering systems, however, manual adjustments may be possible, for example by a department head, who according to his/her own sensitivity can make adjustments to the system reorganization proposal, adjustments that require time since it is necessary to physically go to the aisle in front of the shelf, consequently, in this case too, the reorder time may be non-zero.
- OPT: order preparation time. Time for order preparation It is the time that elapses for the transformation of the quantities to be ordered Q_i of the different items in a structured formal order. We have already seen in Chapt. 4 how a PO_j for a supplier j should be structured; here, we limit ourselves to observing how the order formalization time can vary between zero, in the case of automated reordering systems at the ERP level, to more or less significant times in the event that the order needs to be produced by hand on a Word or Excel template, a common practice in many SMEs.
- OFT: order forwarding time. Once a formal order has been prepared, the document representing the PO must be forwarded by the customer to the supplier. The order forwarding time can assume extremely different values depending on how the order is transmitted. It will certainly be negligible in the case of systems integrated via EDI at the ERP level, in which the transmission of an order takes place automatically through web services (WS), significant if the transmission

takes place manually (e.g. via email, it is necessary that an operator prepares the email and sends it to the supplier).

- *VLT*: Vendor Lead Time. Also called Order Cycle Time (OCT), this is the time needed by a supplier, once a formal document has been received, to send an order confirmation, and fulfil and deliver the order in accordance with the specifications. *VLT* represents the element of the cycle time over which the customer has no control. In the case of Cost and Delivery incoterms (C- and D-type), this time also consists of the order transport time.
- *TRT*: transport time of the order. In the case of ex-works (EXW) or Free Incoterms, once an order has been prepared, the items must be transferred from the supplier to the customer. Since the ex-works (EXW) or Free (F) delivery has been agreed, this operation, and therefore also the time *TRT*, is the customer's responsibility. In Annex A—The INCOTERMS 2020, the INCOTERMS table is shown with the diverse types of transport.
- *ULT*: unloading time. Time required to inbound a delivery from the customer is counted using this term. Typically, this time is represented by the time for unloading the vehicles.
- *RECT*: receiving time. Time needed to verify the correspondence between what was delivered and what was ordered, according to the methods described in detail in the chapter on managing information flows, and to conduct any sample quality checks
- *DET*: data entry time. Once the conformity of the delivery has been verified, the items must be entered in the customer's information system. Only after this operation will the articles be available for use. This time can be encompassed in the *RECT*

Vendor Lead Time—Order Cycle Time

As mentioned, within the cycle time "seen" by the customer there is a term over which the customer has no control, represented by the cycle time of the supplier *VLT*, plus possibly the *TRT* in the event that the transport time of the main leg is a charged to the supplier (C and D INCOTERMS).

This time, of which a summary characterization has already been given, can be defined as follows. The Vendor Lead Time is given by the total time (time interval expressed in units appropriate to the specific situation in question—hours/days/weeks), between

- instant of receiving the order
- instant of making the goods available according to the delivery methods agreed with the end customer.

The Order Cycle Time is critical since it is the time that is crucial from the point of view of the service perceived by the customer, given that, for the customer, this is the only element of the cycle time that depends on the supplier, and on which it then judges the performance of the service received.

Like *BLT*, *VLT* can also be obtained by splitting it into its component times. Considering a case that is as generic as possible, the *VLT* is made up of the procurement time of raw and semi-finished materials (procurement), production time and distribution times. The latter in turn include the storage times of the goods, the transport times between the various levels of the distribution network, and the times for delivery to the customer. In the following discussion, the Make To Stock case is considered (e.g. distribution companies), and therefore it will be assumed that the supplier has the ordered lines available in stock at a certain level of the distribution network. Consequently, we will not consider the case in which certain order lines need to be manufactured, or raw materials have to be procured, since here it is not of interest to enter so much into the merits of the production processes but those that fall more typically within the operations management sphere, it being understood that also production, as a link in the supply chain, nevertheless falls within the scope of SCM. Clearly, this Vendor Lead Time model will need to be reviewed for companies which fulfil on order (ETO, MTO or ATO), and we will try to better understand these implications in the chapter relating to the strategic management of the lead time.

The reference scheme can therefore be the one represented in Fig. 31, in which, against an order from the customer (e.g. the retailer in Fig. 31), the availability of the order lines is checked at the distribution centre of the supplier (S in Fig. 31), which is it is either a full-mix DC or a factory warehouse (W). In the event of availability, the order is prepared and sent to the customer, while in the event that there is no availability, the order is gradually transmitted upstream in the logistics chain, up to a central warehouse. Here, as mentioned, it is assumed that the order lines are present and therefore the order can be fulfilled.

In a general case, however, the order could be transferred even further upstream in the chain, generating a manufacturing or assembly process (O) or the procurement of raw materials and components (P).

Having made these clarifications, the Vendor Lead Time (Order Cycle Time) can be broken down into the sum of three fundamental components, according to the following equation:

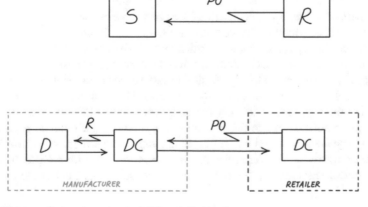

Fig. 31 The manufacturer's system to fulfil a retailer's order

$$OCT(or VLT) = TOP + TOF + TRT$$

where:

- *TOP*: Time for order processing. This is given by the sum of all the times necessary to process the order from an administrative point of view; it ends with the sending of the order confirmation to the buyer.
- *TOF*: Time for order fulfilment. This means the sum of the time needed to physically prepare the order (retrieving/picking, sorting, packing, shipping); it ends with the availability of the order prepared, ready for delivery
- *TRT*: Time for order transport. Time required to perform the operations and the steps required to deliver the order from the supplier to the buyer; it ends with the provision of the goods to the buyer according to the agreed delivery conditions.

Each of the three addenda of the *OCT* is actually in turn the result of a sum of times, since the processing, preparation and delivery of the order are obtained through multiple phases and elementary activities. It is therefore worth analysing each operation in detail, considering its main components.

TIME FOR ORDER PROCESSING TOP

When it comes to the time to process the order *TOP* from an administrative point of view (within the Customer Relationship Management—CRM), then, this can be obtained from the sum of:

$$TOP = TIT + TAC + TIC + TDP + TOC$$

where:

- *TIT*: Time for internal transmission (data entry). Once an order arrives at the supplier (e.g. via email) it must be forwarded within the supplier's organization to those responsible for dealing with it, and who will enter it within an information system. If the transmission takes place via EDI or a web portal, this time is zero.
- *TAC*: Time for administrative checks. Once an order has been taken over by the commercial office, a whole series of administrative checks must be carried out, the duration of which is counted as a whole in *TAC*. Examples of administrative checks could be, for example, verification of whether an order comes from a new customer or not. In the first case, new customer data must be compiled, vice versa, if the customer already exists, it must be checked whether the customer is up to date in terms of previous payments (the customer may be blocked), whether the payment or delivery methods requested by the customer correspond with those agreed, and so on. In both cases, the prices, discounts, payment and billing methods required by the customer must be verified. Also in this case, if the company is structured through a corporate ERP with a CRM (Customer Relationship Management) module, most of these checks can be automated

- *TIC*: Time for inventory checks, i.e. the time required to verify that the order lines requested by the customer are in stock.

This time depends on inventory accuracy, understood as the correspondence between physical stock and stock in the information system (WMS), as seen in detail in the chapter on the management of information flows.

The time for checking for availability can be extremely variable. It can be zero, in the case of WMS systems powered by accurate processes (e.g. BC readings piece by piece in all IN and OUT processes) and/or accurate inventories (e.g. daily RFID inventory counts) it can be extremely onerous if the stock data is not accurate and therefore requires manual verification i to confirm the order.

- *TDP*: Time for delivery planning. Having verified that an order can be fulfilled in terms of the availability of items, it must be checked whether the order can be fulfilled on schedule, in terms of preparation (picking, packing, shipping) and scheduling of deliveries (transport). This therefore means verifying whether it is possible to plan the preparation and shipment by the requested date, for example by combining the delivery with others already scheduled in that area if the deliveries are managed in-house, or by checking the availability of an express courier, where shipments are outsourced.
- *TOC*: Time for order confirmation. The administrative processing phase of the order closes with the time necessary to confirm the order to the customer. The OC (order confirmation) is then sent, a structured document seen in the Chap. 4 on information flows, with which the vendor undertakes to fulfil the customer's order in the manner detailed in the OC.

Ultimately, the *TOP* is different from zero for manual processes, while it is typically zero for automated processes as happens with EDI systems or online channels, so that the order is processed in hidden time during the preparation and forwarding of the order $OPT + OFT$ (see Fig. 32): the order processing time is cancelled since it occurs progressively as the consumer interacts with the web portal which communicates in real time with the vendor ERP.

TIME FOR ORDER FULFILMENT TOF

Once an order has been processed from an administrative point of view, it must be physically prepared, that is, fulfilled. The time to prepare an order *TOF*, *Time*

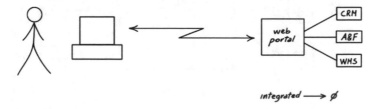

Fig. 32 Order processing time in the case of management via a web portal integrated with CRM, WMS and A&F

for order fulfilment, which belongs to the logistics sphere and therefore depends on how the picking (pick + sort), packing (+ slam), shipping (sorting + loading of the vehicle) processes are organized and structured.

The TOF can also be broken down into the sum of several terms

$$TOF = TLP + TPI + TSO + TCU + TVE + TCP + TSOS + TVL$$

Picking

- *TLP*: Time for picklist generation. This is the time required to enter the order lines of a customer order in the picklists of a picking mission. Depending on whether the picking is organized by order, batch, or zone picking mode, the assignment methods will differ. The process is normally carried out automatically and is optimized to minimize the variable component of the picking time, i.e. the travel time from the WMS, and is therefore zero. Vice versa, in the absence of a WMS and manual processes, the value of a TLP can be significant (see Fig. 33).
- *TPI*: Time for picking items. Once the picklists have been formed and assigned, this is the time physically necessary for picking, then to access the storage areas and selectively pick out the items requested by the customer in the required quantities. This operation can be carried out in various ways, typically "operator to materials" or "materials to operator" as in the case of Amazon where Kiva robots are used to bring the storage pods to the operator, which allows annulment of the variable quota picking times linked to travel times (Fig. 34).

  In fact, the time for picking items TPI can be further broken down into TTR—time for transfer and TPK—time for picking

$$TPI = TTR + TPK$$

The former component is variable, and depends on the location of the reference to be reached, the latter component is fixed for each item picked, and depends on the quantities per order line.

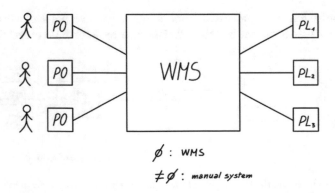

Fig. 33 Time to switch from sales orders to picklists

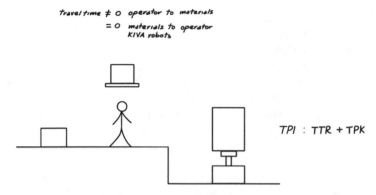

Fig. 34 Picking time—operator to materials versus materials to operator

- *TSO*: this is the sorting time. If the picking takes place in a *batch picking* or *zone picking* mode by grouping customer orders with more than one item per order, it is necessary to proceed with a sorting operation which, starting from the items picked belonging to the same picklist, divides them by customer order. This process, as described extensively in the previous chapters, can be manual or automated by means of automatic sorters. At any rate, the time required to carry out these operations can be significant.

Packing

- *TCU*: Time to customize items. In the event that the customer does not request standard items, it is necessary to carry out customization interventions. Consider, for example, the case in which a reworking of promotional items has to be done (e.g. display packaging, re-packing, application of promotional films, pricing for a particular market, etc.) or the customer requires an aesthetic customization (e.g. a screen-printed shell for an iPhone, addition of paper goods), or a PC with a particular type of software installed as well as a particular type of component, meaning that a certain time for installation and therefore of customization is required. But also activities such as the application of RFID tags for suppliers operating in the fast fashion and luxury or consumer electronics sectors, to comply with retailers mandates like Walmart's or Nordstrom's. All these activities go under the name of VAS—Value Added Services. Orders that require customization must pass through the VAS area, where operators carry out the customizations in a *TCU* time
- *TVE*: Time for verification. Once the picking and customizations have been done, it is necessary at this point to verify that the order being composed corresponds in type of items (mix) and quantity (Q_i) to what has been requested by the customer, before proceeding with the consolidation. It is possible to have sample checks (typical for case level picking in the FMCG) or a complete check at a packing station (the latter is typical of item level picking, for instance for online sales).

- *TCP*: Time for consolidation and packaging. This is the actual packing. The materials collected in different areas and customized in other areas must then be gathered and packaged—in the case of online sales it is at this stage that the verification is made, so that the *TVE* is masked—in one or more loading units in order to be delivered to the customer. This phase includes SLAM operations (Scan, Label, Apply, Manifest) detailed in Chap. 2, with which the parcel label is read in an online channel, the courier's shipping label is printed and applied, and the destination loading plan—*bordereau*—is created

Shipping

- *TSOS*: Time for sorting/staging. Once the shipping unit has been formed, whether it is a palletized shipping unit or a package, it must be brought to the loading dock through which it will be shipped. In some cases it is necessary to sort all the packages destined for shipment from a single dock using a piece of automated equipment—a *sorter* in fact. In the case of palletized loading units, the operator or the Materials Handling/LGV system in the case of industrial automation, will need to physically move the SU to the staging area in front of the shipping dock
- *TVL*: Time for Vehicle loading: Once ready in the staging area, the load units must then be loaded. The loading time and therefore the time necessary for the operator to load the vehicle docked on the quay. This operation can take place by packages (typical if the saturation is by volume) or by pallets (in the case of saturating the load capacity by weight) depending on how the vehicle is to be loaded. In the first case, the loading time is longer; we can try to mitigate the impact by using motorized extendable roller conveyors with which to convey the packages directly inside the load compartment

TIME FOR ORDER TRANSPORT TRT

Once the order has been prepared, the goods must be delivered to the customer. To evaluate TRT, that is the delivery time of the order, a fundamental distinction must be made depending on the delivery method agreed between customer and supplier. In the case of EXW deliveries, the order delivery time is zero, since all transfer operations of the order are the responsibility of the buyer.

In the case of other INCOTERMS (see Annex A), the supplier is also responsible for either some or all the transport operations of the goods to the agreed location, and therefore the order delivery time will be characterized by a certain non-zero value. This time includes not only the time taken to cover the routes, but is the set of all transports, changes of means (road/rail/ship and ship/rail/road) of the times for customs operations, and sorting (crossing of transit point) that an order must undergo through the distribution chain, to get from the supplier to the customer (see Fig. 35, the case of DDP incoterms).

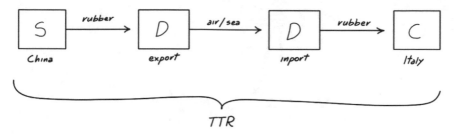

Fig. 35 Transport time in the case of a DDP incoterm

OCT Performance Measurement

There are two measures of performance in terms of lead time. One is a priori, based on process analysis and one is a posteriori, based on field performance. Both can provide complementary information on the supplier's performance in terms of cycle time.

Using the PO issuance scheme as a reference in the figure in the previous paragraph, for a first performance evaluation we can assume as a hypothesis the case in which the order from the store can be fulfilled directly from items in stock in the DC, without the need to then transfer the order further up the chain.

A-Priori Measurement

The a-priori measurement consists in observing how each component of the lead time seen above is actually a random variable, characterized by an average and a standard deviation. It therefore follows that from the relationship with OCT being a sum of all random variables n, it too results in reality as a random variable:

$$
\begin{aligned}
OCT =VLT &= TOP + TOF + TRT \\
&= (TIT + TAC + TIC + TDP + TOC) \\
&+ (TLP + TPI + TSO + TCU + TVE + TCP + TSOS + TVL) \\
&+ (TRT)
\end{aligned}
$$

Assuming the random variables associated with the individual independent component elements, for the central limit theorem (*CLT*), the Vendor Lead Time is therefore a random variable which tends to be normally distributed with mean and variance calculable starting from the mean and variance of the individual components (Fig. 36):

$$
M_{OCT} = \sum M_i \quad \sigma_{OCT} = \sqrt{\sum \sigma_i^2}
$$

This consideration is important to evaluate a priori both the extent of the lead time and its variability, with the aim of having M_{OCT} as low as possible, and $\sigma_{OCT} = 0$.

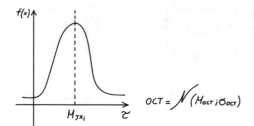

Fig. 36 Order cycle time is a random variable that tends to be normally distributed

Fig. 37 Pareto analysis on
the components of the mean
value of the order cycle time

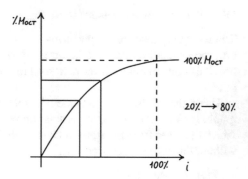

The average value mainly makes it possible to estimate a "presumed" performance and therefore highlight the activities and phases that have the greatest impact on the lead time, in order to optimize and reduce them. It is possible to simply order the addenda from the largest average value to the smallest and thus understand which components to act on, applying a Pareto analysis, as represented in Fig. 37. It turns out, for example, that 20% of the activities are those which generate 80% of the order cycle time, and are therefore the activities on which to act as a priority to try to lower the average value.

For example, if the transport time were the item that had the greatest impact on the average value, the transition to faster modes of transport could be evaluated (obviously taking into account the related costs). These considerations will be further explored in the paragraph "Strategic Lead Time Management".

The standard deviation affects the variability. This too is fundamental information, given that customers not only value the value of the average lead time, but also its reliability, since if the lead time is stochastic, the customer's safety stocks will also need to take this component into consideration. (It is known that SSs depend on the square root of a term in which the square of the standard deviation of the lead time appears). Furthermore, in the event that a customer applies the *DRP*, we saw how a lead time greater than the expected value cannot be absorbed except by means of safety stocks.

For the purposes of customer service, therefore, customers not only evaluate lead time performance in terms of an absolute value but also of reliability, and therefore the analysis of standard deviations allows us to understand which phases are the least reliable (those with the highest standard deviation) and on which to act to reduce the margin of uncertainty.

A-*Posteriori* Measurement

To give a performance evaluation in terms of cycle time, however, it is necessary not to limit oneself to an a priori analysis, but to measure the performance produced and to try and benchmark it against target values, or the best competitors, based on field data.

To do this, according to Ruggeri (1998), the data relating to orders from a cluster of homogeneous customers must first be considered. Orders must therefore first of all be clustered by homogeneous customer, this homogeneity being sought in terms which directly influence the order cycle time, for example: acquired or new customers, order formation methods (it is pointless to compare OCT for customers connected to the supplier via EDI with automatic reordering systems and customers who place their orders manually), routes and delivery distances (it is pointless to compare cycle times for direct orders to customers in a province and customers on other continents), characteristics of demand, etc. (it is pointless to consider in the same cluster customers whose orders that are completely different in terms of average order composition—number of order lines per order and average quantity per order line, so that the picking, consolidation and transportation times could be significantly different).

Once a certain set of orders belonging to a cluster of homogeneous customers has been identified, it is possible to move on to a quantitative evaluation of the performance in terms of cycle time for this homogeneous cluster. To this end, a graphic representation such as the one shown in Fig. 38 can be used, in which the time, expressed in number of WTUs—Work Time Units (hours, days, weeks) elapsed

Fig. 38 A-posteriori measurement of lead time—Lt_{min}, $LT_{xx\%}$ and $LT_{100\%}$

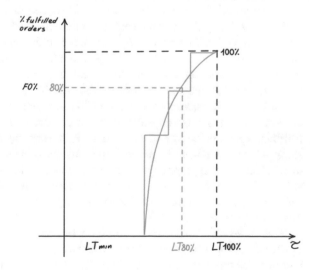

from the moment of receiving an order from a supplier is reported on the abscissa axis up to the point when the last order of the cluster is fulfilled, while on the ordinate there is the percentage of orders completely fulfilled $FO\%$ by the supplier (% fulfilled orders). Typically a stepped curve like the one shown in Fig. 38 is obtained.

This curve should be read in this way. A certain number of orders from a homogeneous cluster (e.g. orders issued during a particular day $WTU = -1$) arrive on time $WTU = 0$; the first orders begin to be fulfilled completely after a certain time interval, then other orders are fulfilled, and so on until 100% of the orders have been fulfilled $FO\% = 100\%$.

Clearly, fulfilment capacity plays a fundamental role in the shape of the curve. The more undersized the capacity is with respect to the cluster of orders to be fulfilled, the greater the number of steps in the curve will be, since not all orders can be fulfilled within the same WTU, but an extended time interval will be required. Conversely, in the case of an oversized fulfilment capacity compared to the cluster to be served, it is likely that all the orders in the cluster can be completely fulfilled within the same work time unit. The same goes for LT_{min}. In the face of an oversized system, it is more than probable (unless there are physiological constraints—e.g. transport time) that the value of LT_{min} is lower than in the case of an undersized system.

Furthermore, other conditions being equal and, in particular, fulfilment capacity and cluster size, the more the orders considered correspond to a homogeneous cluster of customers (in terms of administrative order management, order composition methods, characteristics of the orders, distances from the fulfilment centre, characteristics of the means of transport used, traffic routes), the more the curve will tend to have a reduced number of steps, since all orders will normally be completely fulfilled within a short time interval. Conversely, in the face of an extremely varied cluster of orders, the stepped structure will be particularly accentuated, since some orders will tend to be completely fulfilled after quite dissimilar time intervals.

According to (Ruggeri 1998), on the abscissa axis it is however possible to identify some characteristic performance parameters related to the cycle time.

Firstly, LT_{min}, which represents the minimum lead time in terms of work time units below which the warehouse is unable to fulfil any order. This limit is imposed by technical constraints, which may be linked, for example, to the minimum time required to process an order, prepare it, and transport it to the customer. In the case of the Amazon Prime service, for example, the LT_{min} is equal to 24 h.

There is also a maximum lead time $LT_{100\%}$, which instead represents the number of work time units in which 100% of the orders received at $WTU = 0$ are completely fulfilled. Again, in the case of the Amazon Prime service, $LT_{100\%}$ is equal to 48 h. The stepped curve can then be interpolated by means of a continuous curve, represented in light blue in the graph. This interpolator can be useful for evaluating lead time values LT_{N^*} corresponding to a certain percentage N^* of the orders fulfilled. Thus, for example, $LT_{80\%}$ represents the WTU required to completely fulfil 80% of the orders received at $WTU = 0$.

This curve can also be used to compare the company's lead time values with those of the best competitor if available or with respect to target values (e.g. in the case of e-commerce retailers, the target value is therefore represented by 24–48 h, depending on the cluster, which corresponds to the Amazon Prime benchmark), as represented in Fig. 39. In the specific case, we can see how the target curve is characterized by better performances in terms of cycle time compared to the best competitor, since both the value of LT_{min} and the value of $LT_{100\%}$ are lower.

The representation of the supplier's performance given above in terms of lead time is linked to the scheme assumed: as mentioned, this scheme takes for granted that orders from downstream can be completely fulfilled using the stock available at the DC. This situation is very common in the case of online purchases, in which the available stock is shared between all the stores and therefore the stock service ratio of the individual store is less than one unit. In other words, it is assumed that the DC/store has a stock service ratio define as follow

$$SSR = \frac{\#orders\ fulfilled\ with\ available\ stock}{\#orders\ received}$$

equal to 100%. At the DC/store there are all the items ordered and in the quantities necessary to fulfil the orders (Fig. 40).

In the event that this hypothesis no longer holds, and that some orders therefore require replenishment by upstream warehouses or other stores in order to be fulfilled, a representation of the supplier's performance in terms of cycle time would be the one show in Fig. 41.

To analyse the latter case, we can consider a stock service ratio $SSR = P^*$. It is therefore assumed that only a percentage P^* of the orders coming from the customer

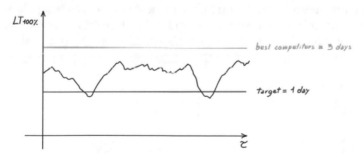

Fig. 39 Benchmarking of lead time KPIs (e.g. $LT_{100\%}$) with respect to target values and values of the best competitors

Fig. 40 Fulfilment of orders when the stock coverage ratio SCR is equal to 100%

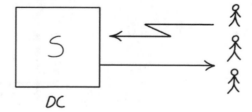

Fig. 41 Fulfilment of orders when the stock service ratio is less than 100%, thus some order lines need to be replenished from upstream

cluster can be fulfilled directly through the stocks present at the DC/store, while the complementary part $100\% - P^*$, in order to be completely fulfilled, needs upstream procurement of order lines from the SC (Distributor or even Manufacturing, or in the case of BOPIS, either replenishment from the DC or transhipment from another store).

As a result of this hypothesis, the supplier's lead time performance curve changes, as shown in Fig. 42.

As we can see, the blue curve represents the time needed to fulfil the P^* by the DC/store using the items in stock. As in the previous case, these orders are fulfilled starting from LT_{min}. As for the remaining orders $(100\% - P^*)$, these, as mentioned, cannot be fulfilled with stocks at the DC/store, and therefore in order to do so they need an additional LT time interval which corresponds to the supply lead time of missing order lines from upstream to the DC, or for transhipment from another store. Once the order lines have been transferred, the $(100\% - P^*)$ orders can be fulfilled according to the red curve. The resulting curve is the green one, obtained by adding the blue curve and the red curve, which corresponds to the actual performance of the DC/store delivered to the cluster.

Also on the green curve it is possible to identify characteristic points, such as $LT_{80\%}$, for example, which corresponds to the time interval required to completely fulfil 80% of orders. As can be seen, there is a general worsening of performance as

Fig. 42 *A-posteriori* measurement of lead time in the case of SSR lower than 1

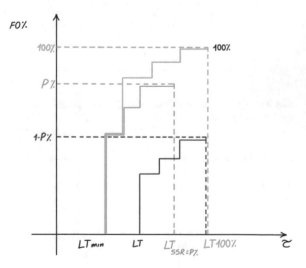

the percentage $(1 - P^*)$ of orders fulfilled through upstream procurement increases, just as performance worsens as the LT between the DC and the upstream rings increases.

If, on the one hand, increasing the stock coverage ratio therefore improves service performance, on the other hand it is evident that increasing the stock coverage has implications on the level of inventories and remaining stocks present at the DC/store and therefore on the one hand on the physical cost of maintaining and managing stock (it is clear that as the number of items kept in stock increases, management complexities increase) but also on the cost of market mediation linked to the risk of obsolescence. The trade-off between stock coverage (and therefore service), capital costs and obsolescence is one of the most typical optimization problems which a supply chain manager has to cope with.

Numerical Example

Here is a simple numerical example of the discussion just concluded (Fig. 43).

Hypothesis

Homogeneous cluster

$SSR = 100\%$

$$LT_{min} = 3gg \quad LT_{50\%} = 1gg \quad LT_{30\%} = 2gg \quad LT_{20\%} = 3gg$$

Let us analyse a case in which:

$$SSR = 70\% \quad LT = 2days \quad LT_{min} = 3days$$

Resolution

A distinction is made between orders completely fulfilled through the stock present in the store $(FO\%_{store})$ and through replenishment from other stores $(FO\%_{rep})$—the

Fig. 43 Retrospective measurement of lead time—numerical example with SCR $= 100\%$

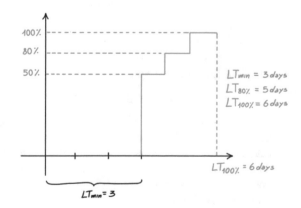

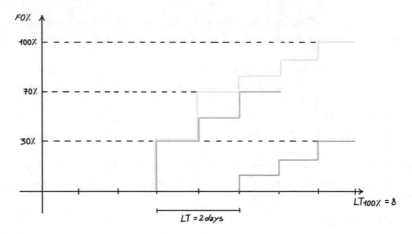

Fig. 44 *A-posteriori* measurement of lead time—numerical example with SCR = 70%

sum *FO%* over the period of these two elements provides the quantity of orders completely fulfilled in that period. The cumulated order fulfilled, i.e. the total of orders fulfilled up to that period *CFO%* is given by the sum *FO%* of all the previous periods (Fig. 44).

days	1	2	3	4	5	6	7	8
$FO\%_{store}$	0	0	0	$0.5*0.7$	$0.3*0.7$	$0.2*0.7$	0	0
$FO\%_{rep}$	0	0	0	0	0	$0.5*0.3$	$0,3*0,3$	$0.2*0.3$
$FO\%_{tot}$	0	0	0	35%	21%	29%	9%	6%
CFO%	0	0	0	35%	56%	85%	94%	100%

As expected, it can be seen that as the stock service ratio *SSR* increases, the $LT_{100\%}$ decreases, on the contrary, if the *SSR* decreases (and therefore the supply from upstream supplier increases) then the $LT_{100\%}$ increases.

Below we will see how other parameters affect the LT, beyond the *SSR*. These are:

- Mode of transportation
- Order formation procedure
- Order fulfilment procedure.

Lead Time—Performance Improvement

After analysing the ways in which to quantitatively measure a posteriori the Vendor Lead Time and how it is possible to provide a representation of the relative performance, the main factors which influence this performance are now introduced and qualitatively analysed.

Four macro-factors can be identified that most influence a supplier's performance in terms of cycle time:

- ICT technologies adopted for order processing
- structure of the distribution network and its organization
- type and mode of transportation
- stock availability ratio SAR.

ICT Technologies and Integrations

As far as the first factor is concerned, clearly the adoption of *ICT* technologies, and in particular the integration between information systems, can streamline the VLT. In particular, we have seen in the previous chapters how tools such as EDI integrations for the transmission/reception of POs and COs between systems (from integration through Web Services to a simple exchange of text files) can reduce transmission times of an order between customer and supplier, while integrated business applications, such as ERP systems, can allow a supplier to reduce certain elements of the time needed to process an order *TOP*, in both cases reducing the value of LT_{min}. Also a WMS, or such automatic identification technologies as BC, but above all RFID, which optimize the and order preparation cycles, play a fundamental role in reducing time for picking *TPI*, time for verification *TVE* as well as the time for vehicle loading *TVL* by acting in this sense towards a reduction in $LT_{100\%}$. As for picking times, the use of RFID technology in-store allows operators in charge of fulfilling online orders to use an RFID hand-held reader in item search mode (Geiger mode); once the SKU to be searched for has been set, the terminal emits a sound when an item corresponding to the search parameters is within the read range of the device. Item search times can even be reduced by as much as 80–90%. With regard to verification times, thanks to RFID smart tables, picked items can be carefully controlled, immediately verifying whether the mix and the quantities picked and/or packed for each order are correct. Finally, RFID technology allows control over masked time single Shipping Units as they are loaded, on the one hand reducing *VLT*, and on the other one avoiding mis-shipments.

Structure of the Distribution Network

As for the distribution network, two sub-factors can be distinguished: the first is the organization of the DC/store network, the second the mode of transportation used.

Clearly, by acting on the conformation of a DC/store network, the value of LT_{N^*} can be influenced, that is the value of the time required to completely fulfil a number N^* of orders. The qualitative link between the number of warehouses/store and LT_{N^*} is shown in the graph in Fig. 45.

As always, the graph, and therefore the performance, must be referred to a given cluster of customers. If the SC is structured with a high number of deposits (e.g. the pharmaceutical case, characterized by distributors and wholesalers) or the e-commerce channel can count on several stores, it will be able to keep the value of LT_{N^*} low for a fairly large cluster of customers. For the same cluster, however, LT_{N^*} will increase as the number of DCs or stores decreases, or vice versa, for the same number of DCs/stores, as the cluster inhomogeneity increases (e.g. if the SC has only national deposits and there are international customers in the cluster, the LT_{N^*}

Fig. 45 Qualitative link
between distribution
structure and lead times

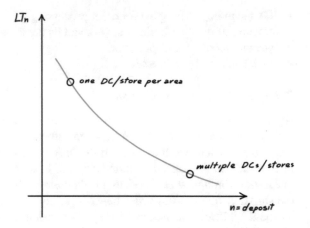

inevitably lengthens; the same happens if online orders need to be fulfilled by few stores). As a result of these considerations, however, it should not be thought that the optimal solution is to have a large number of DCs/stores to guarantee a reduced value of the cycle time for a large cluster of customers. In fact, maintaining a high number of DCs/stores involves costs for the SC, both in terms of Capex costs for infrastructures, and Opex costs on the one hand in terms of management costs of the DCs/stores, and on the other, above all in terms of capitalization of capital for inventories and related risk of obsolescence, which can be significant (it has been seen how the centralization of inventories involves a reduction in stock by a factor of $\sqrt{N}$). As always, the choice by those involved in supply chain management must therefore be made by trying to mediate the two antithetical needs (shortening lead time and improving service level on the one hand, reducing distribution costs on the other), and at any rate in accordance with the strategy pursued by the supply chain itself.

Type of Transportation

Clearly, depending on the mode of transportation adopted, the part of the cycle time relating to the delivery of the order can be compressed, in particular the time to transport the order *TRT* from the supplier to the customer. We therefore speak of a *dedicated courier*, when the shipment is transferred with a specific mission for the customer (e.g. a load entrusted to a cab driver; the rate is fixed according to the quantity transported and the means of transport used, and varies only with the distance). Then there are the express couriers (DHL, UPS), who guarantee the delivery of small packages in a short time (about 48–72 h) at a high cost. And again, the courier forwarder, which can be a transport company which organizes the transfers of consolidated orders between its transit points: in this case the cost varies both with the quantity shipped and with the distance.

Mode of Transportation

Clearly, the faster the mode of transportation, the greater the burden for the supplier in terms of transport costs. Also in this case, it is possible to highlight a qualitative trend of the lead time LT_{N^*} and the cost of transport according to the mode of transportation adopted. Over long distances, therefore, air and ship, and possibly rail, are competitive if the points of origin and destination are connected by land; while on the medium-short distances on the mainland, iron and rubber can prove competitive.

Also in this case, the choice of the optimal mode of the transportation system must be made by the logistician with a view to a trade-off between antithetical needs of service and cost.

Stock Service Ratio SSR—Definitions

We shall now focus in more detail on the fourth factor which influences Vendor Lead Time, namely the *stock service ratio SSR*.

Anywhere in the SC (from store to DC) is said to have stock coverage for a particular item if it is possible to have items immediately available for delivery to the customer. There is therefore stock service if a customer order can be fulfilled directly by a DC with the quantity required for the item. In other words, the availability rate is the % of order lines/SKUs available at a given time T, while the service rate is the ratio of orders fulfilled with available stocks to the total number of orders received at a given time T, as per paragraph "*A-Posteriori* Measurement"

$$SSR = \frac{\#orders\ fulfilled\ with\ available\ stock}{\#orders\ received}$$

From the definition given, we can understand the importance of stock service in determining cycle time. In fact, stock service avoids that for the fulfilment of a certain number of orders, it is necessary to procure from other rings at the same level or further upstream in the supply chain, with an inevitable increase in cycle times, as seen previously.

Stock service can be measured a priori or a posteriori, as detailed in the following paragraphs.

Stock Service Ratio—*A-Priori* Measurement

Ruggeri (1998) observed that the *a-priori* measurement of *SSR* depends on two key dimensions of equal importance:

- a horizontal dimension, linked to the number of order lines/SKUs available in stock at a given time: it is identified with the stock availability ratio *SAR* measured in units [$n°$] or percentages [%].
- a vertical dimension, linked instead to the quantity of the available stock per single order line/SKU *QAS*. The two dimensions are complementary to each other, since neither of them, on its own, can guarantee stock service.

HORIZONTAL DIMENSION—*SAR* STOCK AVAILABILITY RATIO

As for the horizontal dimension, i.e. the number of reference in stock *NRS* means the number of references in the range that are in stock in the DC/store at a given time. The horizontal dimension of stock coverage is usually expressed as a percentage, relating it, for example, to the overall breadth of the range *BR*:

$$SAR = \frac{NRS}{BR} = \frac{number\ of\ references\ in\ stock\,(T)}{breath\ of\ range}[\%]$$

A value of $SAR = 40\%$, for example, indicates that 40% of the items in the range are in stock (at least one piece) and immediately available for delivery. In other cases, especially for comparative evaluations, horizontal coverage can be measured as the ratio between the number of items in stock and the number of items held in stock by a competing supplier. Clearly, the greater the value of the *SAR*, the greater the horizontal stock coverage of the DC/store, i.e. the coverage in terms of the breadth of the range, at the expense of Opex operating costs linked precisely to the stock maintenance costs and the costs of possible shrinkage, always associated with stock management.

To assess the qualitative influence of the horizontal coverage measured by stock availability rate *SAR*, on the Vendor Lead Time measured as usual by means of a value LT_{N*}, we can use the qualitative graph of Fig. 46, where the value of LT_{N*} is reported as a function of the *SAR*. As we can see, the curve has a hyperbolic trend. There is a vertical asymptote, represented by $LT_{N^*_{min}}$ which represents the supplier's minimum cycle time to completely fulfil $N^*_{\%}$ orders in the case of a horizontal coverage equal to 100%. Therefore, it is not possible to go below this value by increasing the number of items in stock, since there are still technical times for the fulfilment of orders that do not depend on the Stock Availability Rate SAR. Conversely, as the value of the *SAR* decreases, the Vendor Lead Time LT_{N*} tends to increase, since the percentage of orders P^* for which it is necessary either to tranship from other stores or procure from an upstream DC also increases.

Fig. 46 Influence of the number of items kept in stock on lead time KPIs

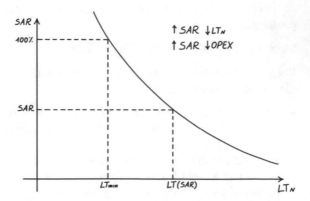

Ultimately, if in a given ring of the SC it is decided to keep a certain percentage SAR^* of the range in stock, this results in a cycle time LT_{SAR^*} to completely fulfil orders N^* which on the one hand is a function SAR^* of choice, on the other it is certainly greater than $LT_{N^*_{min}}$, i.e. minimum cycle time, corresponding to full horizontal coverage. The influence of this aspect has been seen in the exercise relating to the measurement of the LT seen in previous paragraphs.

An FMCG or fashion store tends to keep the entire range in stock, and therefore to have a number of items kept in stock value of 100%, so that customers who enter a store can see and buy the desired product. Unlike the case of capital goods stores such as consumer electronics, in which to avoid excessive immobilization and shrinkage costs, only one model is kept in stock while the other models in the family must be procured from upstream. In the case of pharmaceuticals, even pharmacies tend to keep in stock only the high-rotating items of class A+ or A, in order not to be burdened by the physical costs of stock-keeping, and to take advantage of the particularly short supply lead times to replenish the items on order of class B or C. In the case of fashion stores, the entire range is present only in some flagships, while smaller stores keep in stock only a limited number of lines (models/variants) and stock up from flagships or DCs for those not held. Procurement can take place from other stores even in the case of size gaps, that is, when some sizes are no longer available for a given line, typically the central ones, which are then replenished from other stores or from the DC.

The case of distribution centres is different. Normally there is almost never 100% coverage of the items, but only high rotation ones are kept in stock, or in any case those with low physical costs (immobilization) and market mediation (shrinkage or OOS i.e. dry or non perishable foods). For low-rotation items, or particularly expensive items, and therefore characterized by high physical immobilization costs, or with a high component of market mediation costs, for example shrinkage risk (e.g. fresh or perishable goods) there is a tendency to procure from upstream according to the logic of JIT/cross-docking.

VERTICAL DIMENSION—SQR STOCK QUANTITY PER REFERENCE

In addition to increasing costs, increasing the "number of items in stock", and therefore the horizontal dimension of the stock coverage of a warehouse, is not sufficient in itself to guarantee adequate stock service at the store/DC, since coverage is also determined by the vertical quantity dimension of the inventory per item SQR. Quantity of Available Stock therefore measures how much is held in stock for each reference/SKU. Only if for each item there is an adequate quantity can we speak of stock availability.

As with the horizontal dimension, also in the vertical dimension a qualitative representation of the influence of the size of the stock SQR can be given on the time required to completely fulfil a percentage N^* of orders LT_{N^*}. In this case, it is assumed that the horizontal dimension of the stock coverage has already been fixed, hence also the value SAR^*.

As can be seen Fig. 47, there is also in this case a minimum lead time $LT_{N^*_{min}}$, corresponding to the case of infinite vertical stock coverage: as the amount of stocks

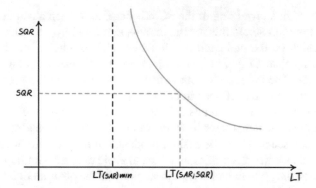

Fig. 47 Influence of the quantity of available stock per item SQR on lead time KPIs

per item decreases SQR, the value of LT_{N*} increases, since as the number of items required to be kept in stock for each item will increase the probability of having to resort to supplies for order fulfilment.

Here too, the choice of the SQR^* to be kept in the store/DC must be made considering that, if high values SQR reduce lead time, at the same time the costs deriving from keeping the products in stock (i.e. Opex) increase, and therefore it is also necessary to determine the value of SQR^*, trying to balance these antithetical needs.

In an FMCG store, the reordering policies (reordering level and reordering interval) determine the necessary quantity for each reference. In the case of a apparel and fashion stores, on the other hand, there is a tendency to think from a row perspective (model/variant) and to adopt a vertical stock coverage which favours the central sizes (M and L) while the quantities of extreme sizes (XXS and XS or XL and XXL) are limited. This situation is illustrated in Fig. 48

Once we start to check the size gaps in the central sizes (the trend of sales in red), the line becomes disassorted. Sometimes to avoid customers not finding the right size, in the case of size gaps, the line may be withdrawn from the sales area, and kept to fulfil online orders, or sent back upstream to the DC which restores the assortments using whatever has been received from all the stores and, if the line is

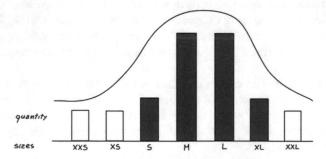

Fig. 48 Amount of stock per line in the case of apparel and fashion retail

still sellable in season, to send it back to the store. Otherwise, the line is kept in stock to fulfil online orders, and the following season will be sent to the outlet channel.

SSR—*A-Posteriori* Measurement

The performance value of any store/DC in terms of stock service ratio can be quantitatively measured a posteriori by following two types of approach: a "commercial" approach and a "technical" approach, depending on whether we wish to respectively emphasize the ability to serve a customer order, or the quantitative aspect in terms of the quantity supplied.

COMMERCIAL Approach

In the case of a commercial approach, the performance can for example be measured in the number of orders SSR_O or order lines SSR_{OL} immediately fulfilled, compared to the number of orders or order lines requested respectively, both parameters evaluated for a given time interval ΔT.

$$SSR_{OL} = \frac{\textit{fulfilled order lines } (\Delta T)}{\textit{requested order lines } (\Delta T)}[\%]$$

$$SSR_O = \frac{\textit{fulfilled orders } (\Delta T)}{\textit{orders received } (\Delta T)}[\%]$$

A measurement is therefore taken for this period of how many orders/order lines have been requested (denominator) and of these, how many orders/order lines have been fulfilled directly using the available stock (numerator).

It is clear that to increase the value of the numerator, and therefore the performance in terms of service, it is necessary to be able to fulfil as many orders as possible, regardless of the size of the order. In other words, all fulfillable orders, whether large or small in terms of quantity, have the same amount of influence on performance.

In choosing between orders or order lines for assessing stock service ratio, it is first of all necessary to avoid considering only one or the other, since the information obtained from the two indices can be complementary. Consider, for example, the case in which, in a given time interval, a series of orders are presented such that only some of these can be completely fulfilled, due to the lack of one or more order lines (which constitute for example the order backlog). The SSR indicator calculated on order lines would give an indication close to unity, suggesting high performance. Conversely, a calculation of the indicator based on orders would give the stock service performance a value close to 0, highlighting that in reality only a few orders were completely fulfilled and that therefore the performance in terms of customer service is low.

The above example also shows that, depending on the type of correlation between the order lines requested by the customer and the product sector, it may be more important to have a high ratio in terms of the number of orders fulfilled or the number of order lines that are immediately fulfillable. When the order lines are extremely independent from one another—as in FMCG retailers, or pharma distributors/wholesalers/pharmacies where in the customer order there is no dependence between the order lines, for example, a hypermarket which buys food, household products, etc.—it is preferable to have high values of the indicator relating to the order lines, since even in the event of an incomplete order, the customer issues an order for the available order lines (frequently, customers either wait for the backlog, or order a replacement product). If, on the other hand, there is a close correlation between a customer's order lines, for example, the purchase of equipment or a kit, in which the various order lines are the different parts of the equipment (software, hardware, manuals, cables and auxiliary materials), it is clear that having high levels of stock coverage in terms of orders that can be completely fulfilled becomes a priority. Ultimately, if the components of an order are highly correlated, the performance measured in terms of orders fulfilled/requested acquires greater importance; if, vice versa, the order lines are not correlated, the measure of the performance through the relationship between fulfillable and requested order lines can prove more significant.

TECHNICAL Approach

If, on the other hand, a technical approach is adopted, the amount of the vertical stock is measured in kg, m^2 or m^3 consistently with the size to be measured, number of units, etc. Similarly to what was done previously, the performance in terms of stock service ratio will still be measured by the ratio between the quantity delivered directly using stock and the total quantity requested, both values evaluated for a reference interval ΔT.

$$SSR_T = \frac{fulfilled\ quantities\ (\Delta T)}{ordered\ quantities\ (\Delta T)}[\%]$$

In this case, however, to increase performance, it is necessary to be able to fulfil orders for which large quantities are required in the first place, while the ability to fulfil a small quantity order exerts a secondary weight on the increase in performance.

Numerical Example

Below, an example summarizes the above, which shows how complementary information can be obtained from the three commercial and technical indicators.

Let us suppose that we have 2 structured orders as in the table below.

Order ID	Total order lines	Status	No. of order lines	Required quantities
A	10	Fulfillable with available stock on hand	9	1
		Not fulfillable with available stock on hand	1	1
B	2	Fulfillable with available stock on hand	1	10
		Not fulfillable with available stock on hand	1	1

The stock service ratio measurements are as follows:

$$SSR_{OL} = \frac{9+1}{10+2} = \frac{10}{12} = \frac{5}{6} = 83,33\%$$

$$SSR_O = \frac{0}{2} = 0\%$$

$$SSR_T = \frac{9+10}{10+11} = \frac{19}{21} = 90,47\%$$

Although SSR_{OL} and SSR_T are very high, the zero value of SSR_O means that actually no order can be fulfilled.

At this point, let's suppose that order B is completely fulfillable. We have then

$$SSR_{OL} = \frac{9+2}{10+2} = \frac{11}{12} = \frac{11}{12} = 91,66\% \quad SSR_O = \frac{1}{2} = 50\%$$

$$SSR_T = \frac{20}{10+11} = \frac{20}{21} = 95,23\%$$

Even in this case, although SSR_{OL} (and SSR_T) are very high and therefore an indication of a high service, only one of the two customers will be satisfied since $SSR_O = 50\%$.

A-Posteriori Measurement: Further Considerations

Having made this due clarification between the parameters of stock service rate of a commercial and technical nature, it is necessary to make a further clarification. Using the expressions given above to evaluate the performance of a store/DC in terms of stock coverage, we could find ourselves in a situation where, despite the indicator assuming values close to unity, the service rate of the warehouse is actually very low. This happens whenever a customer order actually occurs only after the customer has verified the availability in the DC/store of the mix/quantity requested. This is typical of BOPIS orders, when the customer checks for availability online and places an order only for available products.

The relationships given above, if either the technical or commercial approach is used, should therefore be modified, considering the quantities, orders or potential order lines respectively as denominators, i.e. orders not issued but which could have turned into actual orders if there had been stock availability. Again referring to the quantities at the same time interval ΔT, then:

$$SSR_{OL} = \frac{\textit{fulfilled order lines } (\Delta T)}{\textit{requested order lines } (\Delta T) + \textit{POTENTIAL order lines } (\Delta T)}$$

$$SSR_O = \frac{fulfilled\ orders\ (\Delta T)}{requested\ orders\ (\Delta T) + POTENTIAL\ orders\ (\Delta T)}$$

$$SSR_T = \frac{fulfilled\ quantities\ (\Delta T)}{requested\ quantities\ (\Delta T) + POTENTIAL\ quantities\ (\Delta T)}$$

The problem at this point is how to evaluate potential order lines, orders or quantities. In this regard, the following considerations may be made.

In general, in the supply negotiation between customer and supplier, there are a number of orders equal to M that can be completely fulfilled with available stocks, and a number N that either cannot be or can only partially be fulfilled (see Fig. 49). The former are complete fulfilment orders in the sense that all the requested order lines are available in stock in the required quantities. For the remaining orders N, on the other hand, there is the possibility of an incomplete fulfilment of the order, given that for some order lines (backlog) a replenishment from other stores or from the DC may be needed.

Taking into consideration the N orders that can only be partially fulfilled, it can be noted that it is possible to identify:

- a fraction N_1 corresponding to the orders issued, however complete—the customer is willing to await the backlog
- a fraction N_2 of orders issued only for the order lines that can be immediately fulfilled—the order is issued but the missing order lines are cancelled
- a fraction N_3 of missed orders—the whole order is cancelled against one or more non-fulfillable OLs.

This quota N_1 includes all orders that cannot be completely fulfilled due to the lack of certain stock order lines, but for which the customer is willing to wait, generating a backlog. That is, the customer issues a complete order, accepting to initially receive only the order lines immediately available, while those in stock out will be delivered later. The quota N_1 must therefore be ignored for the purpose of calculating potential order lines, potential orders or potential quantities, since these are orders acquired to all intents and purposes, and therefore fall within the term of (*requested orders/order lines/quantities* (ΔT)).

As for the quota N_2 and N_3 this is quite a different matter. These fall into N_2 the fraction of orders issued only for the order lines that can be immediately fulfilled. That is, the customer, instead of issuing a complete order, issues a partial order only for those order lines for which it has checked the availability of stock, while any order lines that are not available are not ordered. Given that, if the DC/store had had

Fig. 49 Orders that can be fulfilled and orders that can only be partially fulfilled

all the order lines in stock, it would have received N_2 complete orders, it could in a sense be said that in this case potential orders have not been lost N_2. On the contrary, the order lines OL_2 (or quantities Q_2) that have not been ordered have certainly been lost.

Finally, the fraction N_3 considers orders that are not issued by the customer as a consequence of having checked the unavailability of some items, in whole or in part. Considering the entire quota N_3 as a quota of potential orders, however, would be incorrect, since a further subdivision must be made. In the fraction N_3 it is in fact possible to distinguish a quota N_{3b} of orders issued by "certain" customers, and a quota N_{3a} issued by potential customers. For the former, it is certain that if the order lines had been available in stock, a complete order would have been received: the quota N_{3a} and the related order lines OL_{3a} (or quantities Q_{3a}) must therefore be counted in all respects among the orders, order lines and potential quantities. As for the quota N_{3b}, relating to orders from "uncertain" customers, this quota includes orders issued by "uncertain" customers, and who therefore would not necessarily have issued an order even if the order lines had been in stock. To calculate the share of potential orders for N_{3b}, we can use the following relationship:

$$potential\ orders = N_{3a} + N_{3b} * market\ share\ [\%]$$

Or in the online case

$$potential\ orders = N_{3a} + N_{3b} * \frac{N°\ orders}{N°\ accesses\ to\ the\ e-commerce\ portal}\ [\%]$$

The same relationship applies, with the appropriate quantities, for order lines and quantities:

$$potential\ order\ lines = OL_{3a} + OL_{3b} * market\ share\ [\%]$$

$$potential\ order\ lines = LO_{3a} \mid LO_{3b}$$
$$* \frac{number\ of\ orders}{no\ accesses\ to\ the\ e-commerce\ portal}\ [\%]$$

$$potential\ quantities = Q_{3a} + Q_{3b} * market\ share\ [\%]$$

$$potential\ quantities = Q_{3a} + Q_{3b} * \frac{number\ of\ orders}{no\ accesses\ to\ the\ e-commerce\ portal}\ [\%]$$

Market share, expressed in terms of volume or turnover, is ultimately assumed to be the probability of receiving potential unissued orders N_{3b}.

Once able to estimate the share of potential orders, in commercial units of measurement (orders or order lines) or technical units (technical units of measurement), it is then possible to correctly evaluate the performance of the DC/store in terms of SSR.

If we wish to measure the SSR_{OL} (for example):

$$SSR_{OL} = \frac{OL\ fulfilled\ \Delta T}{requested\ OL\ \Delta T + potential\ order\ lines\ LO\ \Delta T}$$
$$= \frac{OL\ fulfilled\ \Delta T}{requested\ OL\ \Delta T + OL_{N2} + OL_{N3a} + OL_{N3b} * market\ share}$$

As always, the technical or commercial indicators are not to be seen in contrast, but any correct assessment needs to include a parallel and complementary analysis.

As a further comment, it could be added that in a company it is particularly difficult to trace the quota of potential orders, since there is no systematic and widespread data collection. To correctly calculate the performance in analysis, it is therefore necessary to organize the business processes so that the necessary data are available afterwards, thus keeping track for example also of the unissued parts of orders or missed orders, which have not been billed (POs and invoices are usually the only data available).

In contrast, online systems are structured to track potential quantities, since it is possible to measure how many customers did not purchase a product because it was not available (adding and removing from the cart, or visiting the page and leaving the platform when unavailability of the product has been verified).

Strategic Lead Time Management

After a thorough analysis of lead time, its components and the related KPIs, we complete the examination by introducing how SC lead time can be optimally managed from a strategic perspective, in order to generate a competitive advantage.

The Strategic Lead Management concept was first introduced by Christopher (1992), who then refined it over time. When Christopher talks about *strategic lead time management* he actually means something broader than the order cycle time. He thinks of a coordinated approach at the supply chain level (pipeline management) aimed at optimizing the crossing time of a product unit, from raw material to finished product, modulating and coordinating not only the cycle time in one echelon of the SC, but all the stages of the SC, from procurement to production to distribution, based on the trend in market demand.

Supply Chain Flow Time

In order to strategically manage lead time, it is important to introduce the concept of SC flow time SCFT.

Imagine a unit of product which at the time τ_0 is purchased by the final consumer. If we could wind back the clock of that unit of product, and go back through time throughout its phases—distribution, production and procurement of RM, upstream to the first highest level supplier—we would arrive at a time $\tau_0 - \Delta\tau$ when the first unit of raw material began to be produced.

The flow time of the SC is represented by the value of $\Delta\tau$, that is, the time that a unit of product takes to travel the SC, from raw material until being purchased by the final consumer.

It is important to consider all the processes and phases of the supply chain in the flow time: both the phases in which the raw material, the semi-finished product and the finished product undergo logistical production transformations (transport, handling, transformation, control, etc.), and the waiting phases as stock. Please note that the flow time is given by the following relation

$$SCFT = \sum_{i=1}^{N} LT_i + \tau_{I_i}$$

where.

- LT_i is the order cycle time at the generic echelon
- τ_{I_i} is the time spent as inventory at generic level i.

To better understand some aspects related to flow time, and the implication of flow time on SC stock levels, a hydraulic analogy can be employed (see Fig. 50). Think of the SC as a pipe that carries a fluid: the longer the pipe is, the higher the volume of fluid inside the pipe will be; the slower the fluid velocity, the longer the time it takes for the pipe to empty. Conversely, a short pipe with high fluid velocity will have low volumes of fluid inside, and thanks to the high velocity it will be able to empty quickly. A fluid particle will therefore take little time to cross the pipe from one end to the other.

In the hydraulic analogy, the length of the pipeline is the length of the SC, the velocity of the fluid is related to the LTs of the various processes, thus the product flow, and the volumes of fluid represent the RM, semi-finished and finished products which pass through it, including related stocks, and the emptying time of the pipeline coincides with the flow time of the SC.

Returning to the hydraulic analogy, in a long pipe 100m, under normal conditions, if the SC processes and the related LT cause the fluid to move at 1 m/s, the pipe will take 100 s to empty. Therefore, if downstream of the pipeline the demand for the product changes, it is necessary to empty 100m the pipeline and thus waiting for 100 s before being able to obtain the different product that the market is requesting at the exit point. Similarly, in an SC, if the downstream demand from the market

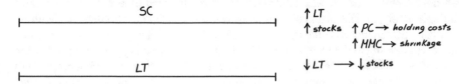

Fig. 50 The supply chain as a pipe carrying a product

changes, a time equal to the flow time will pass before the SC is able to cope with this change.

To reduce the lead time of the SC and improve its reactivity, it is necessary to act along two lines:

- reducing the length of the supply chain
- increasing the speed of the flow inside it (in analogy to the concept of increasing the flow rate, it is necessary to increase the speed), thus reducing the flow time.

In the above hydraulic analogy, if the pipe is shortened to 50 m and the fluid moves at 2 m/s, the pipe will take 25 s to empty, and therefore its responsiveness will be four times the previous one.

Reducing the flow time not only improves responsiveness and therefore the ability to respond to changes in market needs, but also has a very significant impact on profitability. It is clear that there is a proportionality between the duration of the flow time and the amount of stocks in the supply chain. Therefore, the longer and/or slower the supply chain is, i.e. the longer it takes to procure raw materials, produce the product and distribute it to the end customer, the greater the time that the stocks of raw materials, semi-finished products, finished products spend inside it. Consequently, the greater the relative fixed cost (physical cost for holding them) and the risk of obsolescence in the case, for example, of a technological breakthrough (market mediation cost), which instantly renders the stocks present in the SC obsolete.

To reduce these costs, it is necessary to reduce the flow time of the product, and therefore reduce the overall lead time given by the sum of all the lead times plus times spent as inventory, by reducing the quantity of the individual terms (LT_i or τ_{I_i} in the sum above), or by shortening the chain, and thus eliminating some addenda (N in the sum above).

Strategic lead time management therefore means acting according to these two guidelines, and therefore (i) reducing the length and (ii) increasing the speed of the flow. The goal is also to reduce time spent as stock and therefore the associated physical and market mediation costs, up to the limit of passing from a push production to a PULL production in the case of complete closure of the LTG—the lead time gap (understood as the difference between the time that the consumer is willing to wait and the LT of the SC). Thinking of the case of Zara which, thanks to low LTs, has introduced the concept of fast fashion—fast collections introduced in small quantities onto the market which allow it to minimize PCs and especially MMCs.

A fundamental ingredient to be able to effectively shorten the overall flow time is to act according to a global coordinated approach through the involvement of all the players of the SC who contribute to the related lead time. If each player acts independently, the risk is that of compressing phases that are not bottlenecks, without realizing that in any case the overall flow time does not change due to stocks which are accumulating in the warehouses. This is rather like what happens in a production line: if we wish to increase productivity, and therefore the units of product per unit of time, it is pointless to increase the productivity of the other stations unless we increase the productivity of the slowest station. In the supply chain, for instance, it is pointless to reduce transportation costs by shifting from sea to air freight, if the product then spends weeks in stock waiting for an order to arrive.

The following paragraphs illustrate how the SC players can act together in a coordinated manner to reduce the overall flow time of the SC.

Value Adding Versus Non Value Adding

After recognizing that the approach to be adopted is the global one of SC (procurement + operation + distribution) and not the OCT of a single phase, the first step at the basis of strategic lead time management is to distinguish between processes, phases, and single activities which consume time without adding value and those that instead consume time but do add value. In the first case, we speak of *non-value adding time NVAT*, in the second case of *value adding time VAT*. The distinction between non-value activities and value adding time is critical for the management of logistics processes and the optimization of SC lead time.

In general, to understand whether an activity is of a non-value or value adding time type, it is necessary to understand if the activity adds value to the product for which the customer is willing to pay a Delta price. Examples of value adding time are production, transport, picking and packing, but also ripening and seasoning for foods, since they add value by transforming organoleptic properties of a food product; bringing the product closer to the customer, preparing and fulfilling its order, making it available also when raw materials are not available (i.e. farm products). Conversely, activities such as stock-keeping and duplicated activities, are typically ones which generate worthless times, since they only serve to cover inefficiencies of the SC (need to produce/supply economic lots higher than demand due to production or logistic constraints, need to replenish in quantities higher than demand to take advantage of quantity discounts, need to double check at outbound/inbound or at the insert in the production lines because the upstream players cannot be trusted, etc.), and if they were to be eliminated the value of the product perceived by the consumer would be exactly the same. As mentioned before, stock keeping can, on the other hand, be considered value adding time in the case of (i) recurring events, for which it is necessary to produce in advance and therefore generate stocks (ii) the availability of raw materials is limited in time, for which it is necessary to produce in a limited campaign or (iii) the stocks are used to generate particular organoleptic conditions of a product, such as the maturing of cured meat or cheese or the ageing of wine. (iv) the stocks are used to increase the resilience of the SC, and to cope with the risks of interruption.

Just in Time approach synthetizes the non value adding time with the term "Muda", waste. there are several mudas, namely:

- Waiting times: whenever the flow is slowed or halted due to a slowdon or missing materials in the upstream phases
- Internal transport and transhipment times: whenever products need to be moved internally without adding value for the customer (i.e. to free space in a warehouse; to be reworked in a different facility due to a lack of capacity in the main one)

- Duplicated processes: whenever the same process is duplicated, typically at the interface of the different actors in the supply chain
- Inventories: time that the product wastes as inventory

Quantification of Phase Duration and Cost

Once the difference between value adding and non-value adding activities is understood, the second step of strategic lead time management is to map the SC, creating a flowchart of the processes, from upstream to downstream, and highlighting the processes, phases and individual activities. For each of these activities i then:

(1)   Defining the duration either in terms of lead time $LT_i[hordays]$, or inventories τ_{I_i} thus the time spent as inventory at a generic level i. the Question to answer is: "How long does it take for a unit of product to complete this phase?"

(2)   Defining the cost of the activity $c_i\left[\frac{\cancel{\epsilon}}{unit\ of\ product}\right]$. What is the total cost per unit of product of that activity (sum of plant costs—Capex share K, and Opex energy E, manpower/labour L, consumables C, plus the scrap/shrinkage quota S) adding capitalization cost in the case of inventories. The question to be answered here is "How much does that activity cost?"

(3)   Understanding whether it is an activity that adds value or not (the fundamental question to understand whether the activity is of the value adding or non-value adding type is the following: the following question needs to be answered here: "If there were not that activity, would the perceived value of the product by the consumer be the same?") and therefore consequently whether it is value adding $k_i = 1$ or non-value adding time $k_i = 0$, respectively.

It is essential that this step is carried out by involving the various stakeholders responsible for the various processes, trying to reach an agreement on the above classification. It is not easy to reach an agreement, since nobody wants to admit that they are responsible for activities that do not generate value, even if they represent a cost. For example, typically a double check at the vendor's output and at the buyer's input or a double buffer of stocks—FP by the vendor and RM at the buyer—may not add value.

At this point, it is possible to draw a graph like the one in Fig. 51. The overall time of the various activities is shown on the abscissa, while the value adding time is shown on the ordinates (non-value adding activities are parallel to the abscissa axis). The overall extension of the activity axis on the abscissa axis provides the flow time of the product unit, while that on the ordinate axis the value added time. From the relationship between the two values, opportunities for improvement can be highlighted.

Inserire figura

In parallel, a second graph can also be developed, in which the cost of the activity is shown on the abscissa axis and the value added time on the ordinate. The activities parallel to the abscissa axis that do not add value while adding a cost are evident. The overall extension of the activity axis on the abscissa axis provides the total

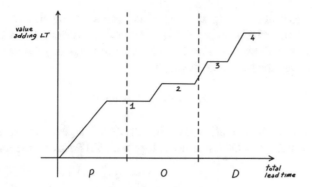

Fig. 51 Value adding time versus SCLT; non value adding time is equal to SCLT-VAT

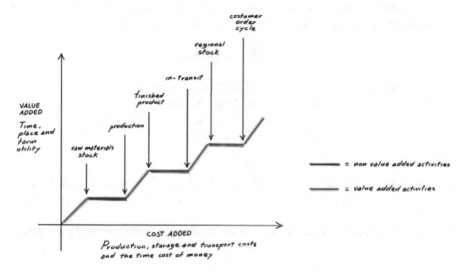

Fig. 52 Cost added versus value added time

logistics cost (procurement, operations and distribution) of the product unit, while that on the ordinate axis the related value added time (Fig. 52).

From the diagram two things are clear:

- the incidence of time which adds value to the overall time. This quantity is measured through the percentage value of value adding time $VAT_\%$, calculated as follows:

$$VAT_\% = \frac{\sum_{i=1}^{n} k_i * \left(LT_i + \tau_{I_i}\right)}{\sum_{i=1}^{n} \left(LT_i + \tau_{I_i}\right)}$$

whether it is value adding $k_i = 1$ or non-value adding time $k_i = 0$, respectively. From the relationship $VAT_\%$, it is possible to obtain in a simple way the percentage of activity that does not add value to the final product and on which to act to reduce the flow time of the SC:

$$NVAT_\% = 1 - VAT_\%$$

- the incidence on the overall cost of the activities that generate value (% of value adding cost $VAC_\%$), calculated in analogy with $VAT_\%$ using the index:

$$VAC_\% = \frac{\sum_{i=1}^{n} k_i * C_i}{\sum_{i=1}^{n} C_i}$$

Also in this case, the complementary of this index represents the percentage of cost borne by activities that do not add value, and therefore ought to be optimized:

$$NVAC_\% = 1 - VAC_\%$$

In some cases, in performing this analysis, it turns out that the value adding time values $VAT_\%$ do not exceed 10–15%. The vast majority of the time, around 85%, the product spends in the SC for non-value generating activities—typically either as stock or in duplicated processes!

Flow Time Optimization

A third step, developed as a logical consequence of the previous one, foresees that improvement opportunities are highlighted, obviously trying to compress those activities that do not generate value.

Once again, it is important to emphasize that the approach must be the integrated one of the SC, avoiding that each function independently reasons by silo. For example, the operations decide to compress the lead time of manufacturing activities to a few days or even hours, through the adoption of automated lines, only to realize that the finished products remain in stock for months downstream in a warehouse in charge of distribution.

Typically, the problems arise:

- In the interfaces between different processes, where duplication of stocks (on both sides) or processes (control in shipment and receipt; quality checks after production at the supplier and before feeding the production/assembly lines at the buyer facility) can be generated.
- Lack of visibility and/or variable lead times—which typically generates stocks on the side of those who do not have visibility or tolerate variability in supply lead time to cover themselves from the uncertainty of demand/lead time, and increases the overall LT in terms of oversizing the pipeline.
- Excessive setup and format change times, which generate overly large economic lots which then turn into stocks.

- Fixed procurement costs or quantity discounts that generate too large economic lots which then turn into stock.
- Bottlenecks which slow the flow by generating upstream inventory.
  - undersized processes.
  - lack of technological resources or skills.

The graph in Fig. 53 shows a qualitative representation of what the impact of having exploited the opportunities for improvement in the activities of the SC could be. With the same convention seen above, the section of the 45° curve identifies value-added activities, the sections parallel to the abscissa axis identify non-value-added activities.

It also highlights the integrated approach that involves Procurement P, Operations O, and Distributions D, which makes it possible to identify the most critical phases.

Another simplified representation is that in which, instead of distinguishing between value adding and non-value adding activities, we make a distinction between stocks (typically not value adding) and therefore τ_{I_i} and other activities related to processes (basically value adding), thus LT_i. With this representation we can see the overall duration of the flow time, given by the sum of the duration of all the phases, and how much time the product spends as stock, therefore doing nothing. Likely figures may show that, compared to a duration of the flow time of 20 days, there is a total stock in the SC of 80 days. Among other things, with this representation it is also possible to highlight at which phase of the supply chain the product remains in stock for the longest time, if as a raw material (and therefore with a lower cost since it is neutral and characterized by more stable demand—a neutral product can be used to produce multiple variants/customized versions) or if as a finished product (with a higher cost since it has already absorbed the costs of previous processing and is customized and therefore more subject to MMCs in the event of a change in demand for that particular model/variant). It is not unusual for most of the time to be spent in stock as a finished product (Fig. 54).

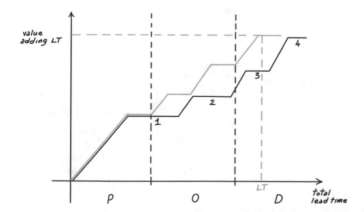

Fig. 53 Review of processes and elimination of non-value adding time

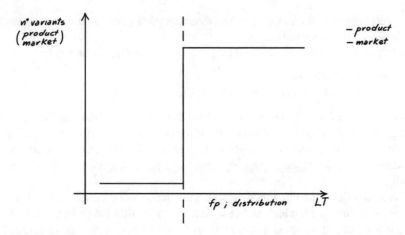

Fig. 54 Stock analysis by number of variants/versions/models as a function of flow time

Introduction to Other Transactional Factors

As an introduction to the second transactional factor of Customer Satisfaction CS analysis, we propose a map of the analysis carried out so far, as a brief summary of what has been discussed and to find the common thread after this long lead time analysis (see Fig. 55).

We began with an analysis of KPIs impacting Customer Service CS as proposed by Ruggeri (1998), and went into detail on the factors supporting sales, having previously analysed the factors supporting marketing. Sales Service Factors are divided, as mentioned, into pre/post/transactional factors, according to (La Londe and Zinszer 1977). Among the transactional factors, the lead time was first analysed in detail, highlighting the different components, the *a-priori* and *a-posteriori* measurement KPIs, and strategic aspects related to optimization of SC flow time (Christopher 1992).

We shall now analyse the other transactional factors, such as (i) frequency, (ii) punctuality, (iii) accuracy, and (iv) delivery flexibility.

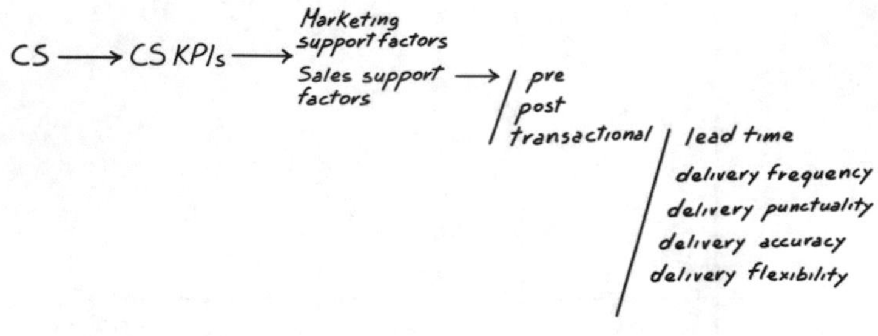

$CS \longrightarrow CS\ KPIs \longrightarrow$ Marketing support factors
Sales support factors $\longrightarrow$ pre / post / transactional
lead time
delivery frequency
delivery punctuality
delivery accuracy
delivery flexibility

Fig. 55 Conceptual scheme of customer satisfaction measurement

Delivery Frequency

Definition

Delivery frequency is usually defined with reference to a cluster of homogeneous customers in terms of average order characteristics, geographical position, etc.

For this sales support service factor the following definition can be given: Delivery frequency DF is defined as the number of scheduled deliveries $DF\left[\frac{1}{t}\right]$, within the unit of time taken as a reference in the product sector in question (hours in pharmaceuticals, days in FMCGs, weeks or months in textiles and clothing). i.e., 2 deliveries per day; 3 deliveries per week, etc.

The inverse obviously represents the time between two deliveries and can also be taken as a service indicator.

Performance Measurement KPIs

According to Ruggeri (1998), measurement of performance in terms of delivery frequency DF can be carried out using an absolute indicator, such as the number of scheduled deliveries NSD in a given time interval T,

$$DF = \frac{NSD}{T}\left[\frac{1}{t}\right]$$

or through a relative indicator, expressed for example by the ratio between the number of actual deliveries NAD and the number of deliveries scheduled NSD in the same time interval ΔT:

$$DF\% = \frac{NAD(\Delta T)}{NSD(\Delta T)}[\%]$$

Optimization

The choice of the minimum delivery frequency offered to the customer with a view to balancing costs. This balancing can be illustrated with reference to Fig. 56. As the delivery frequency varies, there are in fact costs for the customer which vary in contrary ways, so that it is always a question of finding a service/cost trade off with respect to antithetical trends.

In fact, as the minimum delivery frequency increases, transport costs increase for the customer, due to the fact that the number of trips increases, and that for the same route the quantities transported are lower, and therefore also the saturation of the means, which is to say, smaller capacity vehicles are used, the cost of which is higher per kg or m^3 transported.

In addition, the fixed costs of issuing the order increase, since the supply frequency increases. However, in the case of automatic reordering systems with EDI integration, this aspect can be negligible.

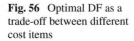

Fig. 56 Optimal DF as a trade-off between different cost items

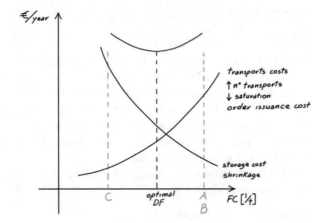

Conversely, as the frequency of delivery and therefore the service increases, the costs associated with the storage (physical costs) and the possible deterioration (market mediation) of the articles tend to decrease for the customer, since as the frequency increases, the average quantity kept in stock by the customer decreases, being reduced to zero for JIT deliveries, compatible with the needs of the final consumer (LTG < 0).

From a minimization of the total cost, the sum of these items, the optimal delivery frequency is determined (*ODF*).

In some product sectors, certain particular factors can contribute to increasing one of the two items of the total cost, making the choice of the delivery frequency practically obligatory. Among these factors, the main ones are:

- the shelf life of the product—thus market mediation costs related to shrinkage
- the customer's willingness to stock items—thus physical holding costs
- the traffic route between customer and supplier—thus physical transport costs.

With regard to the first factor, in the perishable goods sector, the costs deriving from a possible deterioration of a product mean that there is practically an inverse proportionality relationship between the minimum delivery frequency DF_{min} and the shelf life of the item *SL*; it is clear that items with a low shelf life, therefore, highly perishable items, require a high delivery frequency, which is why fresh food departments are replenished 3–4 times a week. Conversely, longer shelf-life items such as dry foods will be able to tolerate lower delivery rates. In the case of products which are simultaneously perishable—in the sense of a short life cycle—and high value as in consumer electronics, there is a tendency not to keep stock and to operate with high delivery frequencies in order to have negative LTGs.

The same qualitative consideration seen with regard to the shelf life of an item can be applied to the cost of immobilization, and therefore to the customer's availability of goods storage. If a customer, due to lack of space, or to avoid significant capital costs, does not have the possibility/availability of storage of sufficient quantities of products (high storage costs, e.g. in the case of drugs, automotive spares), the delivery frequency must be higher, even if the item is not perishable. This is typically the case of pharmacies, or car repair shops, which do not keep minimum quantities of drugs or spare parts in stock, but use JIT and demand a frequency of delivery from distribution centres at least daily or twice a day of whatever has been ordered downstream.

Traffic routes can also play a key role in determining the value of a minimum delivery frequency. If customer and supplier are on a main traffic route, deliveries may take place more frequently; vice versa, in the case of supplies to customers who are far away or difficult to reach, the transport cost will represent the preponderant item of the total cost, consequently the value of the delivery frequency inevitably shifts towards lower values.

"ON TIME" Delivery OTD-Punctuality of Delivery

Definition

Punctuality of deliveries (on time delivery OTD) is defined as a supplier's ability to respect the time window ΔT for delivery of an order agreed with a customer. In other words, if the customer expects delivery of the order to take place within a certain time window ΔT, then the measure of punctuality will refer to the ability of the supplier within that window; neither before nor after (see Fig. 57).

Again with reference to a cluster of homogeneous customers for that window ΔT, typically in terms of order composition (mix quantity), traffic routes, and delivery methods.

In general, the ability to meet delivery dates, and above all to avoid delays, is a significant service factor in almost all product sectors. Punctuality of delivery becomes particularly important when the customer has an appointment, as in the case of online deliveries or express couriers, in which the customer has given the willingness to receive an order in a specific physical place in a specific time slot, and the fact of not respecting this creates considerable inconvenience.

We saw in Chap. 2 how major e-commerce players have organized fixed collection points (lockers, 24/7 shops, gas stations) in order to avoid this constraint.

Performance Measurement KPIs

To analyse the concept of punctuality of delivery and to develop numerical indicators through which to provide a quantitative assessment of performance, the following consideration may be useful. Given the time horizon ΔT—in the case of express couriers or online deliveries typically one day, the totality of orders to be delivered within this window can be considered $TOTD$. Of these orders, a portion $ODIA$ will include orders delivered in advance of the agreed window, a portion ODL will include orders delivered late, a fraction OND will include orders not delivered at all (the so-called "backlog"), and finally a portion $ODOT$ will include orders delivered "on time", i.e. within the agreed time window. Clearly the condition must apply:

Fig. 57 Punctuality of delivery

$$TOTD(\Delta T) = ODIA(\Delta T) + ODL(\Delta T) + OND(\Delta T) + ODOT(\Delta T)$$

Percentage values can be used as quantitative performance indicators relating to the punctuality of deliveries. We therefore have:

$$OIA\%(\Delta T) = \frac{OIA(\Delta T)}{TOTD(\Delta T)}$$

$$ODL\%(\Delta T) = \frac{ODL(\Delta T)}{TOTD(\Delta T)}$$

$$OND\%(\Delta T) = \frac{OND(\Delta T)}{TOTD(\Delta T)}$$

$$OTD = ODOT\%(\Delta T) = \frac{ODOT(\Delta T)}{TOTD(\Delta T)}$$

where the first terms represent respectively the percentage of orders delivered in advance, late, not delivered (backlogs) and delivered on time within the time interval considered. The evolution of each percentage indicator can be observed over time, both in an absolute sense and in relation to a performance target, as represented in Fig. 58.

Alongside these four indicators, it is also necessary to consider two complementary indicators to measure the performance produced by the supplier in terms of on-time delivery.

A first indicator will take into account the fact that in the period of time ΔT in question, the delivery of backlogs and late orders may be scheduled: this indicator is the *NBD*, that is Number of backlogs delivered, thus orders which should have been delivered in the previous period, but which for some unforeseen circumstance were moved to ΔT. This indicator can be given by the ratio between this quota *NBD* and the total number of orders to be delivered *TOTD*.

Fig. 58 Punctuality measurement—temporal evolution with respect to target values of %ODOT

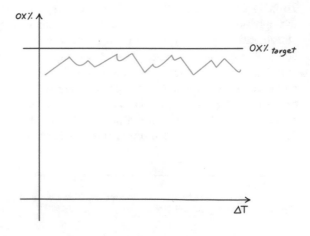

$$BD\%(\Delta T) = \frac{NBD(\Delta T)}{TOTD(\Delta T)}$$

Similarly, a second factor will take into account that in the ΔT period the early delivery of orders could be scheduled NAD, which could have been scheduled in a later period without exceeding the delivery window agreed with the customer, but that, due to economic needs, for example the possibility of combining deliveries along a given route, have been brought forward ΔT.

Also in this case, the indicator related to the Number of Anticipated Deliveries NAD is calculated as the ratio between NAD and $TOTD$.

$$AD\%(\Delta T) = \frac{NAD(\Delta T)}{TOTD(\Delta T)}$$

"IN FULL" Delivery—Accuracy of Delivery

Definition

The fourth sales support service factor of the product delivery area is the delivery accuracy.

Accuracy is one of the key service factors, together with the lead time and timeliness, and means a supplier's ability to comply with the agreed conditions for the delivery of products.

$$A\%(\Delta T) = \frac{accurate\ (orders/order\ lines/quantities)\ fulfilled\ (\Delta T)}{total\ (orders/order\ lines/quantities)\ fulfilled\ (\Delta T)}$$

Delivery accuracy depends on numerous factors and is often identified with the "5Rs" concept where the R stands for "right".

We then speak of the right product in the right quantities—right quantities in the right place—right place in the right state—right state at the right time. This last factor coincides with punctuality, as seen in the previous paragraph. The other 4 components of accuracy will now be analysed in detail: product, quantity, place, state.

Right Product, Right Quantities

With regard to the first two Rs, right product and right quantities, we mean the ability of the SC to deliver to the downstream customer exactly the items ordered in the quantities ordered. In other words, in order for the delivery to be *accurate* in terms of product and quantities, there must be no mix errors (we speak of mix errors to identify the item and therefore the GTIN of the product: e.g. in the spare parts sector for car, instead of the clutch for the xx model, the clutch for the yy model is delivered), nor quantity errors, therefore quantities more or less than those ordered, expressed in items, packages, kg, and so on.

Accuracy indicators related to mix and quantity can be formulated, relating them to orders or order lines fulfilled in a given time interval ΔT. In relation to orders, we therefore speak of orders fulfilled mix or quantity accuracy *OFMA* and *OFQA*; while at the level of order lines we speak of order lines fulfilled mix and quantity accuracy *OLFMA* and *OLFQA*, calculated as follows.

$$OFMA\% = \frac{NOFMA}{NOF}$$

$$OFQA\% = \frac{NOFQA}{NOF}$$

$$OLFMA\% = \frac{NOLFMA}{NOLF}$$

$$OLFQA\% = \frac{NOLFQA}{NOLF}$$

where the numerator shows the number of accurate orders (order lines) by mix or quantity fulfilled in the period, while the denominator is the total of orders and order lines fulfilled in the period (*NOF* and *NOLF* respectively).

Also in this case, the order line indicator is particularly important whenever there is a correlation between the references in the customer's request, and therefore in the event that an error at the order line level compromises the performance perceived by the customer at the level of the entire order. This can happen, for example, in the case of kit sales, where it could happen that if an order line is missing or an order line is wrong, the whole kit becomes unusable. The case of consumer goods is different, and in general the case in which there is no correlation between the order lines. In this case, a mix/quantity error on one reference does not affect the usability of the others.

Complementary information can also be drawn from the two indicators. An index value *OX* % close to 0 and an index value *LOX* % close to unity indicate, for example, that although at the level of order lines it is possible to deliver what has been agreed in terms of mix/quantity, at the level of the orders several errors have been committed since there are no orders without mix/quantity errors. For this reason, it is always a good idea to value and monitor both indicators. The opposite situation, in which the order value is high, and the order line value is low, could mean that an error has been made in a few large orders, while a whole series of smaller orders have been fulfilled in compliance with the specifications.

Right Time

The concept of *right time* is in fact attributable to the concept of punctuality, as seen above, of "ON TIME" punctuality, so please refer to the considerations made above and the related indicators.

Right Place

As regards the concept of *right place*, this means the accuracy of place, that is, the supplier's ability to deliver the goods to the agreed physical location. This in turn depends on the INCOTERMS established with the customer, and therefore lies in a range between the customer's dock door (EXW), the origin dock (Free incoterms), the destination dock (Cost Incoterms), the destination dock (Cost type) or the delivery dock door (delivery Incoterms). Details of the 2020 Incoterms with the different types of returns, the activities that are the responsibility of the vendor and the buyer, and the point of transfer of responsibility from the vendor to the buyer are reported in Annex A.

A possible performance indicator for accuracy of place could be the following:

$$AP\% = \frac{NOPA}{TOTF}$$

where the numerator NOPA (number of orders with place accuracy) is the number of compliant orders fulfilled by place within the period, while the usual denominator TOTF is the total number of orders fulfilled and delivered within the period.

In this case, it does not make much sense to distinguish between orders and order lines since both indicators will have the same trend, since there is a constant of proportionality given by the average number of order lines per order.

Right State

Finally, the concept of *right state*, or accuracy of state, falls within the category of accuracy indicators. In this case, we mean the ability to deliver orders and order lines in suitable conditions with respect to what has been agreed. This is a varied concept which depends on several aspects.

A first aspect which affects the accuracy of the state is damage during transport and handling. In this case, the indicator that will be proposed a little later on may refer to the order or the order lines.

A second aspect, particularly significant for FMCGs and perishable products in general, is that of respecting the expiry dates. Those who receive goods expect a product within the agreed commercial shelf life, which is different from the quality shelf life of the product. Typically, in the FMCG sector, the shelf life of a product is divided by 2/3 to the distributor and 1/3 to the manufacturer. In other words, those who receive the goods know the quality shelf life of the product (time interval between production date and expiration date, not so much in terms of safety but in terms of marketability) and expect to have, at the moment of receipt, at least two thirds of that interval. If, for example, a tray of cooked ham has 30 days of shelf life, upon receipt of the goods the expiry date must not be less than 20 days.

A third aspect relevant to the concept of the right state is that of respect for the cold chain. In food and pharmaceuticals, and in general in all sectors where the product must be kept at a controlled temperature, this aspect is particularly important. Here are some examples. (i) frozen products—e.g. frozen foods are transported at $-27°$, ice cream is transported at $-18°$ and suffer from both too high temperatures but also too low temperatures, as well as defreezing/freezing cycles, which can ruin the original emulsion and the organoleptic aspect, (ii) fresh products (which must be kept between

0 and 4° (e.g. milk, cured meats), or in any case perishable (e.g. chocolate: Ferrero transports products such as *Rocher* and *Pocket Coffee* at a controlled temperature between 8° and 16°, while in summer Ferrero interrupts distribution precisely in order not to risk compromising the organoleptic quality of the product; there's a famous Ferrero advert: "Do you know how many chocolates we sold this summer? Zero!" and then explains the reason why (iii) pharmaceutical products—for about 30% of the drugs distributed, the maintenance of certain environmental conditions is vital. We then speak of products 2–8 or 15–25, depending on the temperature range at which they must be kept and transported. Recently, the distribution of Pfizer's Covid-19 vaccine has highlighted the criticality of this aspect, since the vaccine must be kept at a temperature of −80 °C, the lowest ever touched by an industrial product, which as a result posed enormous logistical problems.

Again in relation to the cold chain aspect, it is equally important to reconstruct the temperature graph T as a function of time t through appropriate sensors (typically BLE or active RFID) which measure the air temperature in the environment where the product is located and transmits it to the legacy systems. This measurement typically takes place for each leg (travel, warehouse) or uninterruptedly (in warehouses and during travel) by positioning the sensor on the product, e.g., on the pallet or inside a package. The objective of this check is not so much to detect temperature peaks (linked for example to the opening of the vehicle doors in the case of a multidrop delivery), but rather to monitor the integral time/temperature and therefore the time that the product passes outside the allowable threshold, as represented in Fig. 59. In fact, this value measures the actual influence of the air temperature on the temperature at the core of the product. The greater the thermal inertia of the product and therefore

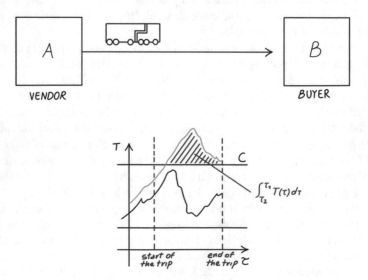

Fig. 59 Management of the cold chain during the transportation of goods

its specific heat and mass, the more it will be able to absorb temperature peaks for long time intervals, and vice versa.

Finally, aspects related to the suitability of the delivered load units and identification systems, the documents accompanying the goods and finally the withdrawal of packaging and/or discarded items fall within the scope of *right state*.

With regard to the suitability of the load units delivered, we mean the supplier's ability to provide handling units which comply with the customer's expectations, from three points of view:

- suitability of size—in the sense that the constraints agreed with the customer regarding the maximum and minimum dimensions of the handling units must be respected; the customer requests the delivery of HUs whose maximum or minimum dimensions may be constrained by the materials handling systems used: shelving, conveyors, doors, etc., so that it is therefore essential to respect in order to avoid laborious operations such as re-palletizing downstream.
- suitability of identification—in the sense that the delivered HUs must have an error-free identification system (standard logistics label and barcode or RFID tag, appropriately coded according to SSCC standards; secondary and primary packaging identified by GTINs, use of RFID tags to identify primary, secondary and/or tertiary packaging
- suitability of packaging—in the sense that the packaging must have the functional characteristics requested by the customer, characteristics which need to be verified on delivery. In some cases, the packaging may lack some particular features requested by the customer, or have lost them as a result of damage (e.g. loss of a vacuum), which in some cases may even have led to damage to the product itself. They must also match the type of preparation required by the customer, e.g., a preparation for customer orders, or single-reference single-lot layers to make the HUs suitable for ventilation in the case of cross-docking

With accuracy of deliveries in terms of documentation, we mean the supplier's ability to issue suitable documentation to accompany the shipment, whether this is related to the economic transaction (accompanying invoice), or the transportation (transport document, BoL), and error-free. Errors in a delivery note take up time and therefore mean administrative management costs and even delays in the goods reaching the system. In tight processes, such as for instance fresh food cross-docking, a few hours of delay in the availability of goods in the system can delay processes downstream and compromise deliveries.

Finally, the accuracy aspect linked to the withdrawal of packaging and/or discarded items is connected with the supplier's ability to meet the customer's needs in terms of withdrawal of packaging and/or discarded and/or expired items, in the face of an agreement to that effect between supplier and customer.

Quantitative state accuracy KPIs can be formulated to express the performance of an SC; depending on the case, these indicators may be more appropriately linked to the orders (O) or to the individual order lines (OL) fulfilled. As ever, the performance values are expressed as percentages and refer to a reference period ΔT.

Table 2 Accuracy KPIs referring to order lines or orders

State indicator	Possible units of measurement	Predominant unit of measurement
Damage	Orders; order lines	Order lines
Shelf life	Orders; order lines	Order lines
Cold chain	Orders	Orders
Handling units	Orders; order lines	Order lines
Type of preparation	Orders; order lines	Order lines
Identification	Orders; order lines	Order lines
Documentation	Orders	Orders

$$AX\%(O) = 1 - \frac{NOEX(\Delta T)}{TOTF(\Delta T)}; AX\%(OL) = 1 - \frac{NOLEX(\Delta T)}{TOLF(\Delta T)}$$

In the formula given above.

- $AX\%$, it is the particular indicator (e.g. accuracy of the documentation, accuracy of the handling units delivered, etc.), either related to orders (O) or order lines (OL)
- while $NOEX and NOLEX$ (number of orders/order lines with errors) is the number of errors made for the particular time interval (ΔT), related to orders or order lines respectively
- and $TOF/TOLF$ is the total order/order lines to deliver in the period.

The following table shows in detail what type of KPIs can be used to measure the state accuracy, and whether both measures (order and order line) can be used, or which one it is more convenient to measure (Table 2).

OTIF—On Time in Full—Order Fulfilment Rate

Ultimately, the overall accuracy indicator, which represents the accuracy of delivery $A\%$, also known as order fulfilment rate OFR or On Time In Full OTIF, is logically given AND by the manufacturers' 5 delivery compliance indicators connected to the 5Rs.

$$OTIF\% = AQ\% \wedge AM\% \wedge AP\% \wedge AT\% \wedge AS\% = \prod_i A_i\%$$

The order fulfilment rate in a given period ΔT (therefore given by the product of $AQ\%$ related to the accuracy of the quantities, $AM\%$ related to the accuracy of the mix, $AP\%$ related to the accuracy of the place, $AT\%$ related to the accuracy of time, and $AS\%$ related to the accuracy of the state. The fundamental hypothesis is the absence of a correlation between the 5 causes of accuracy, a hypothesis which is generally verified, otherwise the terms related to the covariances between the various factors would need to be introduced.

Furthermore, it is important that all indicators are measured consistently, and therefore either all refer to the order lines or all refer to the orders.

The combination of all the indicators in a production logic means that even if each of them is high and close to unity, the OTIF order fulfilment rate and therefore the probability of making a perfect order by the SC may be far from unit values.

If, for example, each of the 5 indicators is worth 0.9, the overall result is equal to $A\% = 0.9^5 = 0.59\%$, and therefore a probability of about 40% of making a mistake; 4 out of 10 incorrect orders is therefore an indication of inferior performance.

Delivery Flexibility

Definition

The last factor of the sales support service belonging to the order delivery area is *flexibility*. Flexibility can be defined as the supplier's ability to meet urgent and/or unforeseen customer demands.

The definition given, of a general nature, leaves room for multiple interpretations of the concept of flexibility, in the sense that it can be evaluated in different terms. Here we analyse the main dimensions through which a customer can evaluate the level of flexibility of a supplier or of the SC as a whole, or through which the supply chain or the supplier itself can measure its performance in terms of flexibility.

Typically, a customer considers flexibility as an ability to vary the mix and/or quantity of items to be included in an order with respect to the quantity agreed in the order (think of an online order, in which you want to add/remove/replace a reference purchased, or vary the quantities of a reference inserted in the cart after checking out) or a supply contract (same thing, but in the case of B2B orders, and therefore you want to vary the mix and/or quantity, delivery address, billing address after the supplier has sent an order confirmation); a supplier is all the more flexible when it can satisfy these requests, increasing or decreasing the articles supplied to the customer in response to an urgent request, up to just a moment before delivery.

Another dimension of flexibility can be considered the ability and/or willingness of the supplier to bring forward or delay deliveries based on a customer's needs with respect to the agreed time window, as well as to change the place of delivery should the customer request it.

KPI of Measurement

As done previously, to give a quantitative assessment of a supplier's performance in terms of flexibility, quantitative percentage indicators are introduced. These indicators refer to a specific time interval ΔT and to a cluster of homogeneous customers j.

For each variation X requested by the customer, i.e. in terms of (i) mix, (ii) quantity, (iii) place of delivery, (iv) delivery date, it is therefore possible to calculate the percentage of requests for variation of quantities satisfied in the period as follows

$$FX\% = \frac{NRVXF(\Delta T)}{NRVXR(\Delta T)}$$

where the denominator is respectively the number of variation requests fulfilled for the generic indicator X (i.e. mix, quantity, place, date), and the numerator is the number of requests for a variation received for the same indicator.

The global indicator will be given by the number of total requests met compared to the number of requests received (again with the same number of ΔT and clusters of customers).

5.5 Evaluation of the Service Produced by the SC

Once the overview of the external analytical indicators has been exhausted, we can move on to analyse how the SC can use the performance indicators relating to the various service factors seen, calculated on the basis of the data in its possession, to obtain a measurement of the quality of the customer service produced and to monitor its evolution over time.

It is important to underline how we are analysing for now, and how it is possible to measure and monitor the service "produced" over time, and not the service "perceived" by the customer. This distinction has been detailed in Sect. 5.3.1.

Having made this fundamental clarification, it is assumed that the SC can, on the basis of the data in its possession relating to the processes, calculate the performance indicators relating to the analytical external service factors seen in the previous paragraphs, for a given cluster of customers J in an interval of time ΔT.

According to (Bryne and Markham 1991), for each indicator that the SC deems relevant with respect to its criteria for that client cluster J, e.g. the lead time LT, the delivery frequency DF, the punctuality of delivery P, the accuracy of the delivery A, the flexibility of the delivery F, etc., can be represented by means of "traffic lights" or "tachometers", which measure, in an absolute or relative scale with respect to the target, the value of the external analytical indicator in question. These "traffic lights" or "speedometers" are typically reported in a business intelligence user interface, a cockpit dashboard with which the manager of the SC process monitors the situation (Bryne and Markham 1991) (Fig. 60).

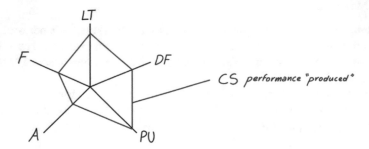

Fig. 60 Summary representation of the service produced by the SC using radar diagrams

Each axis of the diagram relates to a particular performance indicator and the overall service performance "produced" and measured by the supplier is represented by the area of the polygon obtained by combining the values measured by the indicators, each shown on its own axis, with straight lines. This representation is particularly effective since it allows, on the one hand, a summary representation of the product service performance, and on the other because it allows us to compare the performance being measured, for example, with an estimated performance of the best competitor, or with an objective performance to be achieved. In the latter case, it is immediately clear which factors must be acted upon to achieve the intended objectives.

However, a radar diagram representation is a static representation of a service level measured by a supplier. All the values of the indicators (measured and objective), in fact refer to a certain time interval, and the representation therefore gives a static snapshot of the service produced. However, it should be noted that the performance objectives which a supply chain sets itself are generally variable over time, tending in principle to increase with time. In fact, it has been seen how customer service represents a strategic factor on which competitiveness is based in many sectors and therefore the ability to retain acquired customers and acquire new ones, so it is natural that the SC sets service goals that are gradually more ambitious to try to grow and develop by reaching new customers, as well as to catch up with competitors.

Furthermore, according to (Bryne and Markham 1991), the SC is not static but evolves by restructuring both from the point of view of infrastructures and technologies, and from the point of view of the management of processes and human resources. It follows that the comparison between the performance produced and the target performance change over time, precisely because both values vary with time.

To monitor the evolution of performance over time, it may then be interesting to represent in a cockpit the temporal evolution of a generic service indicator, and to relate it, for example, to the evolution of the respective target value over time, as represented in Fig. 61.

In this way, an exact dynamic perception of a product service is obtained. Also in this case, the comparison with a desired target or a reference benchmark supports the evaluation of the service produced by the supplier.

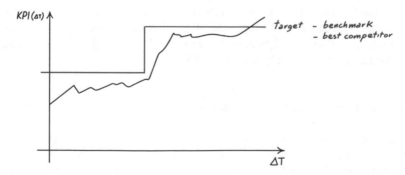

Fig. 61 Evolution over time of the service performance produced and the target value

To obtain a summary representation of a performance, one which provides a snapshot but also an evolution over time, business intelligence programs such as Tableau https://www.tableau.com/it-it, Qlik https://www.qlik.com/it-it/, etc. or applications developed ad hoc through open source software such as Google data studio can be used. In this way, web dashboards can be built—which are therefore accessible anywhere even from smartphones using a simple web browser—based on appropriate views of the ERP system and possibly combined with data from other software systems, which provide an instantaneous value and/or trend over time of the individual external analytical indicators, and/or of the performance produced at an overall level and/or by geographical/customer clusters, or others.

5.6 Evaluation of the Service Perceived by the Customer–Customer Satisfaction

As already mentioned, elevated levels of service "produced" by an SC do not necessarily correspond to high levels of service "perceived" by the customer, and therefore to its satisfaction. On the other hand, given that customer service is in some respects a strategic competitive factor, the evaluation of service performance cannot be limited to the service "produced", but must go further towards the evaluation of the service "perceived" by the customer. This evaluation must also concern both the service perceived as a whole and the service perceived in relation to individual external analytical factors (how the customer or the cluster of customers perceives punctuality, cycle time, etc.).

While the service "produced" is in fact evaluated by an SC on the basis of its own service criteria, the service which the customer "perceives" actually depends on the importance which the customer attaches to the various external analytical factors seen in the previous paragraphs, an importance that does not necessarily coincide with that attributed internally to the SC. Furthermore, as noted previously, buyers and vendors rarely agree when evaluating performance on the same factor. In fact, it may happen that while a vendor considers the performance produced on a factor to be satisfactory, the buyer deems it unsatisfactory, or vice versa, that the vendor tends to underestimate the performance provided on another and undertakes to improve it, while for the vendor this performance is, all in all, quite satisfactory.

There is therefore a need to combine a system for measuring service produced with a system to measure the perceived service.

5.6.1 The Service Vendor Rating process

A first measurement of the Service perceived is the Net Promoter Score seen in Sect. 5.3.1. The main advantage of this is the simplicity of the method and its widespread use, which makes it a benchmarking standard easily understandable for a customer; on the other hand, the limits of this system are the conciseness of its approach, which does not provide detailed information on individual service factors, neither in terms of importance attributed by the customer, nor in terms of perceived performance.

To obtain this kind of information we need to go through a multi-attribute evaluation system based on customer service surveys.

To guarantee the objectivity of the results, it is a good idea that the entire survey is carried out by external consultants rather than by the supplier itself. In this way, the surveys will not be influenced in any way by any pre-established vision, but the problem will be approached with neutrality and without preconceptions.

First and foremost, it is necessary to identify a homogeneous cluster of customers on which to focus the survey, or even a single customer if this is particularly important.

Having defined a generic j-th cluster, the service performance perceived by this cluster can be identified by means of two fundamental steps.

- Identification of the importance that the customer attaches to the various factors
- Measurement of the perceived performance for each factor by the customer itself.

In turn, each of the factors thus identified will depend on how a series of SC processes relevant to that particular factor are organized and managed. It must be pointed out that for SC processes in a broad sense we mean both processes and operations directly managed by the player carrying out the investigation, as well as processes and operations carried out externally, e.g. through third parties or upstream suppliers. On the other hand, being in the perspective of measuring a service perceived by the customer, it would not make sense to distinguish processes that are strictly relevant to one or another player of the supply chain. In fact, for each external analytical factor, the customer perceives a performance of the SC upstream without knowing and without caring whether any deterioration of the performance itself is due to the supplier itself, to one of its Level I or II suppliers, or to its subcontractor, and in any case attributes the responsibility directly to the one with which it has established the commercial relationship, the focal company with which it identifies the SC. It will therefore be up to the vendor to verify that the business processes in a broad sense, extended to the whole SC, do not compromise the performance in terms of external analytical factors relevant to that particular cluster of customers. Finally, for each business process, it is possible to identify the individual phases that can be improved, through which, for example, to improve, rationalize and stream-line business processes. These improvements have a cascade impact according to a cause/effect principle, first on external analytical factors and then on the quality perceived by the cluster.

The analysis takes place through administration of a questionnaire, asking the customer directly what are the most important factors of marketing support and sales support (pre/post/transactional) for it n, through the attribution of a weight of importance w_i to each factor $i = 1, \ldots, n$, and to attribute a measure of perceived performance to each factor identified as significant p_i.

There are appropriate techniques to structure this survey, both as regards the attribution of weights (e.g. fuzzy logic, multi attribute AHP Analytic Hierarchy

Process) and as regards the scales to be used to detect the judgement on performance (e.g. numerical, or fuzzy logic), which, however, are only cited as a possible reference for an in-depth analysis of specific in-depth texts.

5.6.2   Customer Satisfaction Measurement KPIs

Once the questionnaires have been returned, the perceived service, and therefore the judgements on individual weights w_i and performance p_i are available attributed to each factor $i = 1, \ldots, n$—i.e. the customer satisfaction CSj—will be given for example by the weighted sum:

$$CSj = \sum_{i}^{n} w_{i,j} * p_{i,j}$$

CSj which will depend on a series of analytical external service factors (cycle time, breadth of range, etc.) which change according to the cluster/customers J being analysed. Therefore, for a given a cluster/customer j, there will be an appropriate and particular combination of external analytical factors on the basis of which the Customer Satisfaction CS_j for that cluster J will be determined.

Again, the opinion expressed by the different customers could be attributed a different importance, based on, for example, the importance of the customer and gauged against the sales percentage of that customer $\frac{S_j}{S_{tot}}$. The above formula then changes in this way

$$CSj = \sum_{i}^{n} w_{i,j} * \frac{S_j}{S_{tot}} * p_{i,j}$$

This observation highlights a concept which will be dealt with extensively in the concluding paragraphs of this chapter, namely the fact that, like all things also service costs, and therefore it is necessary to concentrate resources on strategic elements, while the less strategic ones can be ignored. In terms of service, this means that customers are not all the same, and the more important the customer is in terms of turnover or business growth potential, the more important it is to ensure a high service level.

5.7   Strategic Management of Customer Service

We have seen in the previous paragraphs and chapters how, in the vast majority of mature sectors, the product tends to be undifferentiated due to the growth of competition and the spread of technologies; the intangible value expressed by a brand, if not supported by solid foundations of quality and service, tends to lose value in the eyes of the consumer unless there are singular cases, and the competition is now played out in most product sectors at the service level (Customer Satisfaction) and price. Today's leading companies cannot ignore excellent service standards, because in the current competitive scenario the product is a prerequisite to sit at the table of competition, and the winner is whoever manages to provide the best service at the lowest possible price. The Amazon case, with its breath of range, lead time and home delivery, teaches this.

In the previous paragraphs, it was described how it is possible to quantitatively measure the service produced, and, above all, the service perceived by the customer, that is, the one on the basis of which the SC is able to win new customers and retain existing ones. The need now arises to combine these two aspects: if the service level is the strategic factor on which to base the competition, what information can be obtained from evaluation of a perceived service? Again, what strategic actions must be taken as a consequence of the surveys carried out?

From a conceptual point of view, we have seen how service level largely depends on how an SC is designed and managed. Therefore, the strategic intervention and performance calibration procedure which the SC delivers must be strictly connected with the monitoring of the perceived service level performance.

The strategic management process of the service can be formalized according to the flow chart in Fig. 62, revised from (Ruggeri 1998; Bryne and Markham 1991).

The inputs from the external environment are represented on the one hand by the needs of the market and on the other by the behaviour of competitors.

The company perceives market needs in terms of service level through tools like those seen previously. The elaboration of questionnaires and the subsequent analysis, as seen, allows us to derive significant external analytical factors, the service priorities and the judgement which the customer has of the performance provided by a company.

A simple analysis of a market and the relationship that a company has with it would still give a short-sighted view of the strategic context, since other competitors also operate on the market together with the company in question. Therefore, the initiatives implemented by the latter in terms of customer service must be constantly monitored and evaluated (in the transportation sector, for example, the reduction of a competitor's cycle time can shift significant market shares). Amazon, through its Prime service, has become the number one online retailer in the world precisely because it was the first to be able to provide a service that no one had thought of before, considered by many to be a risky move—on the one hand it was thought that few consumers would be willing to spend €40 per year regardless, and on the other hand it was considered dangerous to guarantee zero shipping costs even for a myriad of small shipments. In reality, the move was a huge success. By 2020, there were 50 million Prime subscribers around the world. To this Amazon has added other services such as Prime Videos or Prime Books which have pushed Prime subscribers to over 200 million in 2021, thanks also to the Covid-19 pandemic and the explosion of online purchases, a record figure which allows Amazon to count on 200 million loyal customers who, having subscribed to Prime, then buy on Amazon and not on

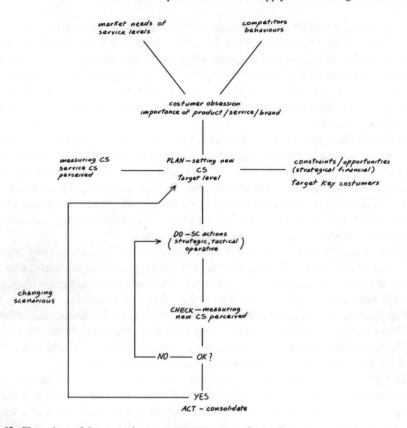

Fig. 62 Flow chart of the strategic management process of a service

other portals, and a fixed and anticipated cash flow of 8 billion in advance thanks to the subscriptions (Tommasi 2021).

Through these two inputs (what the market demands in terms of service and what the competitors offer), an SC can objectively evaluate the relative importance of the service/brand/product offered to develop and grow within its own competitive scenario. As mentioned, in most cases and in general in mature sectors, the space for companies who do not perceive service level as something strategic is gradually reducing, so this can have a general value for the most diverse product sectors.

Some companies, like Amazon or Dell, have raised the level of strategic importance of customer satisfaction to brand-new levels, coining the terms "customer obsession" and "customer delight". With these terms is meant on the one hand an ability to anticipate the service needs of a consumer by putting oneself in his/her shoes (if I were my customer, what would I expect from me?), and on the other, the ability to satisfy these needs, or even go beyond the expectations, to a level of "delight" in fact. Customer obsession also means putting the consumer first, and guiding every decision using his/her satisfaction as a litmus test. In other words, to understand when faced with a dubious situation whether a decision is right or wrong, the question to ask is: "what impact will this decision have on customer satisfaction?". And it will be correct if it goes in the direction of satisfying the customer.

Once a company has become aware of the strategic value of its service level, the first thing to do is to evaluate its position by operationally measuring the *CS* as perceived by its customers. This process, in addition to providing a quantitative assessment of the current perceived service, allows evaluation of the margins of intervention for an improvement of the service, through sensitivity analyses of the various factors. These margins must be constantly related to any existing constraints (financial, technological, organizational-union), or opportunities linked to the specific operational context, however an assessment must also be made if the interventions concern the entire client portfolio, or if they need to be specific for a cluster of target customers.

Traditionally, any interventions and repositioning of the SC at all levels, i.e. strategic—in the long term (multi-year impact), tactical—in the medium term (the horizon is the year), and operational—in the short term (the horizon is order fulfilment) must in fact always be evaluated with a view to cost/benefit trade-offs, balancing the costs associated with the intervention and the possible economic or strategic advantages that could derive from it. In fact, as a service level varies, also the turnover will vary, since the service level affects the ability to retain existing customers or acquire new ones, as will the costs incurred by the company to provide this service.

As regards turnover, customers tend to use different suppliers to whom they usually entrust proportional market shares based on the product/price/service mix they receive. If the first two terms are aligned among various competitors, then the discriminating factor to determine market share is precisely the service.

An "S" trend can therefore reasonably be considered plausible, with the turnover, as represented in Fig. 63 which initially grows very slowly since it is improving the CS on low values does not entail substantial benefits, since the customer always perceives a lower service than its expectations. This happens up to a "lower" threshold value from which the assigned turnover actually grows as the CS grows, due to the turnover distribution mechanism described above. As the service grows further, the

Fig. 63 Turnover and costs according to the CS: excellent service level

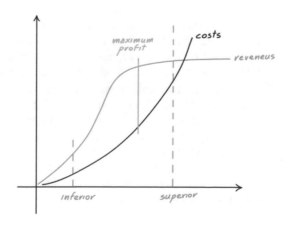

turnover settles into an asymptote, starting from the "superior" value. From this point onwards, improvements to the service level do not posit proportional benefits, since the customer is already satisfied with the service they are receiving, and therefore a further increase in the service does not lead to further business growth.

Service costs typically include:

(i)   the cost of the distribution network (first level and second level full mix warehouses and transit points)—to try and bring the product closer to the customer and thereby improve the service, the network must be increasingly widespread

(ii)   the cost of stocks (not only finished products, but also semi-finished products and raw materials)—to increase a service it is necessary to increase the stock service ratio, in order to have the requested product available when ordered, to be delivered within the requested timeframe

(iii) the direct costs of order management, in order to reduce the time associated with administrative management

(iv)   the cost of the logistic structure (workforce), to reduce processing times

(v)   the production cost linked to the size of the lots, in order to provide fast responses to a customer order thanks to production flexibility

(vi)   the cost of transportation (inter-plant but above all to the end customer and "the last mile", which becomes critical especially in sparsely populated areas), also in this case, dedicated deliveries increase the service level but tremendously increase costs.

These costs tend to increase exponentially with the increase in CS, as represented in Fig. 63, since once a high service level has been reached, further increasing the CS involves more than proportional investments especially in terms of Capex (a widespread distribution network to be able to reach even the most isolated areas, plus duplication of structures) and operating costs in terms of Opex, mainly on transportation and stock maintenance costs deriving from the high value of the SSR (Stock Service Ratio).

To summarize, while the turnover tends to have an "S-shaped" trend with the CS, the logistic costs dependent on the CS typically assume an exponential trend, as represented in the figure below.

From the two curves it is theoretically possible to determine the optimal service level to be provided for a given cluster of customers—which maximizes profit.

Depending on the sectors, the two curves could then be moved further the right or left, moving the optimal point correspondingly. In any case, from an analysis of Fig. 63, a crucial element emerges: regardless of the shape of the turnover trend as the service level varies, if the cost curve is shifted to the right, the profit increases for any value of the service level. The shift to the right of the cost curve typically occurs with technological innovations (SC collaboration portals, integrations between WMS through EDI, sharing of stock data, automated sorters, autonomous robots such as drones for delivery, picking robots, Internet of Things such as RFID) or process innovations (e.g. cross-docking, synchronization of flows and lean optimization of process times and methods, adoption of CPFR programmes) which make it possible to reduce the cost for the same service or, vice versa, to offer a superior service for the same cost.

Similarly, if, instead of assuming an S-shaped trend, the turnover curve continues to grow as the service level increases, the optimal point shifts to the right. Companies like Amazon believe that continuing to increase service levels by pushing them, through "customer obsession" to "customer delight" levels, is the basis for this result. Only through "customer delight" is it possible to retain the customers acquired, making them grow in terms of CLV (Customer Lifetime Value) and thus making them promoters; therefore, a source of positive advertising which brings in new customers and which, unlike investments in marketing, costs nothing.

Returning to the flow chart of Fig. 62, once the interventions have been implemented, a phase of experimentation and verification of the results obtained follows, measured again through NPS or CS, depending on the critical nature of the customer, which may be satisfactory or not. If they are not, then it is necessary to proceed iteratively by incremental adjustments, until an acceptable situation is reached, again measured by the service level as perceived by the customer.

At this point comes a phase of adjustment and consolidation of the results obtained. This process of strategically calibrating the logistics system according to the perceived performance of the service is in any case dynamic, in the sense that it must be constantly reconsidered due to the continuous modification of the surrounding conditions.

5.7.1 The Service Turnover Matrix

At this point, it is legitimate to wonder what the right service is for each customer/cluster of customers, in the sense that since customer service generates service costs, the service level cannot be the same for all customers, but customers of greater importance will theoretically be provided with a higher service level, while customers of lesser importance may receive a lower one. In other words, the CS (as well as the NPS) is specific according to the importance of a customer or a cluster of customers.

In reality, as we saw in Sect. 5.3.1 on the Net Promoter Score—NPS, retaining customers with a low service level is always dangerous, since a dissatisfied customer is potentially a detractor and therefore a source of negative publicity, so it is better to abandon them.

To understand whether the *right service* is being provided to the *right customer*, it is convenient to map customers (or clusters) according to turnover and perceived service, evaluated, for example, by means of CS or NPS. In this way, it is possible to understand the right strategy to be adopted in terms of service but also of product/price in relation to that particular customer/cluster.

This analysis is reported in reference to Fig. 64, in which customers are entered in a four-quadrant matrix (the medians or averages could be used for the division into quadrants) according to the turnover generated and the perceived service level.

Fig. 64 The service
turnover matrix for strategic
management of customer
service

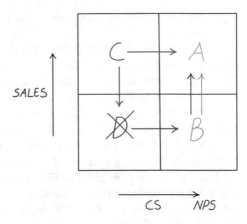

In the proposed figure, Customer A is a one for whom the perceived service level justifies the turnover guaranteed to the SC. In other words, a customer who generates high turnover and who is guaranteed a high service; indeed it is very likely thanks to the guaranteed high service that the customer generates a high turnover. The key word in this case is "MAINTAIN", in the sense that the high standard must be maintained to keep the customer who generates business loyal.

At the antipodes, is Customer D, who perceives a low service level and generates little turnover. A customer in this quadrant can prove dangerous, since we have seen how detractors are a source of bad publicity, so, given the negligible turnover, it is necessary to decide (i) whether this is a customer to be abandoned permanently (and in this case the key word is therefore "ABANDON") (ii) or to grow in terms of turnover, typically according to the path D → B → A. It is therefore necessary to initially guarantee a service level that is perceived to be better than the current one D → B, which prompts the customer to increase the assigned business quota, changing its positioning in the matrix from B to A. the key word is therefore "EVOLVE THROUGH SERVICE", and this evolution must be monitored over time.

Customer B, on the other hand, is a customer for whom a lot is spent in terms of service, but who is not very significant in terms of turnover since it guarantees little business. In this case, it is necessary to understand whether the leverage on this customer is not so much the service, with which the customer says it is satisfied, as the product and/or the price. Customers in this quadrant will then be treated differently from a commercial point of view by repositioning their offer in terms of the products, by offering discounts, and so on. In this case, however, care must be taken to decrease the service level, to avoid losing a promoter customer, who was satisfied and who is therefore a source of good publicity, and who, if push comes to shove, could become a detractor. The key word in this case is: "EVOLVE THROUGH PRODUCT/PRICE".

Finally, Customer C is one who, despite perceiving poor service, still guarantees a high turnover. This situation can occur in rare circumstances, typically: (i) there is a monopoly situation, so that the customer has no alternative solutions from which to obtain supplies; this is a rare situation, limited to the case of particularly innovative products, patents, etc. or in which the intangible value represented by the brand also compensates for the low service level which the SC provides (e.g. in order to take possession of a fashion product the customer is willing to wait for the very long delivery times of the SC). This is a situation of competitive advantage that will not last over time, since, when a competitor is able to supply the same product or the same intangibles along with a superior service, it is going to absorb whole slices of the market (ii) there are different supply alternatives for the customer, but a satisfactory product is proposed at such an advantageous price that despite the low service level, the customer guarantees a large slice of business. This is an equally unstable situation, since also in this case a competitor who becomes aware of the potential volumes may strategically decide to lower prices, perhaps even working for a certain period at a loss, and supply the same product at the same price but with better service performances, to snatch the slice of B customers that generates business anyway. This is the case of a retailer's brand or that of certain online players, such as Amazon itself, who, in the face of competitor products that are handled through a store or an online marketplace and which see a lot of sales and generate turnover even with high delivery lead times, decide to offer their white-label or Amazon Basic product at the same or a lower price but with faster deliveries, precisely to move the substantial slice of business over to their own product. In both situations, therefore, it is a good idea to have Customer C evolve towards A, by improving the service level. The keyword is therefore once again "EVOLVE THROUGH SERVICE".

6 Annex A

INCOTERMS 2020

In the discussion, the terms delivered ex works and delivered ex destination were introduced, with which it is possible to identify the two fundamental methods of order delivery between customer and supplier.

In the international trade of goods, delivery methods are in force which are regulated by the international chamber of commerce, through documents, the so-called INCOTERMS (INternational COmmercial Terms) which are periodically updated. The latest revisions date back to 2020.

The 11 terms governed by the 2020 edition of the Incoterms®, can be classified by an accumulation of obligations for the seller:

- Group E: EXW (term which identifies the minor obligations of the seller)
- Group F: FCA—FAS—FOB (main transport to be paid by the buyer)
- Group C: CPT—CIP—CFR—CIF (the seller pays the freight, but the risk lies with the buyer)
- Group D: DAP—DPU—DDP (the seller delivers to destination. Transport and risks borne by seller).

The general scheme of transport responsibility is represented in Fig. 65.

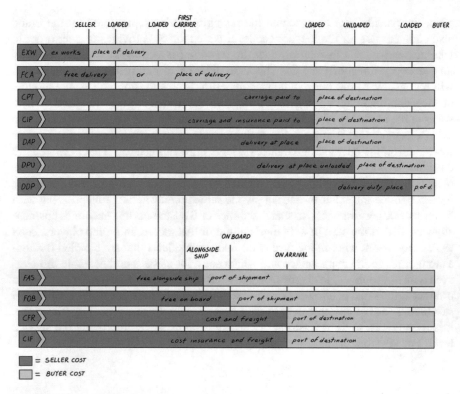

Fig. 65 General scheme of Incoterms 2020

Below is a brief analysis of each individual term:

EXW—"EX WORKS"

The seller delivers when it makes the goods available to the buyer at the seller's premises (or at another designated location, for example, factory, warehouse, etc.). The seller is not required to load or clear the goods for export. The parties must precisely specify the point within the indicated delivery location, as all costs and risks up to that time are borne by the seller. The buyer subsequently bears all costs and risks of the transportation.

FCA "FREE CARRIER"

The seller delivers the goods to the buyer in two ways:

(1)   When the place indicated is the seller's headquarters, the goods are delivered when they are loaded on the means of transport made available by the buyer.

(2)   When the place of delivery is another place than the seller's premises, the goods are delivered when, after being loaded on the seller's means of transport, they reach the other indicated place and are ready for unloading from the means of

transport of the seller. seller and at the disposal of the carrier or another person appointed by the buyer.

FAS—"FREE ALONGSIDE SHIP"

With the term FAS, the seller delivers when the goods are placed alongside the ship (e.g. on a quay or barge) nominated by the buyer at a specified shipping port.

FOB—"FREE ON BOARD"

With the term FOB, the seller delivers the goods by placing them on board the ship nominated by the buyer at the indicated shipping port or when it "procures the goods already so delivered".

CPT—"CARRIAGE PAID TO"

With the term CPT, the seller delivers the goods—and transfers the risk—to the buyer by delivering them to the first carrier appointed by the seller or when it "procures the goods already so delivered". The seller is obliged to conclude the transport contract by paying the transport costs necessary to take the goods to the indicated destination.

CIP—"CARRIAGE AND INSURANCE PAID TO"

With the term CIP, as with CPT, the seller delivers the goods—and transfers the risk—to the buyer by delivering them to the first carrier appointed by the seller or when it "procures the goods already so delivered". The seller is obliged to conclude the transport contract by paying the transport costs necessary to take the goods to the indicated destination. The seller is also obliged to conclude a contract for insurance coverage against the buyer's risk of loss or damage to the goods during transport. The seller is required to take out insurance coverage to cover risks such as Institute Cargo Clauses (A) or similar (maximum coverage), also taking into account the mode of transportation.

CFR—"COST AND FREIGHT"

With this deadline, the seller delivers the goods on board the ship or when it "procures the goods already so delivered". The risk of loss or damage to the goods passes when the goods are on board the ship. The seller must also enter into a transport contract and pay the related costs to transport the goods to the indicated port of destination.

CIF—"COST, INSURANCE AND FREIGHT"

As in CFR, with this deadline the seller delivers the goods on board the ship or when it "procures the goods already so delivered". The seller must also enter into a transport contract and pay the related costs to transport the goods to the indicated port of destination. The risk of loss or damage to the goods passes when the goods are on board the ship. The seller is also obliged to conclude a contract for insurance coverage against the buyer's risk of loss or damage to the goods during transport. The seller is required to take out insurance coverage to cover risks such as Institute Cargo Clauses (C) or similar (minimum coverage), If the buyer wishes to have greater insurance protection, it must expressly agree with the seller or take out additional insurance contracts.

DAP—"DELIVERED AT PLACE"

With this deadline, the seller delivers—and transfers the risk—to the buyer when the goods, ready to be unloaded from the arriving means of transport, are made available to the buyer at an agreed destination or at the agreed point at the interior of that place, if that point is agreed. The seller therefore assumes all the risks associated with the transport of the goods to the place of destination indicated or to the agreed point within that place.

DPU—"DELIVERED AT PLACE UNLOADED"

With this deadline, the seller delivers—and transfers the risk—to the buyer when the goods are made available to the buyer unloaded at an agreed place of destination or at the agreed point within that place, if that point is agreed. The seller therefore assumes all the risks associated with the transport of the goods to the place of destination indicated or to the agreed point within that place.

DDP—"DELIVERED DUTY PAID"

With this deadline, the seller delivers—and transfers the risk—to the buyer when the goods, ready to be unloaded from the arriving means of transport, are made available to the buyer at an agreed destination or at the agreed point at the interior of that place, if that point is agreed. The seller therefore assumes all the risks associated with the transport of the goods to the place of destination indicated or to the agreed point within that place. Furthermore, the seller is obliged to clear the goods, not only for export but also for import.

Furthermore: in all terms, except EXW, the seller is obliged to carry out the customs clearance operations for the export. In all terms, except for DDP, the buyer is required to carry out the import customs clearance operations. In all terms, except EXW, the seller is required to load the goods on board the means of transport. In all terms, except for DPU, the buyer is obliged to unload the goods.

Conclusions—Can Technology and Industry 4.0 Really Change the Structure of the Entire Supply Chain and Society Too?

The Forth Industrial Revolution

We have entered a period of great change, a period which is going to substantially change not only the supply chain and therefore the way we produce, distribute and consume goods and services, but the whole of society. How we work, what we do, how long and in what way we live.

This is the fourth industrial revolution. A process we have been involved in for at least 10 years, and which we must become acquainted with as quickly as possible in order to make the most of it. As Darwin said, *in the struggle for survival, the fittest win out at the expense of their rivals because they succeed in adapting themselves best to their environment*. Which is why it is vital to understand what we are living through, and how scenarios are changing in a move towards even higher levels of productivity, abundance, and prosperity.

To talk about the fourth industrial revolution it is necessary to start from two basic premises.

An industrial revolution means the introduction of a product or process innovation which radically changes productivity, and therefore the output available to the consumer for the same amount of resources consumed. Every industrial revolution has been characterized by this property. The steam engine, the electric motor, oil and Taylorism, information technology and electronics, industrial automation...

Industry 4.0—or the fourth industrial revolution—is a term which was coined for the first time in Germany, when a working group dedicated to Industry 4.0, chaired by Siegfried Dais, the manager of Bosch GmbH, and Henning Kagermann of Acatech (the German Academy of Sciences and of Engineering) submitted a series of recommendations for its implementation to the German federal government. On 8 April 2013, at the annual Hanover Fair, the final report of the working group was released, a project for the development of the German manufacturing sector, the "Zukunftsprojekt Industrie 4.0", which was intended to help the country's industry enjoy a leading role in the world again. Subsequently, the German model inspired

© The Editor(s) (if applicable) and The Author(s), under exclusive license to Springer Nature Switzerland AG 2022
A. Rizzi, *Supply Chain*,
https://doi.org/10.1007/978-3-030-95707-0

numerous European initiatives and the term "Industry 4.0" began to spread to other countries around the world (Dais 2014)

What characterizes the fourth industrial revolution is the fact that machines and products have acquired two characteristics which they never used to display. Machines and products have become "intelligent" and are "connected", i.e. they are able to communicate with one another via the Internet.

If in the third industrial revolution machines simply automated people's work in a repetitive manner, with the fourth industrial revolution they have begun to behave like people. People process information coming from the five senses through the brain and generate coherent responses or actions. Similarly, 4.0 machines receive input from the outside world through increasingly sophisticated sensors; since the machines are "connected" to a greater or lesser extent via the Internet, they can send this information to a "brain" which represents a virtual model of the system, the so-called "digital twin". This model can be on the same machine or on the other side of the world; can comprise a single machine or multiple machines in multiple factories. In any event, the parameters coming from the field are processed, simulating different possible scenarios in fractions of a second, and generating an output that is consistent with certain objective functions. An output which is sent back to the machines, just like the human brain does! The talk now is of "Autonomous Robots", autonomous machines, connected to one another and to the outside world, capable of replacing and supporting people in carrying out not only repetitive tasks but also increasingly complex tasks, thanks to the ability to make decisions at the changing boundary conditions. Consequently, it is not only forklift drivers in warehouses who are being replaced by robots which carry pallets or boxes around, or truck drivers for the distribution of goods that will be replaced by drones or self-driving vehicles. But also call-centre employees or those who carry out assistance activities which in a 4.0 world will be replaced by Siri, Google Assistant, or chatbots. In June came the news that IKEA had "hired" Vera, a robot launched in 2017 by a Russian start-up, capable of carrying out 1,500 job interviews a day (Ansa 2018).

What is a 4.0 Product?

So, thanks to the fourth industrial revolution, products have also—and above all—become "intelligent" and "connected".

Another premise. In a 4.0 scenario, products can first and foremost be made in a totally innovative way. It is no longer necessary to operate through traditional, rigid production or assembly cycles, in which the always-identical products of a production batch are created in a physical place called a "factory". 4.0 products are unique, a one-off batch, made-to-order. But above all, they may be generated from files, through Additive Manufacturing. 3D printing opens up hitherto unexplored production scenarios. The manufacturing cycle of the products, as we have said, potentially always unique and made to measure, is generated automatically each time on a computer starting off from a technical drawing. The file can then be sent anywhere in the world via the Internet, and the product then comes to life by means of a 3D printer, with the same ease with which a document is edited, e-mailed, and hard-copied using a printer. In a 4.0 scenario, not only will there be no need to go to a store to buy a 4.0 product, a store where traditional products physically arrive through the supply chain, but there will also be no need to buy it online. Unlike traditional products purchased on a B2C portal which are sent to the home by courier, the 4.0 product materializes and comes to life directly in our own home by purchasing a file, exactly as now happens for music.

In addition, in a 4.0 scenario, products become "intelligent" and "connected" through the Internet of Things.

The products are "smart" because they are unique and can communicate automatically with the outside world. Smart products are thus alive. First, a "licence plate" is assigned to each single 4.0 object. Each item, through this plate, can be uniquely recognized anywhere in the world. This plate is then coded and, thanks to automatic identification technologies such as RFID (Radio Frequency IDentification), 4.0 products talk, meaning they can be recognized in a totally automated way and then communicated to the outside world. In this way, that physical object can be unified to its, digital twin, the invisible cloud of data that surrounds it.

For example, machines can identify products and decide whether to process that particular single item or not. Is the product authentic? Is the product authorized to be processed? Has the product expired? Is the product suitable? Or, alternatively, the machines can be set automatically by varying certain process parameters without the need for a human operator. Think for example how a 4.0 food kit and a 4.0 oven would work. The products are 4.0 because an RFID tag gives each single kit its own identity; the tag also allows the kit to be automatically identified by the 4.0 kitchen oven, an autonomous robot which through a digital twin recognizes the product, checks its authenticity, verifies that the product is safe, and consequently decides the recipe with which to process it, sending to electric motors, grills, microwaves, water dispensers, etc. the machine instructions to cook a first-rate risotto or knead a fragrant pizza.

By scanning the RFID tag of an item, end users will be able to get things like the instruction manual or the way to use it, verify product authenticity or to find out product recalls, software updates. They could check warranties and expand after sales services, or automatically reorder them by throwing an empty product in a 4.0 trash bin. In a 4.0 scenario there is a hybrid of physical product identity and digital identity, that can be brought together thanks to the internet of things, and will bring in radical changes.

But not only that. Thanks to RFID technology, the 4.0 product can be automatically identified anywhere along the supply chain; both when it is moving and when it remains stationary. RFID readers in the form of portals and tunnels identify raw materials and semi-finished products through a manufacturing process on a production line, or finished products in a logistics process as they move from one department to another or from one plant to another. In this scenario, even when the product is stationary in a warehouse, or in an inter-operational buffer, or on the shelf of a store, it can be read automatically using fixed or mobile readers. In other words,

4.0 products leave *traces* throughout the supply chain, a sort of "seen to arrive, stop and leave" among the various processes and actors. These traces are collected and shared in a standardized way in regular, secure recordsets via the Internet. In other words, anyone in the world, if authorized, can interpret them and understand who (which actor, which reader) has seen a certain *event* (e.g. has seen something "born" "arrive", "stop", "leave" or "die") *what* (as a single 4.0 object, with all its attributes such as batch number, expiry date in the case of a food product, model, size, colour, for a textile and clothing product, explanatory leaflet in the case of a drug, etc.), *when* (date, minute, second), *where* (plant, department, store), and *why* (which business process).

Thus far, according to (Das 2019) there are around 20 billion 4.0 objects which, thanks to an RFID tag, can be uniquely recognized and automatically identified throughout the supply chain. Starting from 2021, over the next three years this will grow to 25 billion. Every year.

It is also RFID technology, at least from a numbers point of view, which creates the Internet of Things. A network of billions of physical objects born every year and equipped with sensors which make them unique, able to exchange data automatically and share their status and position. These data are the "Big Data" with which various stakeholders can build value. A value that becomes greater the more this data is shared. The supply chain becomes completely transparent, and stakeholders have information that is *selective* (related to a single item), *accurate* (with a minimal margin of error) and *timely* (available in real time) regarding what is happening here and now and is going to happen. Thanks to the ability to share, analyse, and process this data through advanced algorithms, neural networks and machine learning which we call "Artificial Intelligence", and make decisions in real time, or even in advance, machines can optimize processes and software programs can minimize inventories, eliminate waste, obsolete and overdue items, and clear stock outages and unfulfilled orders.

The Impact of 4.0 Technologies on the SC

4.0 technologies are essential to support the evolution of the supply chain currently used in many competitive contexts. The last decades of the last century were characterized by a relocation of production. Factories were moved to and concentrated by large multinationals in developing countries, especially in the East, in search of economies of scale in production and low-cost labour. Factories which produced large volumes of standard products for a world market based on demand forecasts, and therefore in advance, to allow low-cost transport. The distribution was worldwide and the contact with the consumer was the exclusive prerogative of groups of retailers and took place exclusively in a place which the consumer had to physically visit to buy the product. The challenge of the new millennium is to offer the consumer a product that is always new and personalized—even made to measure—and available with shorter and shorter waiting times. To have this product, the consumer no longer needs to physically go to a store to see it, touch it and buy it, but can buy it online and have it delivered to the home. Moreover, unpredictable events like pandemics, wars, or even a ship stuck in the Suez canal, highlighted how these model may be utterly inadequate since they suffer of low resilience. For this reason, in many industrial sectors, the *lean supply chain*, able to produce and distribute standard products and working rigidly to demand forecasts, no longer functions in nor need evolve towards an agile supply chain, but is able to change rapidly with the changing tastes of the consumer. The result is one-off, ever-different products that are renewed as swiftly as possible. A constantly evolving supply chain, flexible from the points of view of procurement, production and distribution, which can also function with minimum stocks, not in a Push logic but being pulled by market demand.

But how to do so without having the costs of procurement, production and distribution explode?

All possible thanks to 4.0 technologies.

- RFID and the Internet of Things, Big Data, and Cloud Computing ensure that information relating to consumer demand, the availability of suppliers, raw materials, semi-finished and finished products in the supply chain are accessible in real time and empower the concepts of supply chain integration, pull supply chain, and agile supply chain;
- Technologies such as Simulation, Autonomous Robots, Additive Manufacturing and System Integration, mean that factories, distribution centres, and transportation become more and more flexible and rapid, with a low content of unskilled labour. There is no longer a need to ship raw materials from the other side of the world, to produce large quantities in advance to take advantage of economies of scale or low-cost labour, and then re-transport the finished goods back to the other side of the world! In the new millennium we will be returning to producing locally, in small, semi-autonomous factories, capable of producing whatever the market is seeking at that time in unit batches. The real musts will be speed and flexibility.

The Impact of 4.0 Technologies on the Society

The effect of the fourth industrial revolution on our society is the exaltation of the concept of abundance to the maximum. That is, the ability to supply the final consumer—namely all of us—with products of ever-higher quality and of ever-higher variety at lower prices. The ability to deliver exactly the product we require, when required, where required, as required, at the lowest possible price.

It is thanks to the abundance generated by the industrial revolutions that the world population has exploded, letting us live in the best possible era with the longest life expectancy. With the first industrial revolution, for the first time since the appearance of humankind on the planet, the world population in 1804 reached one billion people. Before then, the number of inhabitants of the planet had always fluctuated below this threshold, with wars, famines and epidemics at regular intervals lowering the number of inhabitants to around 500 million people. It took 10,000 years to achieve this important milestone. With the second industrial revolution, thanks to progress in the various sectors of technology and medicine, by 1927 the world population had reached two billion people. It took only 123 years to add the second billion to the world population, compared with 10,000 years for the first. From a demographic point of view, the third industrial revolution marked an exponential growth in the planet's inhabitants, testifying to the increased level of wellbeing and available wealth created "by industrial abundance". The number of people soared from 2 billion in 1927 to 3 billion in 1959, in just 34 years and net of a world war, and after another thirty years, the world population grew to 6 billion in 1999. With the fourth industrial revolution, we went from 6 billion in 1999 to 7.5 billion in 2018, and we are now moving towards the milestone of 9 billion.

It is also thanks to the concept of abundance that in the last 100 years life expectancy has increased by +250%, going from 30 before the first industrial revolution, to 40 at the beginning of the 1900s, to 69.12 in 1960, to the current 83.49 years. And some visionaries like Ray Kurzweil foresaw a future of immortality around 2045, thanks to the union of human and machine, in which the knowledge and skills embedded in our brains and bodies will be powered with exponential growth (meaning that in ten years progress and performances multiply 1024 times, and not ten times like in linear growth) in electronic devices performances and miniaturization (Kurzweil 2005)

In 100 years, the world's gross domestic product has increased by +300%, the cost of food has decreased 13 times, that of energy 30 times, that of transport 100 times, and the cost of communication 100,000 times. The cost of access to knowledge has gone to zero, if it is true that anyone in the world, even in the poorest region of the planet, can access the Internet from a smartphone.

The first two industrial revolutions multiplied the strength of humans and animals. Thanks to the steam engine first, the electric motor, and then the internal combustion engine, physical activity was delegated to a machine, leaving people with the task of control and steering, freeing them from physically tiring, repetitive work. The third, but above all the fourth industrial revolution have been even more radical. In effect, they no longer have an impact on people's physical strength but on their intellectual capacity. Electronic and computer technologies in the '80s and '90s first, and then technologies enabling industry 4.0 in the 2000s (Big Data and Cloud Computing, Artificial Intelligence, Simulation, the Internet of Things, Autonomous Robots) multiply by several orders of magnitude the ability of our brains and minds to analyse, process information and make consequent decisions. Machines are able to manage information of thousands of orders of magnitude higher and made available for the first time thanks to the Internet of Things and integrated systems, taking exact decisions in times of thousands of orders of magnitude smaller, entrusting people with the role of programming intelligence, control and validation of decisions.

I truly believe that also in this case, the fourth industrial revolution will continue to bring prosperity, like the first three, if not even more, thanks on the one hand to the already seen concept of increasing abundance but above all thanks to the fact that it will free people from repetitive, boring and low-added-value or strenuous jobs, developing more professional trades and professions at a higher intellectual and creative level. On the other hand, it is thanks to the concept of abundance that we have gone from working 40 hours a week against approximately double the hours of strenuous work carried out by one of our ancestors two centuries ago. There are those who think that technology will steal jobs. I think it is instead a great opportunity to reinvent better and slower work. 100 years ago in the United States about 50% of the workforce worked in agriculture; today is less than 2%, and this has not caused famine or the lack of food. With the fourth industrial revolution, new professions and opportunities will emerge that did not exist before, especially in the fields of engineering and information technology, biomedicine and nanotechnology, but also entertainment and leisure, personal services, culture and high class manufacturing. On the other hand, who 50 years ago would have imagined jobs such as AI Engineer, gamification designer, Youtuber, Influencer, Data Scientist, Web Analyst, Sentiment Analyst, Social Media Manager, Yoga personal trainer, Wedding planner, etc…

There is still one thing which remains peculiar to our mind and that machines have not been able to do: on the one hand, our ability to conceive, design and program them, of course. But most of all the ability to paint the *Mona Lisa*, to discover the Theory of Relativity or a vaccine capable of defeating cancer, the ability to compose Beethoven's Fifth or write *The Betrothed*.

Innovation, intuition, creative imagination, the ability to inspire a dream and generate emotions remain the exclusive prerogative of humankind. Only for the moment? I don't think so.

References

Al Volante (2015) Ma la Volkswagen ha bisogno di una nuova Phaeton?. https://www.alvolante.it/news/volkswagen-phaeton-2017-18-339015. Accessed 30 Sept 2021

Ansa (2018) Vera, il robot assunto all'Ikea per i colloqui di lavoro. https://www.ansa.it/canale_scienza_tecnica/notizie/tecnologie/2018/04/27/vera-il-robot-assunto-allikea-per-i-colloqui-di-lavoro_5df1664e-c325-422c-b636-a4ba4816e37e.html, accessed 05 Oct 2021

Beck A (2002) shrinkage in Europe: stock loss in the fast moving consumer goods sector. Secur J 15(4):25–40

Bertolini M, Ferretti G, Montanari R, Rizzi A, Vignali G (2012) A quantitative evaluation of the impact of the RFID technology on shelf availability. Int J RF Technol: Res Appl 3(3):159–180

Bottani E, Montanari R, Rizzi A (2009) The impact of RFID technology and EPC system on stock-out of promotional items. Int J RF Technol: Res Appl 1(1):6–22

Bottani E, Bertolini M (2009) Technical and economic aspect of RFID implementation for asset tracking. Int J RF Technol: Res Appl 1(3):169–193

Bryne MP, Markham WJ (1991) Improving quality and productivity in the logistics process—Achieving customer satisfaction breaktrhoughts. Council Logist Manag Edn

Chen F, Drezner Z, Ryan JK, Simchi-Levi D, The bullwhip effect: managerial insights on the impact of forecasting and information on variability in a supply chain. In: Quantitative models for supply chain management

Christopher M (2000) The agile supply chain: competing in volatile markets. Ind Market Manag 29:37–44

Christopher MG (1992) Logistics and supply chain management. Pitman Publishing, London, UK

Dais S (2014) Industrie 4.0—Anstoß, vision, Vorgehen. In: Bauernhansl T, TenHompel M, Vogel-Heuser B (eds) Industrie 4.0 in produktion, automatisierung und logistik—Anwendung, technologien, migration. Springer, Wiesbaden, pp 625–634

Das R (2019) RFID forecasts, players and opportunities 2019–2029. In: The complete analysis of the global RFID industry. IdTechEx

Delen D, Hardgrave BC, Sharda R (2007) RFID for better supply-chain management through enhanced information visibility. Prod Oper Manag 16(5):613–624

Distribuzione Moderna (2019). https://distribuzionemoderna.info/notizia-del-giorno/amazon-schiera-piu-di-300-marchi-esclusivi. Accessed 30 Sept 2021

Drucker PF (1999) Managing for results: economic tasks and risk-taking decisions. Routledge

Fisher ML (1997) What is the right supply chain for your product? Harv Bus Rev 75(2):105–116

Forrester J (1961) Industrial dynamics. MIT Press

Gale T, Rajamani D, Sriskandarajah C (2009) The impact of RFID on supply chain performance. Technol Oper Manag 2

A. Rizzi, *Supply Chain*,
https://doi.org/10.1007/978-3-030-95707-0

Ganeshan R, Harrison TP (1992) An introduction to supply chain management. In: Department of Management. Science and information systems. Penn State University, US

Gibson (2019) Amazon Warehouses trash millions of unsold products, media reports says, CBS News

Green K, Zelbst P, Sower V, Bellah J (2017) Impact of radio frequency identification technology on environmental sustainability. J Comput Inf Syst 57(3):1–9

GS1 Italy (2021) GTIN—Global trade item number. https://gs1it.org/assistenza/standard-specif iche/gtin/. Accessed . Accessed 30 Sept 2021

GS1 Italy (2021b) GS1 EANCOM: lo standard per l'EDI. https://gs1it.org/assistenza/standard-spe cifiche/gs1-eancom-standard-edi/. Accessed 30 Sept 2021

GS1 Italy (2021c) GLN—Global location number. https://gs1it.org/assistenza/standard-specifiche/ gln/. . Accessed 30 Sept 2021

GS1 Italy (2021d) Euritmo: lo standard EDI italiano. https://gs1it.org/assistenza/standard-specif iche/euritmo-standard-edi-italiano/. Accessed 30 Sept 2021

GS1 Italy (2021e) SSCC—Serial shipping container code. https://gs1it.org/assistenza/standard-spe cifiche/sscc/. Accessed 30 Sept 2021

Lee HL, Billington C (1992) Managing supply chain inventory: pitfalls and opportunities. Sloan Manag Rev 33(3):65–73

Hammer M (1990) Reengineering work: don't automate, obliterate. Harv Bus Rev 1–8

Handley R (2021) The convenience store that visits you. RFID Journal. https://www.rfidjournal. com/the-convenience-store-that-visits-you. Accessed 28 Feb 2022

Hardgrave BC, Waller M, Miller R (2006) RFID's impact on out of stocks: a sales velocity analysis. RFID Research centre Sam Walton College of Business University of Arkansas

Kodali S (2019) Retailers, reduce the pain of online returns. https://www.forrester.com/report/Ret ailers-Reduce-The-Pain-Of-Online-Returns/RES157782. Accessed 04 Feb 2022

Kraljic P (1983) Purchasing must become supply management. Harv Bus Rev 61(5):109–117

Kurzweil R (2005) The singularity is near. Penguin Publishing Group

La Londe BJ, Zinszer PH (1977) Customer service: meaning and measurement. In: National council of physical distribution management

La Londe BJ, Masters JM (1994) Emerging logistics strategies: blur print for the next century. Int J Phys Distrib Logist Manag 24(7):35–47

La Repubblica (2013). https://www.repubblica.it/cronaca/2013/02/19/news/nestle_carne_cavallo-52940673/. Accessed 30 Sept 2021

Lee H (2002) Alligning supply chain strategies with product uncertainties. Califor Manag Rev 44(3):105–119

Lee H, Padmanabhan V, Seunjin W (1197) The bullwhip effect in supply chains. Sloan Manag Rev 38(3):93–102

Lewis (1996) When cultures collide. Nicholas Bradley pub

MaRS (2020) Case study: dell—Distribution and supply chain innovation. https://learn.marsdd. com/article/case-study-dell-distribution-and-supply-chain-innovation/. Accessed 01 Oct 2021

Mason T, Knights M (2017) Omnichannel retail: How to build winning stores in a digital world. Koogan

Mentzer JT, DeWitt W, Keebler J, Ming S, Nix N, Smith C, Zacharia Z (2001) Defining supply chain management. J Bus Logist 22(2):1–25

Meti (2020) Demonstration tests to be held for food waste reduction taking advantage of electronic tags (RFID). https://www.meti.go.jp/english/press/2020/1028_003.html.. Accessed 28 Feb 2022

Moroni L (2020) Lo store Rfid non è più un miraggio, GDOweek. https://www.gdoweek.it/lo-store-rfid-non-piu-miraggio. Accessed 30 Sept 2021

OC&C (2021) The OC&C FMCG global 50. https://www.occstrategy.com/en/about-occ/news-and-media/article/id/3262/2016/07/the-occ-fmcg-global-50. Accessed 01 Oct 2021

Pagnobianchi IM (2004) Merchandising strategico pianificare il merchandising per i prodotti di largo consume e misurare l'efficacia, Hoepli

Parunak HVD, VanderBok R (1998) Modeling the extended supply network. In: ISA-Tech'98 Houston, Texas, USA, Industrial Technology Institute

Peters TJ, Waterman RH, Jones I (1982) In search of excellence: lessons from America's best-run companies. Harper & Row, New York

Porter M (1985) Competitive advantage: creating and sustaining superior performance. Free PR

PWC Consulting (2002) Focus on retail: applying auto-ID to improve product availability at the retail shelf. Auto-ID centers

Reichheld FF (2003) The one number you need to grow. Harv Bus Rev 81(12):46–55

Richardson (1993) Parallel sourcing and supplier performance in the Japanese automobile industry. Strat Manag J 14(5):339–350

Rizzi A, Montanari R, Bertolini M, Botttani E, Volpi A (2011) Logistica e tecnologia RFID: creare valore nella filiera alimentare e nel largo consumo. Springer

Ruggeri R (1998) Logistica industriale. Cusl, Milan

Sicurauto (2014). https://www.sicurauto.it/news/toyota-richiama-ancora-le-stesse-auto-per-difetti-allairbag/. Accessed 30 Sept 2021

Simchi-Levi D, Kaminsky P, Simchi-Levi E (2000) Designing and managing the supply chain. McGraw-Hill Higher Education

Sinek S (2019) The infinite game, Ed Portfolio

Sollish F, Semanik J (2011) Strategic global sourcing best practices. Wiley

Stephens D (2017) Reengineering retail: the future of selling in a post-digital world. Figure 1 Pub

Strickland T (1999) Strategic management, concepts and cases. McGraw Hill College Division, New York

Sweeberg C (2022a) Walmart recomits to RFID. RFID J. https://www.rfidjournal.com/walmart-re-commits-to-rfid-with-supplier-mandates. Accessed 28 Feb 2022

Sweeberg C (2022b) RFID tracks inventory, sales, security at unmanned modular smart store. RFID J. https://www.rfidjournal.com/rfid-tracks-inventory-sales-security-at-unmanned-modular-smart-store. Accessed 28 Feb 2022

Tellkamp (2006) The impact of auto-ID technology on process performance—RFID in the FMCG supply chain. Master thesis. https://www.academia.edu/8203055/The_impact_of_Auto_ID_technology_on_process_performance_RFID_in_the_FMCG_supply_chain. Accessed 01 Oct 2021

Tommasi D (2021) Amazon prime: 200 milioni di abbonati nel mondo. https://www.punto-informatico.it/amazon-prime-200-milioni-abbonati. Accessed 05 Oct 2021

Verde Azzurro notizie (2018) Chi produce i prodotti a marchio Conad?. https://www.verdeazzurronotizie.it/prodotti-marchio-conadchi-produce-i-prodotti-conad/. Accessed 30 Sept 2021

VICS (2004) Collaborative planning, forecasting and replenishment CPFR. https://www.gs1us.org/DesktopModules/Bring2mind/DMX/Download.aspx?Command=Core_Download&EntryId=492&language=en-US&PortalId=0&TabId=134. Accessed 30 Sept 2021

You tube (2016) Introducing Amazon go and the world's most advanced shopping technology. https://www.youtube.com/watch?v=NrmMk1Myrxc

Youtube (2011) A day in the life of a Kiva Robot. https://www.youtube.com/watch?v=6KRjuuEVEZs. Accessed 01 Oct 2021

Youtube (2012) Use of the serial shipping container code (SSCC) in the retail sector. https://www.youtube.com/watch?v=R2pNIbI6kAs. Accessed 30 Sept 2021

Youtube (2014) CNET news—Meet the robots making Amazon even faster. https://www.youtube.com/watch?v=UtBa9yVZBJM. Accessed 01 Oct 2021

Youtube (2021) How Amazon's super-complex shipping system works. https://www.youtube.com/watch?v=2qanMpnYsjk&t=671s. Accessed 01 Oct 2021

Printed in the United States
by Baker & Taylor Publisher Services